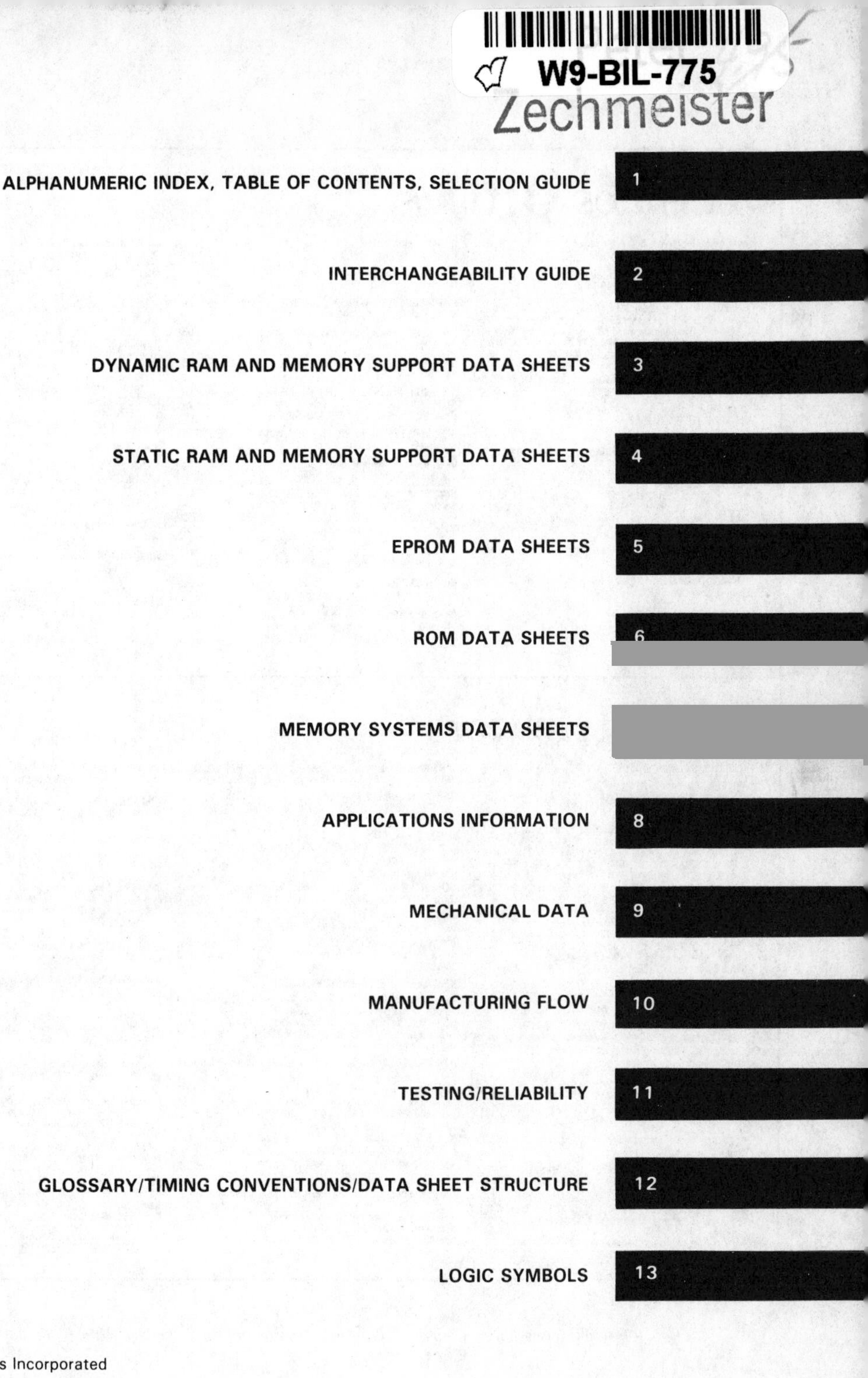

Texas Instruments Incorporated
MOS Memory Division, M/S 6946
P. O. Box 1443
Houston, Texas 77001

TI Sales Offices

ALABAMA: Huntsville, 500 Wynn Drive, Suite 514, Huntsville, AL 35805, (205) 837-7530.

ARIZONA: Phoenix, P.O. Box 35160, 8102 N. 23rd Ave., Suite A, Phoenix, AZ 85069, (602) 249-1313.

CALIFORNIA: El Segundo, 831 S. Douglas St., El Segundo, CA 90245, (213) 973-2571; **Irvine,** 17620 Fitch, Irvine, CA 92714, (714) 545-5210; **Sacramento,** 1900 Point West Way, Suite 171, Sacramento, CA 95815, (916) 929-1521; **San Diego,** 4333 View Ridge Ave., Suite B., San Diego, CA 92123, (714) 278-9600; **Santa Clara,** 5353 Betsy Ross Dr., Santa Clara, CA 95051, (408) 748-2300 **Woodland Hills,** 21220 Erwin St., Woodland Hills, CA 91367, (213) 704-7759.

COLORADO: Denver, 9725 E. Hampden St., Suite 301, Denver, CO 80231, (303) 695-2800.

CONNECTICUT: Wallingford, 9 Barnes Industrial Park Rd., Barnes Industrial Park, Wallingford, CT 06492, (203) 269-0074.

FLORIDA: Clearwater, 2280 U.S. Hwy. 19 N., Suite 232, Clearwater, FL 33515, (813) 796-1926; **Ft. Lauderdale,** 2765 N.W. 62nd St., Ft. Lauderdale, FL 33309, (305) 973-8502; **Maitland,** 2601 Maitland Center Parkway, Maitland, FL 32751, (305) 646-9600.

GEORGIA: Atlanta, 3300 Northeast Expy., Building 9, Atlanta, GA 30341, (404) 452-4600.

ILLINOIS: Arlington Heights, 515 W. Algonquin, Arlington Heights, IL 60005, (312) 640-2934.

INDIANA: Ft. Wayne, 2020 Inwood Dr., Ft. Wayne, IN 46805, (219) 424-5174; **Indianapolis,** 2346 S. Lynhurst, Suite J-400, Indianapolis, IN 46241, (317) 248-8555.

IOWA: Cedar Rapids, 373 Collins Rd. NE, Suite 200, Cedar Rapids, IA 52402, (319) 395-9550.

MARYLAND: Baltimore, 1 Rutherford Pl., 7133 Rutherford Rd., Baltimore, MD 21207, (301) 944-8600.

MASSACHUSETTS: Waltham, 504 Totten Pond Rd., Waltham, MA 02154, (617) 890-7400.

MICHIGAN: Farmington Hills, 33737 W. 12 Mile Rd., Farmington Hills, MI 48018, (313) 553-1500.

MINNESOTA: Edina, 7625 Parklawn, Edina, MN 55435, (612) 830-1600.

MISSOURI: Kansas City, 8080 Ward Pkwy., Kansas City, MO 64114, (816) 523-2500; **St. Louis,** 11861 Westline Industrial Drive, St. Louis, MO 63141, (314) 569-7600.

NEW JERSEY: Clark, 292 Terminal Ave. West, Clark, NJ 07066, (201) 574-9800.

NEW MEXICO: Albuquerque, 5907 Alice NSE, Suite E, Albuquerque, NM 87110, (505) 265-8491.

NEW YORK: East Syracuse, 6700 Old Collamer Rd., East Syracuse, NY 13057, (315) 463-9291; **Endicott,** 112 Nanticoke Ave., P.O. Box 618, Endicott, NY 13760, (607) 754-3900; **Melville,** 1 Huntington Quadrangle, Suite 3C10, P.O. Box 2936, Melville, NY 11747, (516) 454-6600; **Poughkeepsie,** 201 South Ave., Poughkeepsie, NY 12601, (914) 473-2900; **Rochester,** 1210 Jefferson Rd., Rochester, NY 14623, (716) 424-5400.

NORTH CAROLINA: Charlotte, 8 Woodlawn Green, Woodlawn Rd., Charlotte, NC 28210, (704) 527-0930.

OHIO: Beachwood, 23408 Commerce Park Rd., Beachwood, OH 44122, (216) 464-6100; **Dayton,** Kingsley Bldg., 4124 Linden Ave., Dayton, OH 45432, (513) 258-3877.

OKLAHOMA: Tulsa, 3105 E. Skelly Dr., Suite 110, Tulsa, OK 74105, (918) 749-9547.

OREGON: Beaverton, 6700 SW 105th St., Suite 110, Beaverton, OR 97005, (503) 643-6758.

PENNSYLVANIA: Ft. Washington, 575 Virginia Dr., Ft. Washington, PA 19034, (215) 643-6450.

TENNESSEE: Johnson City, P.O. Drawer 1255, Erwin Hwy., Johnson City, TN 37601, (615) 461-2191.

TEXAS: Austin, 12501 Research Blvd., P.O. Box 2909, Austin, TX 78723, (512) 250-7655; **Dallas,** P.O. Box 225012, Dallas, TX 75265, (214) 995-6531; **Houston,** 9100 Southwest Frwy., Suite 237, Houston, TX 77036, (713) 778-6592; **San Antonio,** 1938 N.E. Loop 410, San Antonio, TX 78217, (512) 828-9101.

UTAH: Salt Lake City, 3672 West 2100 South, Salt Lake City UT 84120, (801) 973-6310.

VIRGINIA: Fairfax, 3001 Prosperity, Fairfax, VA 22031, (703) 849-1400; **Midlothian,** 13711 Sutter's Mill Circle, Midlothian, VA 23113, (804) 744-1007.

WISCONSIN: Brookfield, 205 Bishops Way, Suite 214, Brookfield, WI 53005, (414) 784-3040.

WASHINGTON: Redmond, 2723 152nd Ave., N.E. Bldg 6, Redmond, WA 98052, (206) 881-3080.

CANADA: Ottawa, 436 McClaren St., Ottawa, Canada, K2P0M8, (613) 233-1177; **Richmond Hill,** 280 Centre St. E., Richmond Hill L4C1B1, Ontario, Canada, (416) 884-9181; **St. Laurent,** Ville St. Laurent Quebec, 9460 Trans Canada Hwy., St. Laurent, Quebec, Canada H4S1R7, (514) 334-3635 O

TI Distributors

ALABAMA: Hall-Mark (205) 837-8700.

ARIZONA: Phoenix, Kierulff (602) 243-4101; Marshall (602) 968-6181; R.V. Weatherford (602) 272-7144; Wyle (602) 249-2232; **Tucson,** Kierulff (602) 624-9986.

CALIFORNIA: Los Angeles/Orange County, Arrow (213) 701-7500, (714) 851-8961; JACO (714) 540-5600, (213) 998-2200; Kierulff (213) 725-0325, (714) 731-5711; Marshall (213) 999-5001, (213) 686-0141, (714) 556-6400; RPS (213) 744-0355, (714) 521-5230; R.V. Weatherford (714) 634-9600, (213) 849-3451, (714) 623-1261; Time (213) 320-0880; Wyle (213) 322-8100, (714) 641-1611; **San Diego,** Arrow (714) 565-4800; Kierulff (714) 278-2112; Marshall (714) 578-9600; R.V. Weatherford (714) 695-1700; Wyle (714) 565-9171; **San Francisco Bay Area,** Arrow (408) 745-6600; Kierulff (415) 968-6292; Marshall (408) 732-1100; Time (408) 734-9888; United Components (408) 496-6900; Wyle (408) 727-2500; **Santa Barbara,** RPS (805) 964-6823; R.V. Weatherford (805) 465-8551.

COLORADO: Arrow (303) 758-2100; Diplomat (303) 740-8300; Kierulff (303) 371-6500; R.V. Weatherford (303) 428-6900; Wyle (303) 457-9953.

CONNECTICUT: Arrow (203) 265-7741; Kierulff (203) 265-1115; Marshall (203) 265-3822; Milgray (203) 795-0714.

FLORIDA: Ft. Lauderdale, Arrow (305) 973-8502; Diplomat (305) 971-7160; Hall-Mark (305) 971-9280; Kierulff (305) 652-6950; **Orlando,** Arrow (305) 725-1480; Diplomat (305) 725-4520; Hall-Mark (305) 855-4020; Milgray (305) 647-5747; **Tampa,** Diplomat (813) 443-4514; Kierulff (813) 576-1966.

GEORGIA: Arrow (404) 449-8252; Hall-Mark (404) 447-8000; Marshall (404) 923-5750.

ILLINOIS: Arrow (312) 893-9420; Diplomat (312) 595-1000; Hall-Mark (312) 860-3800; Kierulff (312) 640-0200; Newark (312) 638-4411.

INDIANA: Indianapolis, Arrow (317) 243-9353; Graham (317) 634-8202; **Ft. Wayne,** Graham (219) 423-3422.

IOWA: Arrow (319) 395-7230; Deeco (319) 365-7551.

KANSAS: Kansas City, Component Specialties (913) 492-3555; Hall-Mark (913) 888-4747; **Wichita,** LCOMP (316) 265-9507.

MARYLAND: Arrow (202) 737-1700, (301) 247-5200; Diplomat (301) 995-1226; Hall-Mark (301) 796-9300; Milgray (301) 468-6400.

MASSACHUSETTS: Arrow (617) 933-8130; Diplomat (617) 429-4120; Kierulff (617) 667-8331; Marshall (617) 272-8200; Time (617) 935-8080.

MICHIGAN: Detroit, Arrow (313) 971-8200; Diplomat (313) 477-3200; Newark (313) 967-0600; **Grand Rapids,** Newark (616) 241-6681.

MINNESOTA: Arrow (612) 830-1800; Diplomat (612) 788-8601; Hall-Mark (612) 854-3223; Kierulff (612) 941-7500.

MISSOURI: Kansas City, LCOMP (816) 221-2400; **St. Louis,** Arrow (314) 567-6888; Hall-Mark (314) 291-5350; Kierulff (314) 739-0855.

NEW HAMPSHIRE: Arrow (603) 668-6968.

NEW JERSEY: North, Arrow (201) 797-5800; Diplomat (201) 785-1830; JACO (201) 778-4722; Kierulff (201) 575-6750; Marshall (201) 340-1900; **South,** Arrow (609) 235-1900; General Radio (609) 964-8560; Hall-Mark (609) 424-0880; Milgray (609) 424-1300.

NEW MEXICO: Arrow (505) 243-4566; International Electronics (505) 345-8127.

NEW YORK: Long Island, Arrow (516) 231-1000; Diplomat (516) 454-6400; JACO (516) 273-5500; Milgray (516) 546-5600, (800) 645-3986; QPL (516) 467-1200; **Rochester,** Arrow (716) 275-0300; Marshall (716) 235-7620; Rochester Radio Supply (716) 454-7800; **Syracuse,** Arrow (315) 652-1000; Diplomat (315) 652-5000; Marshall (607) 754-1570.

NORTH CAROLINA: Arrow (919) 876-3132, (919) 725-8711; Hall-Mark (919) 832-4465; Kierulff (919) 852-6261.

OHIO: Cincinnati, Graham (513) 732-1661; **Cleveland,** Arrow (216) 248-3990; Hall-Mark (216) 473-2907; Kierulff (216) 587-6558; **Columbus,** Hall-Mark (614) 846-1882; **Dayton,** Arrow (513) 435-5563; ESCO (513) 226-1133; Marshall (513) 236-8088.

OKLAHOMA: Component Specialties (918) 664-2820; Hall-Mark (918) 835-8458; Kierulff (918) 252-7537.

OREGON: Almac/Stroum (503) 641-9070; Kierulff (503) 641-9150; Wyle (503) 640-6000.

PENNSYLVANIA: Arrow (412) 856-7000.

TEXAS: Austin, Arrow (512) 835-4180; Component Specialties (512) 837-8922; Hall-Mark (512) 258-8848; Kierulff (512) 835-2090; **Dallas,** Arrow (214) 386-7500; Component Specialties (214) 357-6511; Hall-Mark (214) 341-1147; International Electronics (214) 233-9323; Kierulff (214) 343-2400; **El Paso,** International Electronics (915) 778-9761; **Houston,** Arrow (713) 491-4100; Component Specialties (713) 771-7237; Hall-Mark (713) 781-6100; Harrison Equipment (713) 879-2600; Kierulff (713) 530-7030.

UTAH: Diplomat (801) 486-4134; Kierulff (801) 973-6913; Wyle (801) 974-9953.

WASHINGTON: Almac Stroum (206) 643-9992; Arrow (206) 575-0907; Kierulff (206) 575-4420; United Components (206) 643-7444; Wyle (206) 453-8300.

WISCONSIN: Arrow (414) 764-6600; Hall-Mark (414) 761-3000; Kierulff (414) 784-8160.

CANADA: Calgary, Cam Gard Supply (403) 287-0520; Future (403) 259-6408; Varah (403) 230-1235; **Downsview,** CESCO (416) 661-0220; **Edmonton,** Cam Gard Supply (403) 426-1805; **Halifax,** Cam Gard Supply (902) 454-8581; **Hamilton,** Varah (416) 561-9311; **Kamloops,** Cam Gard Supply (604) 372-3338; **Moncton,** Cam Gard Supply (506) 855-2200; **Montreal,** CESCO (514) 735-5511; Future (514) 694-7710; **Ottawa,** CESCO (613) 226-6905; Future (613) 820-8313; **Quebec City,** CESCO (418) 687-4231; **Regina,** Cam Gard Supply (306) 525-1317; **Saskatoon,** Cam Gard Supply (306) 652-6424; **Toronto,** Future (416) 663-5563; **Vancouver,** Cam Gard Supply (604) 291-1441; Future (604) 438-5545; Varah (604) 873-3211; **Winnipeg,** Cam Gard Supply (204) 786-8481; Varah (204) 633-6190. AH

Texas Instruments invented the integrated circuit, microprocessor and microcomputer. Being first is our tradition.

TEXAS INSTRUMENTS
INCORPORATED

Printed in U.S.A.

The MOS Memory Data Book

1982

TEXAS INSTRUMENTS
INCORPORATED

LCC 7061
72488-52-CSI

Printed in U.S.A.

IMPORTANT NOTICE

Texas Instruments reserves the right to make changes at any time in order to improve design and to supply the best product possible.

TI cannot assume any responsibility for any circuits shown or represent that they are free from patent infringement.

Information contained herein supersedes previously published data on these devices from TI.

ISBN 0-89512-112-3

Library of Congress Catalog Number: 82-60332

Alphanumeric Index, Table of Contents, Selection Guide

ALPHANUMERIC INDEX TO DATA SHEETS

TABLE OF CONTENTS

WORDS	BITS PER WORD		
	1	4	8
1K		(4K) SRAMs TMS 2114 TMS 2114L TMS 2149	(8K) EPROMs TMS 2708 TMS 27L08
2K			(16K) SRAMs: TMS 4016 EPROMs: TMS 2516, TMS 2716
4K	(4K) SRAMs TMS 2147H TMS 4044 TMS 40L44	(16K) SRAMs TMS 2168 TMS 2169	(32K) ROMs: TMS 4732 EPROMs: TMS 2532, TMS 25L32
8K			(64K) ROMs: TMS 4764 EPROMs: TMS 2564
16K	(16K) SRAMs: TMS 2167 DRAMs: TMS 4116	(64K) DRAMs TMS 4416	
64K	(64K) DRAMs TMS 4164		

(Numbers in parenthesis indicate overall complexity)

Interchangeability Guide

PART I — ALTERNATE VENDOR PART NUMBERING (EXAMPLES)

TEXAS INSTRUMENTS (TI)

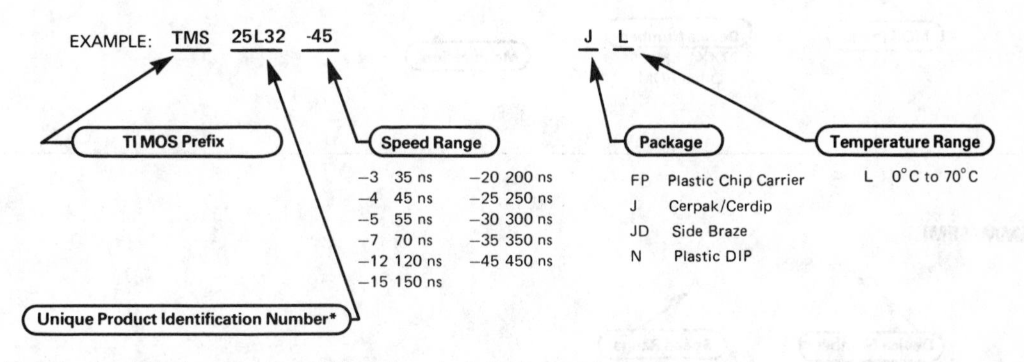

* Inclusion of an "L" in the product identification indicates the device operates at low power.

ADVANCED MICRO DEVICES (AMD)

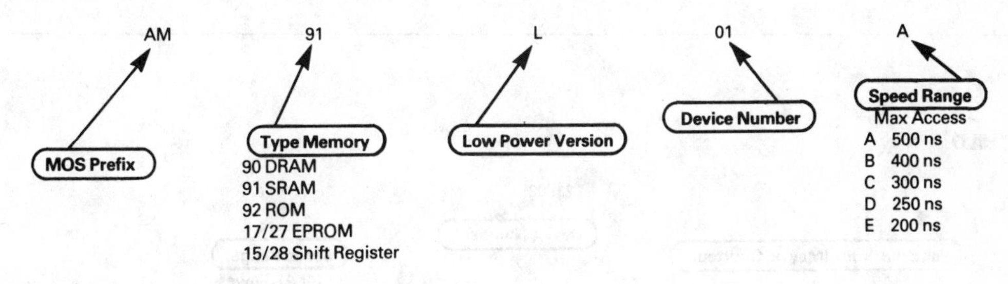

AMERICAN MICROSYSTEMS, INC. (AMI)

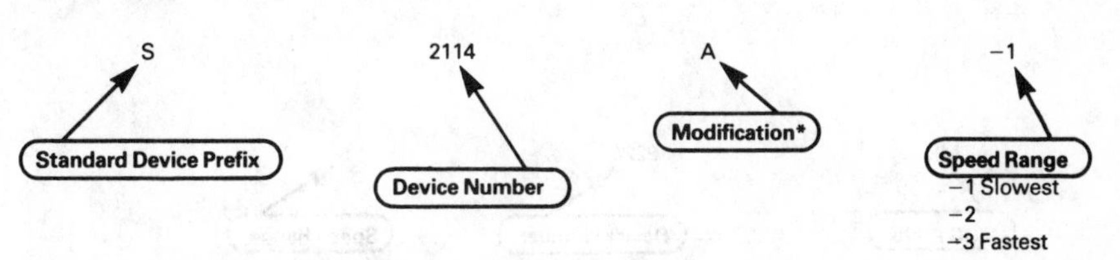

*Can also be "L", usually indicates lower power and/or improved speed

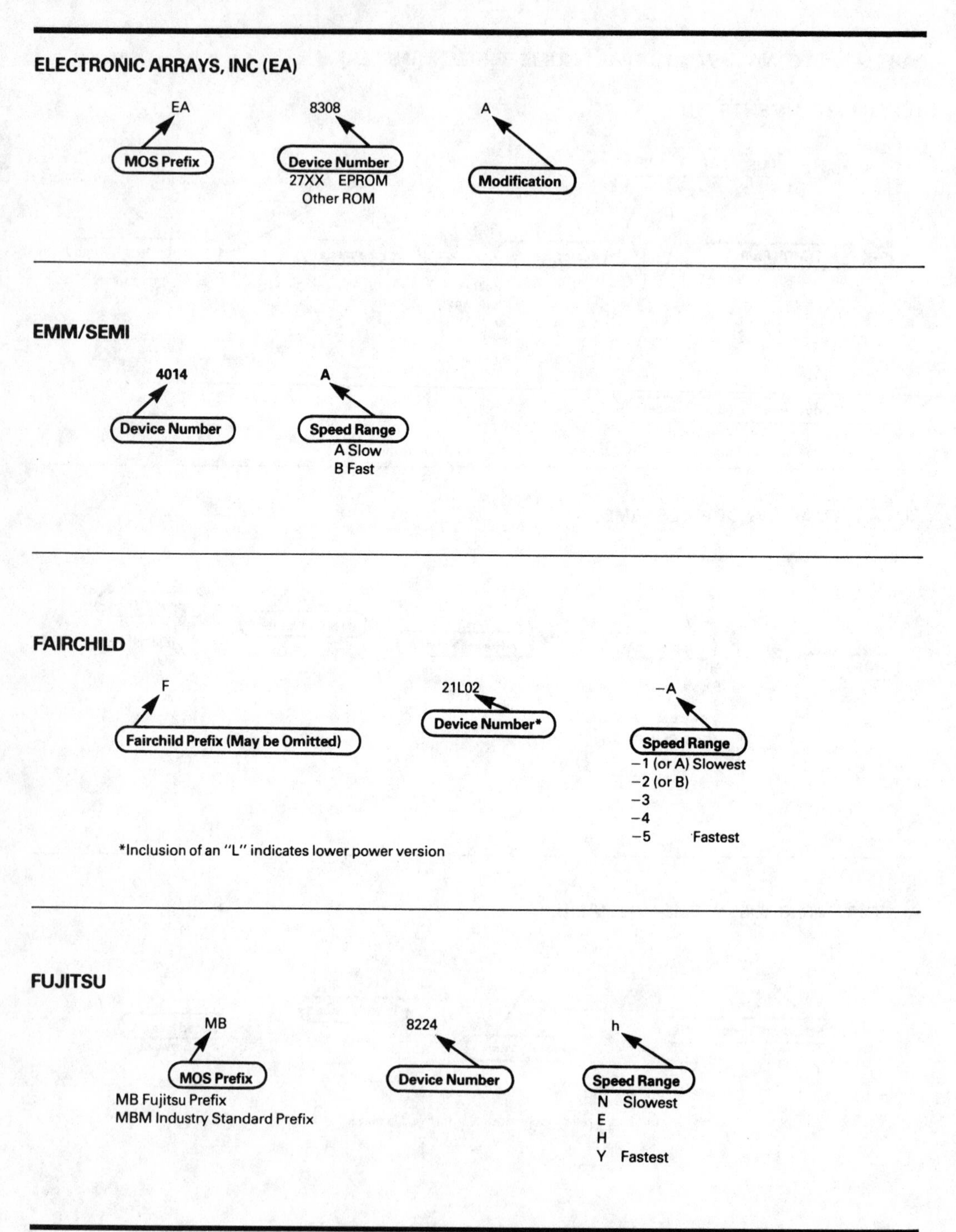
ELECTRONIC ARRAYS, INC (EA)
EA
8308
A
MOS Prefix
Device Number
27XX EPROM
Other ROM
Modification
EMM/SEMI
4014
A
Device Number
Speed Range
A Slow
B Fast
FAIRCHILD
F
21L02
–A
Fairchild Prefix (May be Omitted)
Device Number*
Speed Range
–1 (or A) Slowest
–2 (or B)
–3
–4
–5 Fastest
*Inclusion of an "L" indicates lower power version
FUJITSU
MB
8224
h
MOS Prefix
MB Fujitsu Prefix
MBM Industry Standard Prefix
Device Number
Speed Range
N Slowest
E
H
Y Fastest

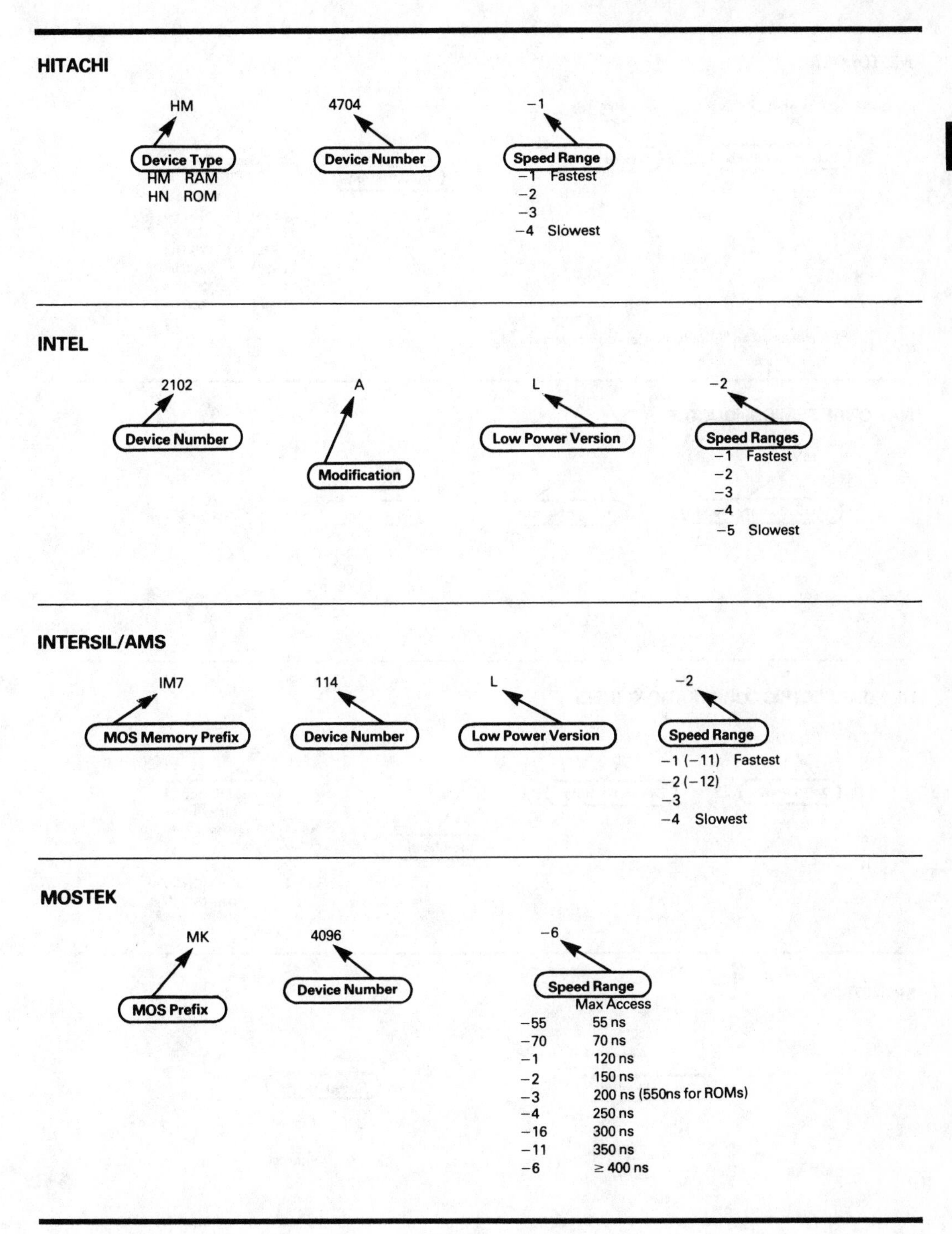
HITACHI
HM
4704
−1
Device Type
HM RAM
HN ROM
Device Number
Speed Range
−1 Fastest
−2
−3
−4 Slowest
INTEL
2102
A
L
−2
Device Number
Modification
Low Power Version
Speed Ranges
−1 Fastest
−2
−3
−4
−5 Slowest
INTERSIL/AMS
IM7
114
L
−2
MOS Memory Prefix
Device Number
Low Power Version
Speed Range
−1 (−11) Fastest
−2 (−12)
−3
−4 Slowest
MOSTEK
MK
4096
−6
MOS Prefix
Device Number
Speed Range
Max Access
−55 55 ns
−70 70 ns
−1 120 ns
−2 150 ns
−3 200 ns (550ns for ROMs)
−4 250 ns
−16 300 ns
−11 350 ns
−6 ≥ 400 ns

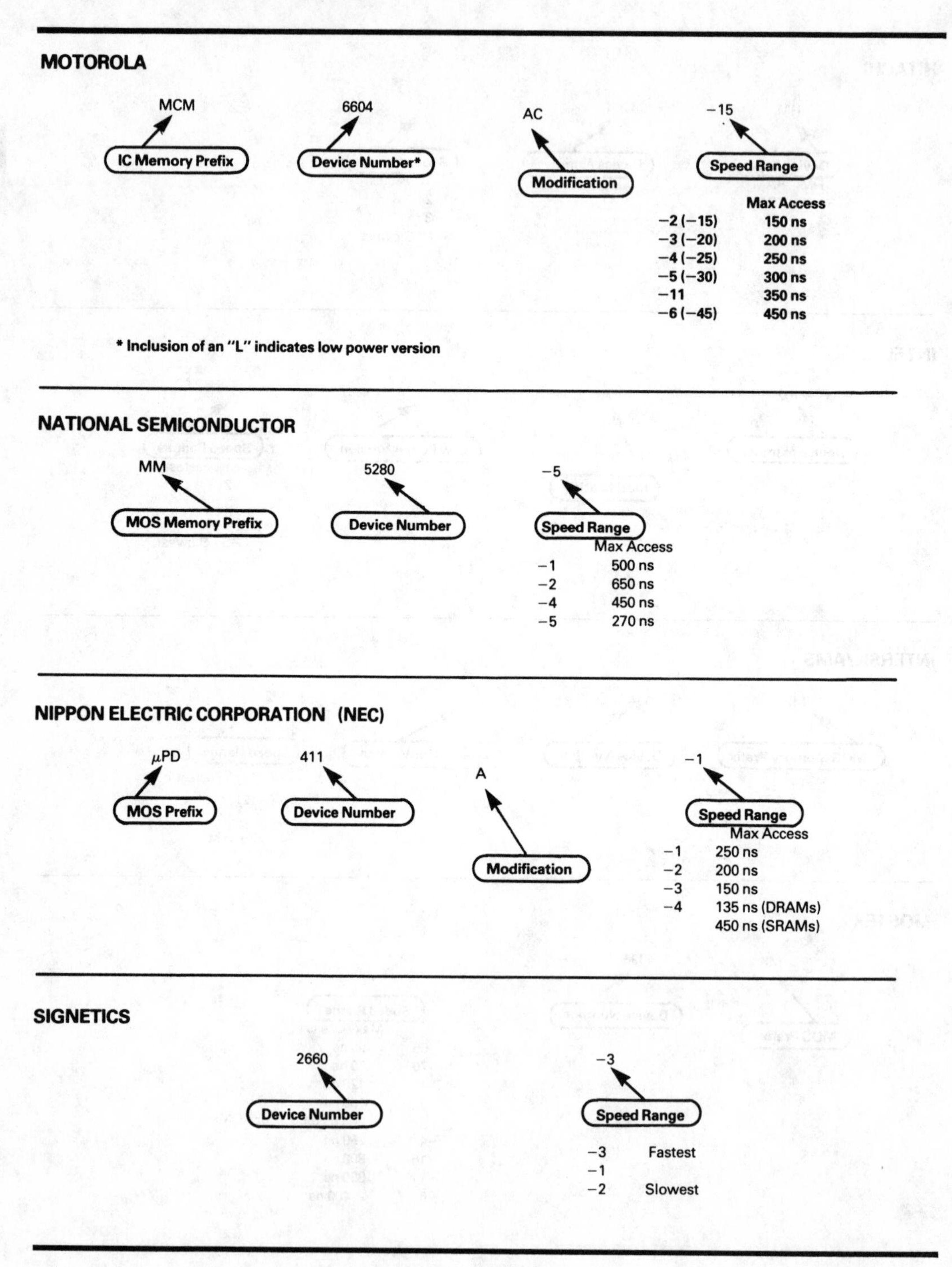

TEXAS INSTRUMENTS
INCORPORATED
POST OFFICE BOX 225012 • DALLAS, TEXAS 75265

SYNERTEK

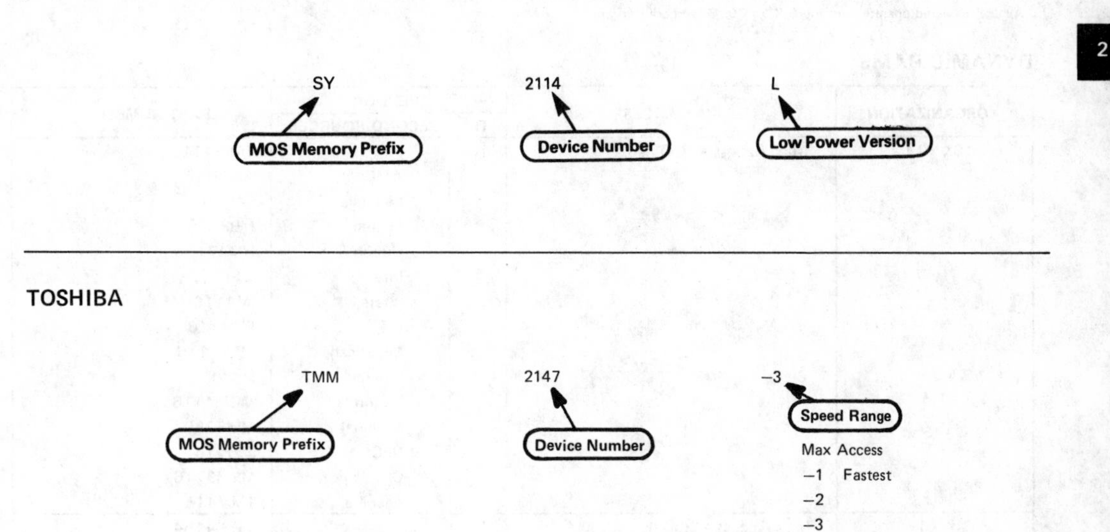
SY
2114
L
MOS Memory Prefix
Device Number
Low Power Version
TOSHIBA
TMM
2147
–3
MOS Memory Prefix
Device Number
Speed Range
Max Access
–1 Fastest
–2
–3
–4
–5 Slowest

PART II – SECOND SOURCES*

*Based on available published data. (Official second sourcing agreements not necessarily implied.)
All devices listed operate over the 0 °C to 70 °C temperature range.

DYNAMIC RAMS

ORGANIZATION	MAX ACCESS	VENDOR		PART NUMBER
		TI	SECOND SOURCES	
16K × 1	Max Access = 250 ns	TI		TMS 4116
			AMD	AM9026
			Fairchild	F4116
			Fujitsu	MB8116
			Hitachi	HM4716
			Intel	2117
			Intersil	IM4116
			ITT	ITT4116
			Mitsubishi	M5K4116
			Mostek	MK4116
			Motorola	MCM4116
			National	MM5290
			NEC	μPD416
			OKI	MSM3716
			Toshiba	TMM416
64K × 1 (5 V)	Max Access = 250 ns	TI		TMS 4164
			Fairchild	F64K
			Fujitsu	MB8264
			Hitachi	HM4864
			Intel	2164
			Mitsubishi	M5K4164S
			Mostek	MK4164
			Motorola	MCM6664
			National	NMC4164
			NEC	μPD4164
			OKI	MSM3764
			Toshiba	TMM4164

STATIC RAMS

ORGANIZATION	MAX ACCESS	VENDOR		PART NUMBER
		TI	SECOND SOURCES	
4K × 1	Max Access = 120 ns	TI		TMS 4044/TMS 40L44
			AMD	4044
			GTE Micro	2141/L2141
			Intersil	IM7141/IM7141L
			Intel	2141/2141L
			National SC	MM2141
			Mitsubishi	M5T4044
			Mostek	MK4104
			NEC	μPD4104
			Synertek	SY2141/SY2141L

TEXAS INSTRUMENTS
INCORPORATED
POST OFFICE BOX 225012 • DALLAS, TEXAS 75265

STATIC RAMS (continued)

ORGANIZATION	MAX ACCESS	VENDOR		PART NUMBER
		TI	SECOND SOURCES	
1K × 4	Max Access = 150 ns	TI		TMS 2114/TMS 2114L
			AMD	9114/91L14
			EA	EA2114L
			EMM/SEMI	2114
			Fairchild	F2114
			Hitachi	HM472114A
			Intel	2114A/2114AL
			Intersil	IM2114/IM2114L
			Mitsubishi	M5L2114L
			Motorola	MCM2114/MCM21L14
			National SC	MM2114/MM2114L
			NEC	μPD2114/μPD2114L
			OKI	MSM2114/MSM2114L
			Synertek	SY2114/SY2114L
4K × 1	Max Access = 70 ns	TI		TMS 2147
			AMD	9147
			AMI	S2147
			Fujitsu	MBM2147
			Hitachi	HM4847
			Intel	2147/2147L
			Mostek	MK2147
			Motorola	MCM2147
			National SC	MM2147/MM2147L
			NEC	μPD2147
			Toshiba	TMM315
1K × 4	Max Access = 70 ns	TI		TMS 2149
			Hitachi	HM6148/HM6148L
			Intel	2149H/2148H
			Motorola	MCM2149
			National	NMC2148
			NEC	μPD2149
			Synertek	SY2149
2K × 8	Max Access = 120 ns	TI		TMS 4016
			Fairchild	F3528
			Fujitsu	MB8128
			Mitsubishi	M58725
			Mostek	MK4802
			OKI	MSM2128
			Toshiba	TMM2016
16K × 1	Max Access = 70 ns	TI		TMS 2167
			Fujitsu	MB8167
			Hitachi	HM6167
			Intel	2167
			Mitsubishi	M58757
			NEC	μPD2167

EPROMS

ORGANIZATION	MAX ACCESS	VENDOR		PART NUMBER
		TI	SECOND SOURCES	
1K × 8 (3 Supply)	Max Access = 450 ns	TI		TMS 2708/TMS 27L08
			AMD	2708
			EA	EA2708
			Fairchild	F2708
			Fujitsu	MB8518
			Intel	2708/2708L
			Mitsubishi	M5L2708
			Motorola	MCM2708
			National SC	MM2708
			OKI	MSM2708
			Signetics	2708
			Toshiba	TMM322
2K × 8 (3 Supply)	Max Access = 450 ns	TI		TMS 2716
			Motorola	TMS2716/TMS27A16
2K × 8 (5 V)	Max Access = 450 ns	TI		TMS 2516
			AMD	2716
			Fairchild	F2716
			Fujitsu	MBM2716
			Hitachi	HN462716
			Intel	2716
			Mitsubishi	M5L2716
			Mostek	MK2716
			Motorola	MCM2716/MCM27L16
			National	MM2716
			NEC	μPD2716
			OKI	MSM2716
			Synertek	SY2716
			Toshiba	TMM323
4K × 8 (5 V)	Max Access = 450 ns	TI		TMS 2532/TMS 25L32
			Hitachi	HN62532
			Motorola	MCM2532/MCM25L32
			National	MM2532
8K × 8 (5 V)	Max Access = 450 ns	TI		TMS 2564
			Motorola	MCM68764

TEXAS INSTRUMENTS
INCORPORATED
POST OFFICE BOX 225012 • DALLAS, TEXAS 75265

Dynamic Ram and Memory Support Data Sheets

TMS 4116 NL
16,384-BIT DYNAMIC RANDOM-ACCESS MEMORY

OCTOBER 1977 – REVISED MAY 1982

- 16,384 X 1 Organization
- 10% Tolerance on All Supplies
- All Inputs Including Clocks TTL-Compatible
- Unlatched Three-State Fully TTL-Compatible Output
- 3 Performance Ranges:

	ACCESS TIME ROW ADDRESS (MAX)	ACCESS TIME COLUMN ADDRESS (MAX)	READ OR WRITE CYCLE (MIN)	READ, MODIFY-WRITE† CYCLE (MIN)
TMS 4116-15	150 ns	100 ns	375 ns	375 ns
TMS 4116-20	200 ns	135 ns	375 ns	375 ns
TMS 4116-25	250 ns	165 ns	410 ns	515 ns

- Page-Mode Operation for Faster Access Time
- Common I/O Capability with "Early Write" Feature
- Low-Power Dissipation
  - Operating 462 mW (max)
  - Standby 20 mW (max)
- 1-T Cell Design, N-Channel Silicon-Gate Technology
- 16-Pin 300-Mil (7.62 mm) Package Configuration

16-PIN PLASTIC DUAL-IN-LINE PACKAGE (TOP VIEW)

Signal	Pin	Pin	Signal
V_{BB}	1	16	V_{SS}
D	2	15	$\overline{CAS}$
$\overline{W}$	3	14	Q
$\overline{RAS}$	4	13	A6
A0	5	12	A3
A2	6	11	A4
A1	7	10	A5
V_{DD}	8	9	V_{CC}

PIN NOMENCLATURE			
A0-A6	Address Inputs	$\overline{W}$	Write Enable
$\overline{CAS}$	Column address strobe	V_{BB}	–5-V power supply
D	Data input	V_{CC}	+5-V power supply
Q	Data output	V_{DD}	+12-V power supply
$\overline{RAS}$	Row address strobe	V_{SS}	0 V ground

description

The TMS 4116 series is composed of monolithic high-speed dynamic 16,384-bit MOS random-access memories organized as 16,384 one-bit words, and employs single-transistor storage cells and N-channel silicon-gate technology.

All inputs and outputs are compatible with Series 74 TTL circuits including clocks: Row Address Strobe $\overline{RAS}$ (or $\overline{R}$) and Column Address Strobe $\overline{CAS}$ (or $\overline{C}$). All address lines (A0 through A6) and data-in (D) are latched on chip to simplify system design. Data out (Q) is unlatched to allow greater system flexibility.

Typical power dissipation is less than 350 milliwatts active and 6 milliwatts during standby (V_{CC} is not required during standby operation). To retain data, only 10 milliwatts average power is required which includes the power consumed to refresh the contents of the memory.

The TMS 4116 series is offered in a 16-pin dual-in-line plastic (NL suffix) package and is guaranteed for operation from 0°C to 70°C. Package is designed for insertion in mounting-hole rows on 300-mil (7.62 mm) centers.

operation

address (A0 through A6)

Fourteen address bits are required to decode 1 of 16,384 storage cell locations. Seven row-address bits are set up on pins A0 through A6 and latched onto the chip by the row-address strobe ($\overline{RAS}$). Then the seven column-address bits are set up on pins A0 through A6 and latched onto the chip by the column-address strobe ($\overline{CAS}$). All addresses must be stable on or before the falling edges of $\overline{RAS}$ and $\overline{CAS}$. $\overline{RAS}$ is similar to a chip enable in that it activates the sense amplifiers as well as the row decoder. $\overline{CAS}$ is used as a chip select activating the column decoder and the input and output buffers.

†The term "read-write cycle" is sometimes used as an alternative title to "read-modify-write cycle".

write enable ($\overline{W}$)

The read or write mode is selected through the write enable ($\overline{W}$) input. A logic high on the $\overline{W}$ input selects the read mode and a logic low selects the write mode. The write enable terminal can be driven from standard TTL circuits without a pull-up resistor. The data input is disabled when the read mode is selected. When $\overline{W}$ goes low prior to $\overline{CAS}$, data-out will remain in the high-impedance state for the entire cycle permitting common I/O operation.

data-in (D)

Data is written during a write or read-modify write cycle. Depending on the mode of operation, the falling edge of $\overline{CAS}$ or $\overline{W}$ strobes data into the on-chip data latch. This latch can be driven from standard TTL circuits without a pull-up resistor. In an early write cycle $\overline{W}$ is brought low prior to $\overline{CAS}$ and the data is strobed in by $\overline{CAS}$ with setup and hold times referenced to this signal. In a delayed write or read-modify write cycle, $\overline{CAS}$ will already be low, thus the data will be strobed in by $\overline{W}$ with setup and hold times referenced to this signal.

data-out (Q)

The three state output buffer provides direct TTL compatibility (no pull-up resistor required) with a fan-out of two Series 74 TTL loads. Data-out is the same polarity as data-in. The output is in the high-impedance (floating) state until $\overline{CAS}$ is brought low. In a read cycle the output goes active after the enable time interval $t_{a(C)}$ that begins with the negative transition of $\overline{CAS}$ as long as $t_{a(R)}$ is satisfied. The output becomes valid after the access time has elapsed and remains valid while $\overline{CAS}$ is low; $\overline{CAS}$ going high returns it to a high-impedance state. In an early write cycle, the output is always in the high-impedance state. In a delayed write or read-modify-write cycle, the output will follow the sequence for the read cycle.

refresh

A refresh operation must be performed at least every two milliseconds to retain data. Since the output buffer is in the high-impedance state unless $\overline{CAS}$ is applied, the $\overline{RAS}$ only refresh sequence avoids any output during refresh. Strobing each of the 128 row addresses (A0 through A6) with $\overline{RAS}$ causes all bits in each row to be refreshed. $\overline{CAS}$ remains high (inactive) for this refresh sequence, thus conserving power.

page mode

Page mode operation allows effectively faster memory access by keeping the same row address and strobing successive column addresses onto the chip. Thus, the time required to setup and strobe sequential row addresses on the same page is eliminated. To extend beyond the 128 column locations on a single RAM, the row address and $\overline{RAS}$ is applied to multiple 16K RAMs $\overline{CAS}$ is decoded to select the proper RAM.

power-up

V_{BB} must be applied to the device either before or at the same time as the other supplies and removed last. Failure to observe this precaution will cause dissipation in excess of the absolute maximum ratings due to internal forward bias conditions. This also applies to system use, where failure of the V_{BB} supply must immediately shut down the other supplies. After power up, eight $\overline{RAS}$ cycles must be performed to achieve proper device operation.

TEXAS INSTRUMENTS
INCORPORATED
POST OFFICE BOX 225012 • DALLAS, TEXAS 75265

logic symbol†

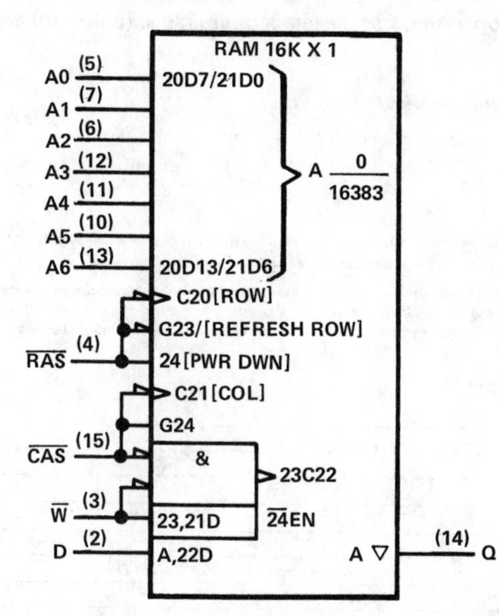

† This symbol is in accordance with IEEE Std 91/ANSI Y32.14 and recent decisions by IEEE and IEC. See explanation on page 289.

funtional block diagram

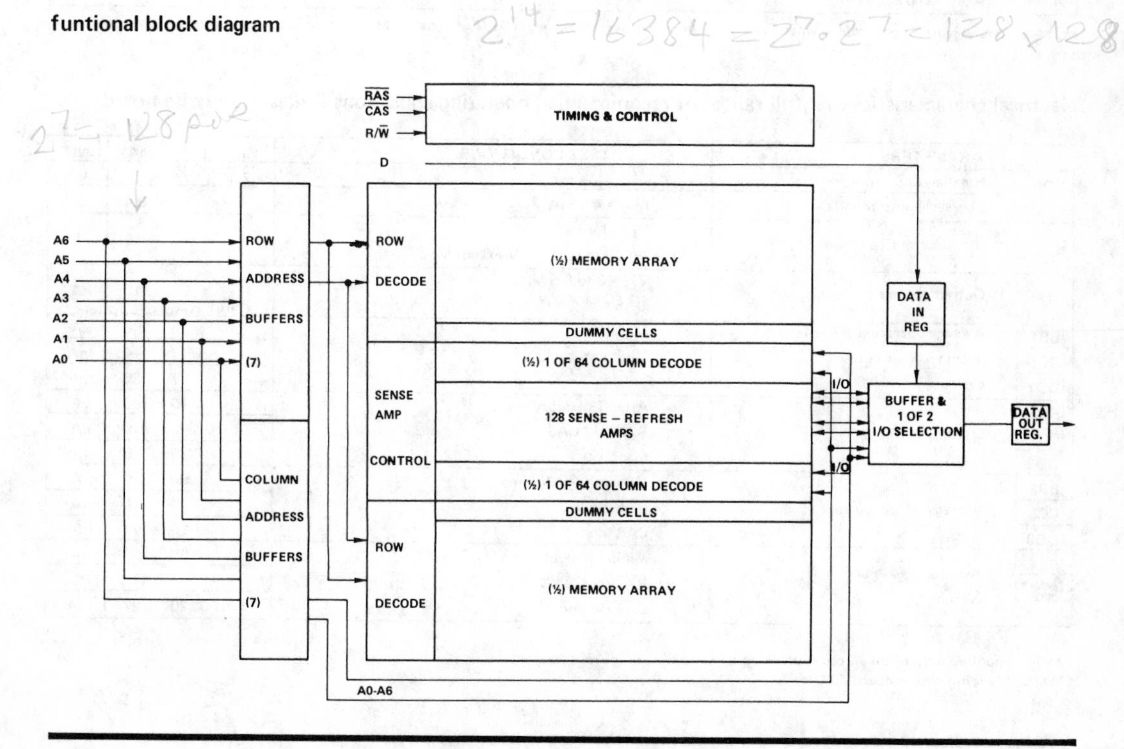

absolute maximum ratings over operating free-air temperature range (unless otherwise noted)*

Voltage on any pin (see Note 1)	−0.5 to 20 V
Voltage on V_{CC}, V_{DD} supplies with respect to V_{SS}	−1 to 15 V
Short circuit output current	50 mA
Power dissipation	1 W
Operating free-air temperature range	0°C to 70°C
Storage temperature range	−65°C to 150°C

NOTE 1: Under absolute maximum ratings, voltage values are with respect to the most-negative supply voltage, V_{BB} (substrate), unless otherwise noted. Throughout the remainder of this data sheet, voltage values are with respect to V_{SS}.

*Stresses beyond those listed under "Absolute Maximum Ratings" may cause permanent damage to the device. This is a stress rating only and functional operation of the device at these or any other conditions beyond those indicated in the "Recommended Operating Conditions" section of this specification is not implied. Exposure to absolute-maximum-rated conditions for extended periods may affect device reliability.

recommended operating conditions

PARAMETER		MIN	NOM	MAX	UNIT
Supply voltage, V_{BB}		−4.5	−5	−5.5	V
Supply voltage, V_{CC}		4.5	5	5.5	V
Supply voltage, V_{DD}		10.8	12	13.2	V
Supply voltage, V_{SS}			0		V
High-level input voltage, V_{IH}	All inputs except $\overline{RAS}$, $\overline{CAS}$, $\overline{WRITE}$	2.4		7	V
	$\overline{RAS}$, $\overline{CAS}$, $\overline{WRITE}$	2.7		7	
Low-level input voltage, V_{IL} (see Note 2)		−1	0	0.8	V
Operating free-air temperature, T_A		0		70	°C

NOTE 2: The algebraic convention, where the more negative (less positive) limit is designated as minimum, is used in this data sheet for logic voltage levels only.

electrical characteristics over full ranges of recommended operating conditions (unless otherwise noted)

PARAMETERS		TEST CONDITIONS	MIN	TYP†	MAX	UNIT
V_{OH}	High-level output voltage	I_{OH} = −5 mA	2.4			V
V_{OL}	Low-level output voltage	I_{OL} = 4.2 mA			0.4	V
I_I	Input current (leakage)	V_I = 0 V to 7 V, All other pins = 0 V except V_{BB} = −5 V			10	μA
I_O	Output current (leakage)	V_O = 0 to 5.5 V, $\overline{CAS}$ high			±10	μA
I_{BB1}	Average operating current during read or write cycle	Minimum cycle time		50	200	μA
I_{CC1}*					4**	mA
I_{DD1}				27	35	mA
I_{BB2}	Standby current	After 1 memory cycle $\overline{RAS}$ and $\overline{CAS}$ high		10	100	μA
I_{CC2}					±10	μA
I_{DD2}				0.5	1.5	mA
I_{BB3}	Average refresh current	Minimum cycle time $\overline{RAS}$ cycling, $\overline{CAS}$ high		50	200	μA
I_{CC3}					±10	μA
I_{DD3}				20	27	mA
I_{BB4}	Average page-mode current	Minimum cycle time $\overline{RAS}$ low, $\overline{CAS}$ cycling		50	200	μA
I_{CC4}*					4**	mA
I_{DD4}				20	27	mA

*V_{CC} is applied only to the output buffer, so I_{CC} depends on output loading.
**Output loading two standard TTL loads.

582

TEXAS INSTRUMENTS
INCORPORATED
POST OFFICE BOX 225012 • DALLAS, TEXAS 75265

capacitance over recommended supply voltage range and operating free-air temperature range, f = 1 MHz

PARAMETER		TYP†	MAX	UNIT
$C_{i(A)}$	Input capacitance, address inputs	4	5	pF
$C_{i(D)}$	Input capacitance, data input	4	5	pF
$C_{i(RC)}$	Input capacitance, strobe inputs	8	10	pF
$C_{i(W)}$	Input capacitance, write enable input	8	10	pF
C_o	Output capacitance	5	7	pF

switching characteristics over recommended supply voltage range and operating free-air temperature range

PARAMETER		TEST CONDITIONS	ALT. SYMBOL	TMS 4116-15		TMS 4116-20		TMS 4116-25		UNIT
				MIN	MAX	MIN	MAX	MIN	MAX	
$t_{a(C)}$	Access time from $\overline{CAS}$	C_L = 100 pF, Load = 2 Series 74 TTL gates	t_{CAC}		100		135		165	ns
$t_{a(R)}$	Access time from $\overline{RAS}$	t_{RLCL} = MAX, C_L = 100 pF Load = 2 Series 74 TTL gates	t_{RAC}		150		200		250	ns
$t_{dis(CH)}$	Output disable time after $\overline{CAS}$ high	C_L = 100 pF, Load = 2 Series 74 TTL gates	t_{OFF}	0	40	0	50	0	60	ns

†All typical values are at $T_A = 25°C$ and nominal supply voltages.

timing requirements over recommended supply voltage range and operating free-air temperature range

PARAMETER		ALT. SYMBOL	TMS4416-15		TMS4416-20		TMS4116-25		UNIT
			MIN	MAX	MIN	MAX	MIN	MAX	
$t_{c(P)}$	Page mode cycle time	t_{PC}	170		225		275		ns
$t_{c(rd)}$	Read cycle time	t_{RC}	375		375		410		ns
$t_{c(W)}$	Write cycle time	t_{WC}	375		375		410		ns
$t_{c(rdW)}$	Read, modify-write cycle time	t_{RWC}	375		375		515		ns
$t_{w(CH)}$	Pulse width, $\overline{CAS}$ high (percharge time)	t_{CP}	60		80		100		ns
$t_{w(CL)}$	Pulse width, $\overline{CAS}$ low	t_{CAS}	100	10,000	135	10,000	165	10,000	ns
$t_{w(RH)}$	Pulse width $\overline{RAS}$ high (precharge time)	t_{RP}	100		120		150		ns
$t_{w(RL)}$	Pulse width, $\overline{RAS}$ low	t_{RAS}	150	10,000	200	10,000	250	10,000	ns
$t_{w(W)}$	Write pulse width	t_{WP}	45		55		75		ns
t_t	Transition times (rise and fall) for $\overline{RAS}$ and $\overline{CAS}$	t_T	3	35	3	50	3	50	ns
$t_{su(CA)}$	Column address setup time	t_{ASC}	−10		−10		−10		ns
$t_{su(RA)}$	Row address setup time	t_{ASR}	0		0		0		ns
$t_{su(D)}$	Data setup time	t_{DS}	0		0		0		ns
$t_{su(rd)}$	Read command setup time	t_{RCS}	0		0		0		ns
$t_{su(WCH)}$	Write command setup time before $\overline{CAS}$ high	t_{CWL}	60		80		100		ns
$t_{su(WRH)}$	Write command setup time before $\overline{RAS}$ high	t_{RWL}	60		80		100		ns
$t_{h(CLCA)}$	Column address hold time after $\overline{CAS}$ low	t_{CAH}	45		55		75		ns
$t_{h(RA)}$	Row address hold time	t_{RAH}	20		25		35		ns
$t_{h(RLCA)}$	Column address hold time after $\overline{RAS}$ low	t_{AR}	95		120		160		ns
$t_{h(CLD)}$	Data hold time after $\overline{CAS}$ low	t_{DH}	45		55		75		ns
$t_{h(RLD)}$	Data hold time after $\overline{RAS}$ low	t_{DHR}	95		120		160		ns
$t_{h(WLD)}$	Data hold time after $\overline{W}$ low	t_{DH}	45		55		75		ns
$t_{h(rd)}$	Read command hold time	t_{RCH}	0		0		0		ns
$t_{h(CLW)}$	Write command hold time after $\overline{CAS}$ low	t_{WCH}	45		55		75		ns
$t_{h(RLW)}$	Write command hold time after $\overline{RAS}$ low	t_{WCR}	95		120		160		ns
t_{RLCH}	Delay time, $\overline{RAS}$ low to $\overline{CAS}$ high	t_{CSH}	150		200		250		ns
t_{CHRL}	Delay time, $\overline{CAS}$ high to $\overline{RAS}$ low	t_{CRP}	−20		−20		−20		ns
t_{CLRH}	Delay time, $\overline{CAS}$ low to $\overline{RAS}$ high	t_{RSH}	100		135		165		ns
t_{CLWL}	Delay time, $\overline{CAS}$ low to $\overline{W}$ low (read, modify-write-cycle only)	t_{CWD}	70		95		125		ns
t_{RLCL}	Delay time, $\overline{RAS}$ low to $\overline{CAS}$ low (maximum value specified only to guarantee access time)	t_{RCD}	20	50	25	65	35	85	ns
t_{RLWL}	Delay time, $\overline{RAS}$ low to $\overline{W}$ low (read, modify-write-cycle only)	t_{RWD}	120		160		200		ns
t_{WLCL}	Delay time, $\overline{W}$ low to $\overline{CAS}$ low (early write cycle)	t_{WCS}	−20		−20		−20		ns
t_{rf}	Refresh time interval	t_{REF}		2		2		2	ms

TEXAS INSTRUMENTS
INCORPORATED
POST OFFICE BOX 225012 • DALLAS, TEXAS 75265

read cycle timing

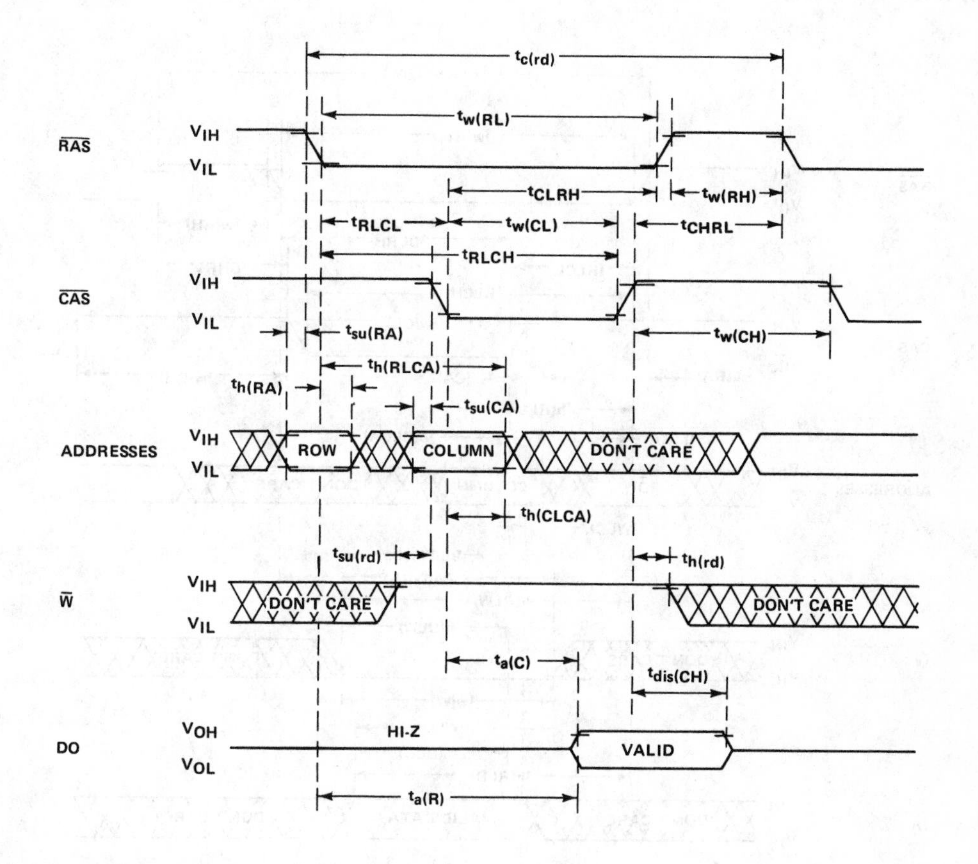

TEXAS INSTRUMENTS
INCORPORATED
POST OFFICE BOX 225012 • DALLAS, TEXAS 75265

early write cycle timing

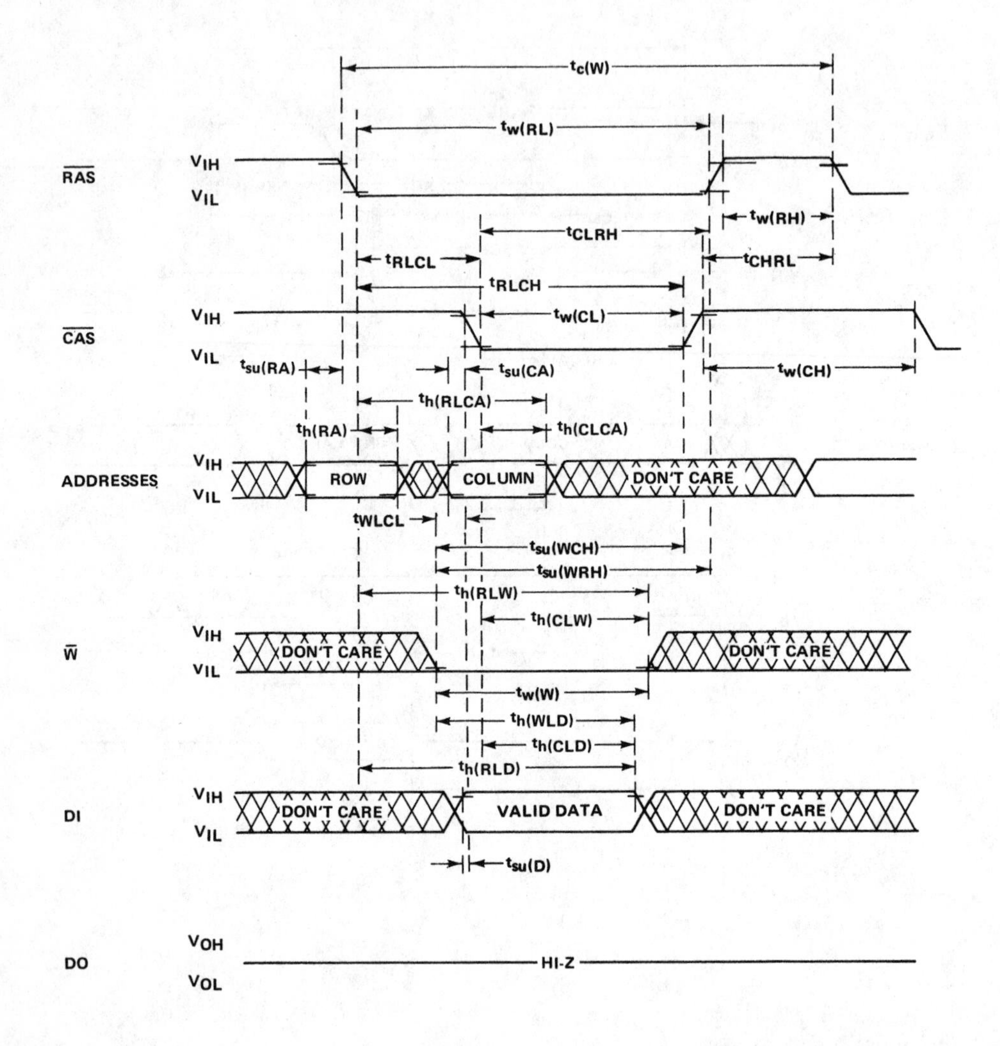

TEXAS INSTRUMENTS
INCORPORATED
POST OFFICE BOX 225012 • DALLAS, TEXAS 75265

write cycle timing

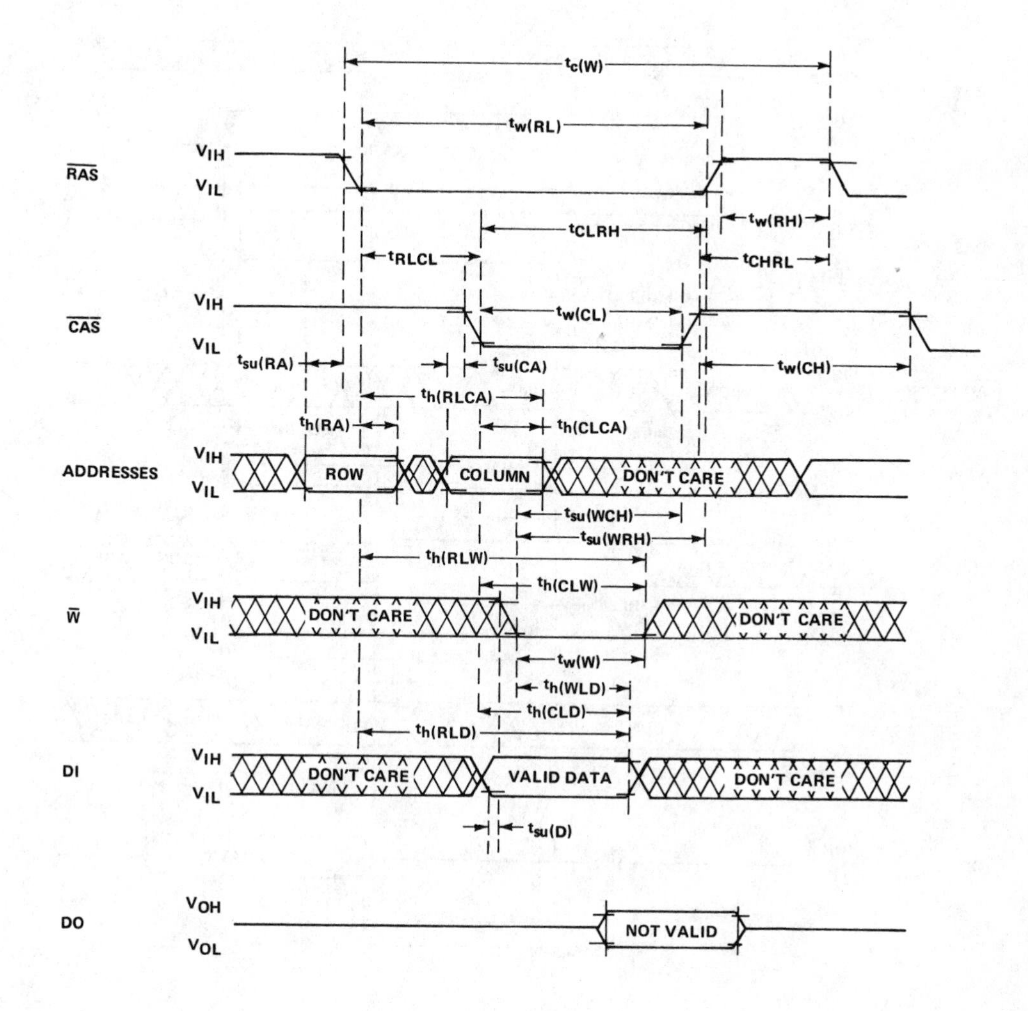

read-write/read-modify-write cycle timing

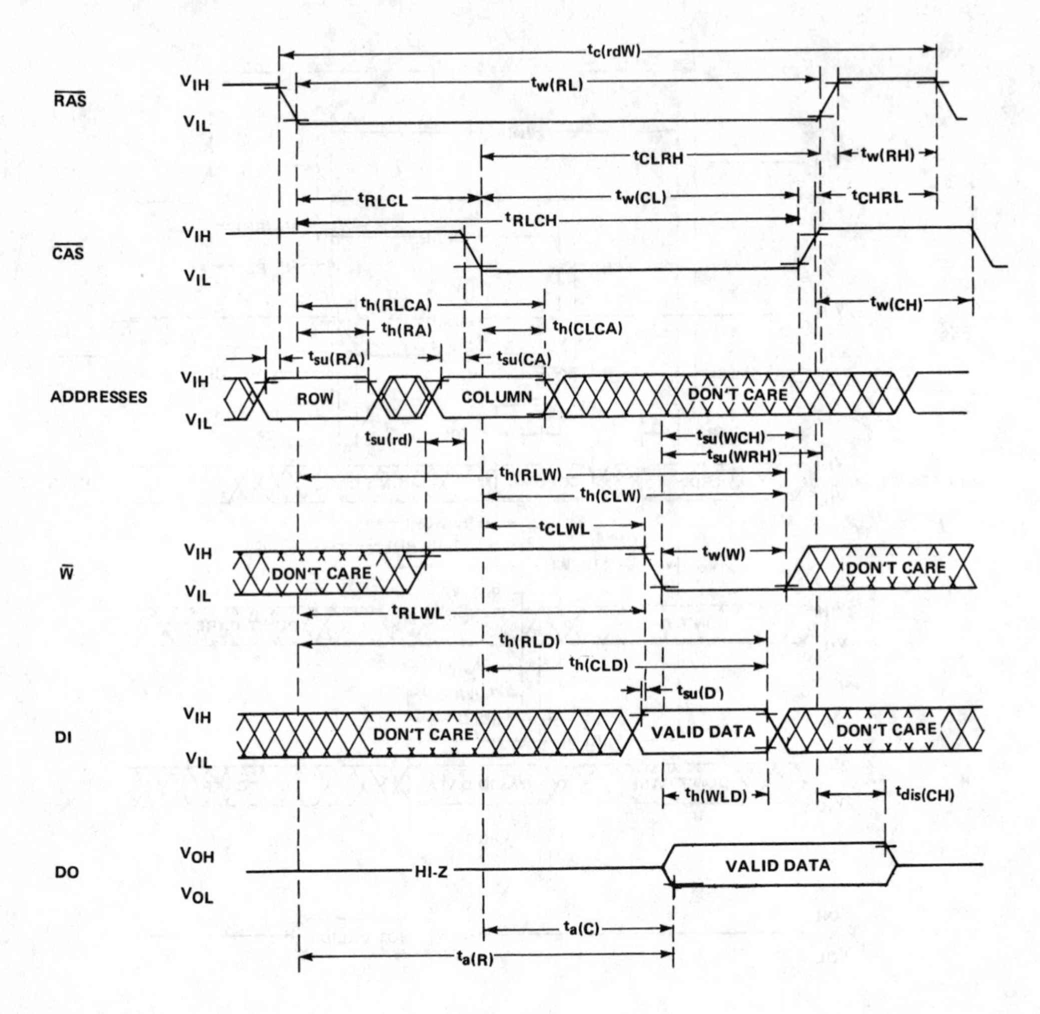

TEXAS INSTRUMENTS
INCORPORATED
POST OFFICE BOX 225012 • DALLAS, TEXAS 75265

page-mode read cycle timing

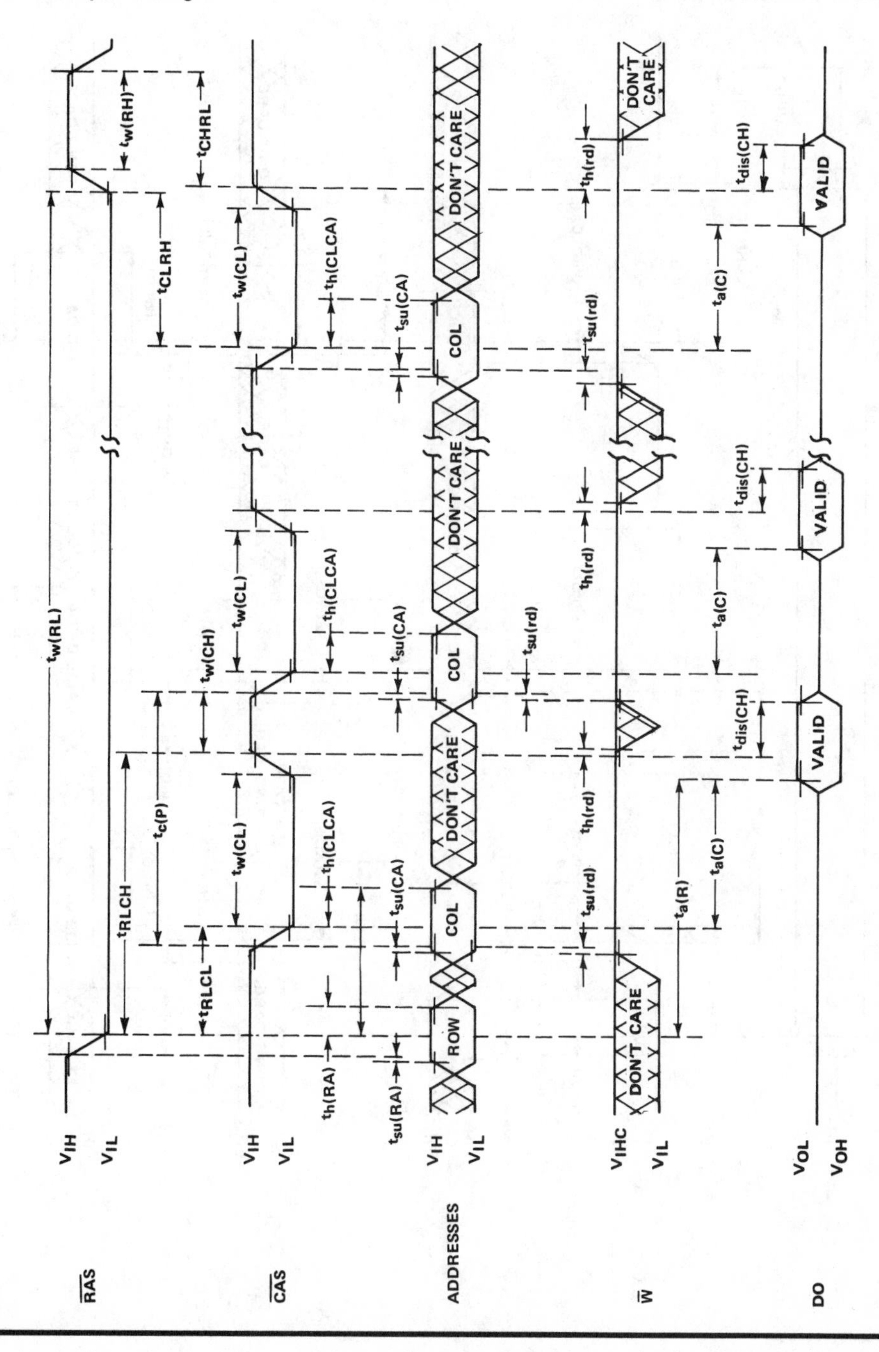

TEXAS INSTRUMENTS
INCORPORATED
POST OFFICE BOX 225012 • DALLAS, TEXAS 75265

page-mode write cycle timing

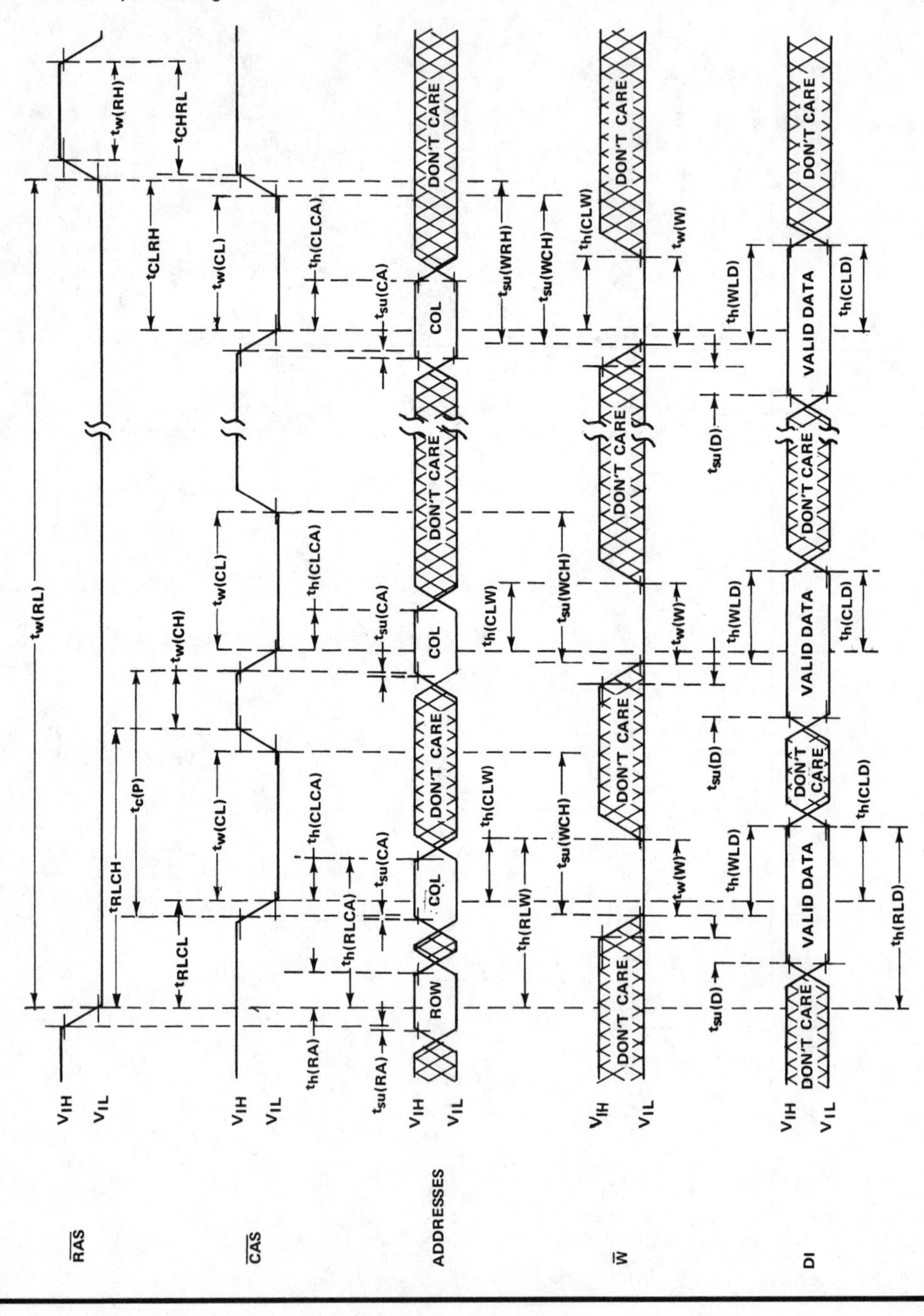

TEXAS INSTRUMENTS
INCORPORATED
POST OFFICE BOX 225012 • DALLAS, TEXAS 75265

$\overline{RAS}$-only refresh timing

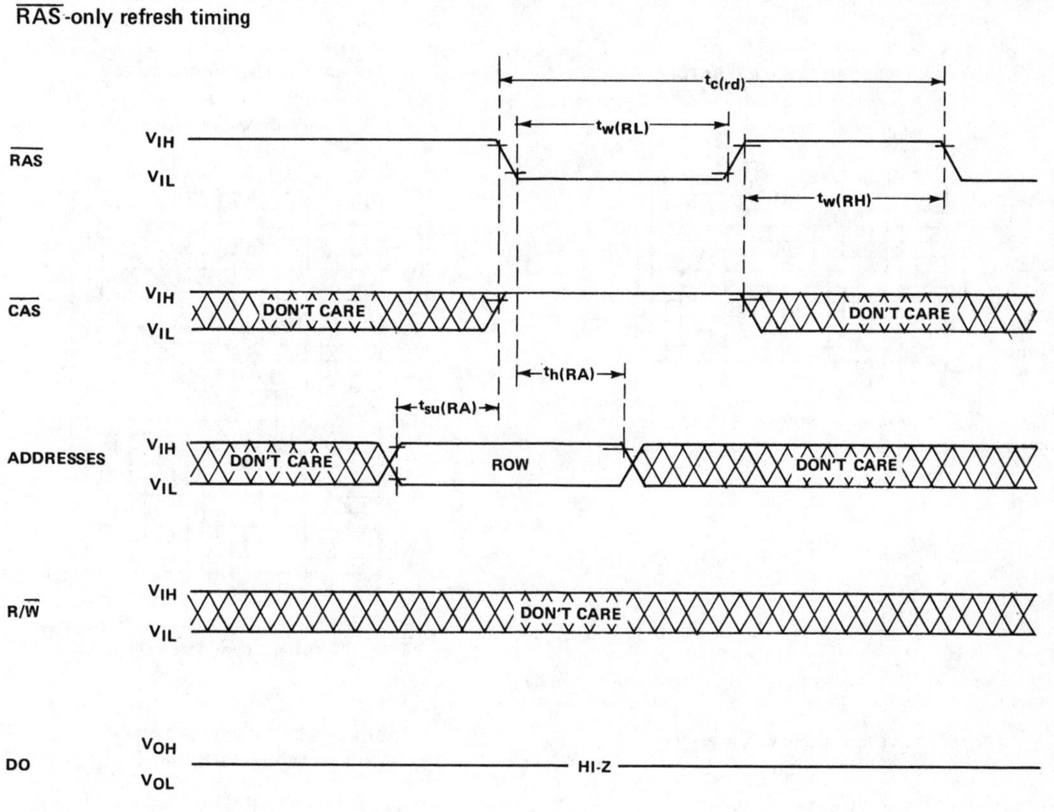

CYCLE RATE (& TIME) VS TEMPERATURE

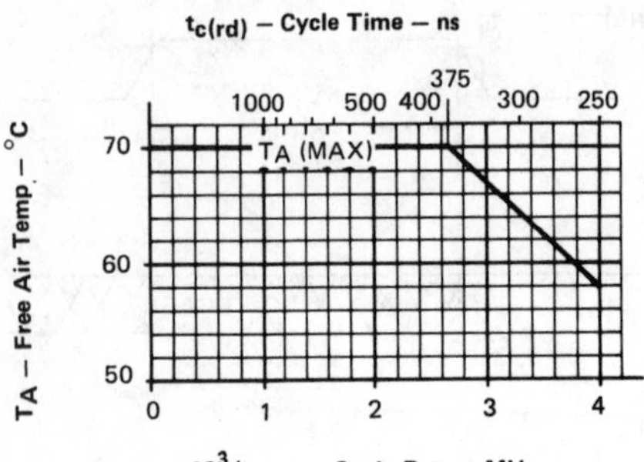

CYCLE RATE (& TIME) vs MAX SUPPLY CURRENT, I_{DD1}

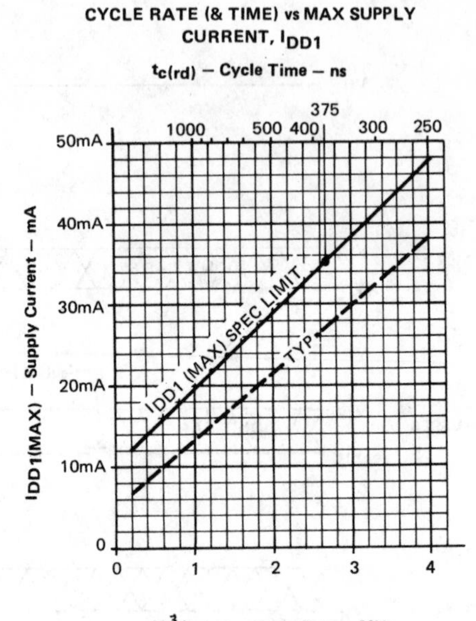

CYCLE RATE (& TIME) vs MAX SUPPLY CURRENT, I_{DD3}

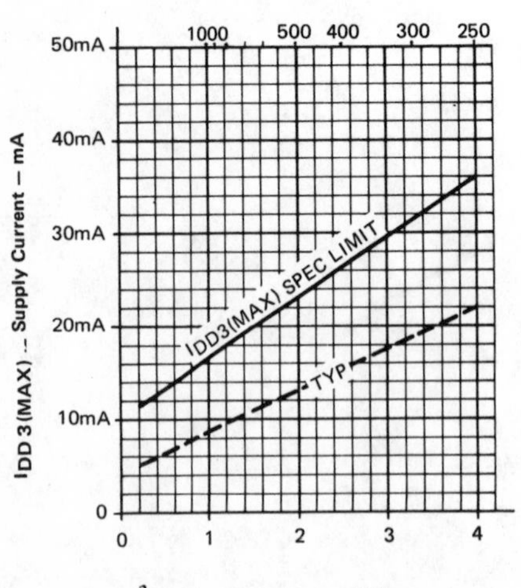

PAGE-MODE CYCLE RATE (& TIME) vs MAX SUPPLY CURRENT, I_{DD4}

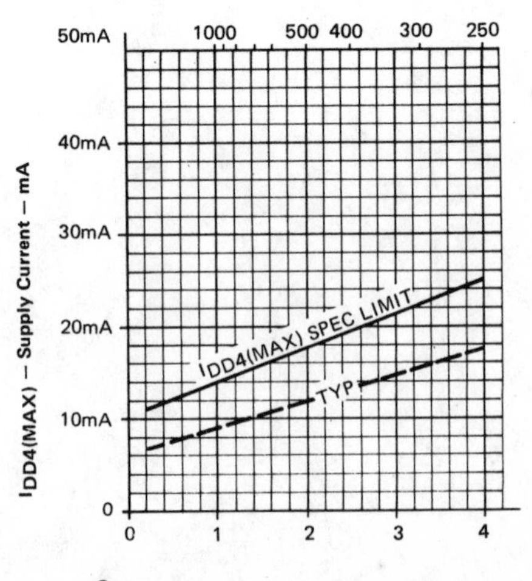

582

MOS LSI

TMS 4164 JDL, NL, FPL
65,536-BIT DYNAMIC RANDOM-ACCESS MEMORY

JULY 1980 — REVISED MAY 1982

- 65,536 X 1 Organization
- Single +5 V Supply (10% Tolerance)
- JEDEC Standardized Pin Out in Dual-In-Line Packages
- Upward Pin Compatible with TMS 4116 (16K Dynamic RAM)
- Performance Ranges:

	ACCESS TIME ROW ADDRESS (MAX)	ACCESS TIME COLUMN ADDRESS (MAX)	READ OR WRITE CYCLE (MIN)	READ, MODIFY, WRITE CYCLE (MIN)
TMS 4164-12	120 ns	75 ns	230 ns	260 ns
TMS 4164-15	150 ns	100 ns	260 ns	285 ns
TMS 4164-20	200 ns	135 ns	330 ns	345 ns
TMS 4164-25	250 ns	165 ns	410 ns	455 ns

- Long Refresh Period . . . 4 milliseconds
- Low Refresh Overhead Time . . . As Low As 1.8% of Total Refresh Period
- All Inputs, Outputs, Clocks Fully TTL Compatible
- 3-State Unlatched Output
- Common I/O Capability with "Early Write" Feature
- Page-Mode Operation for Faster Access
- Low Power Dissipation
  - Operating . . . 125 mW (typ.)
  - Standby . . . 17.5 mW (typ.)
- New SMOS (Scaled-MOS) N-Channel Technology

16-PIN CERAMIC AND PLASTIC DUAL-IN-LINE PACKAGES (TOP VIEW)

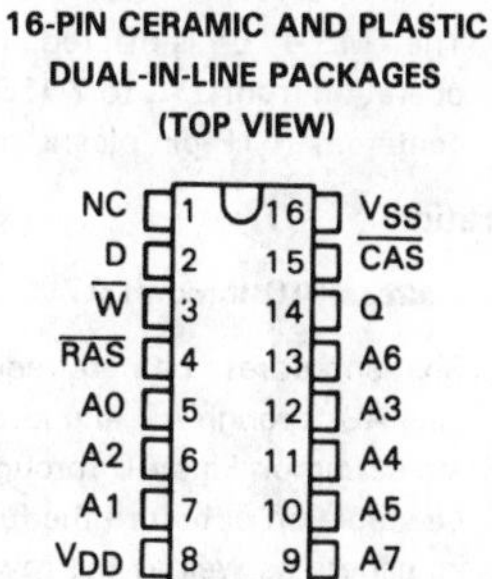

18-PIN PLASTIC CHIP CARRIER PACKAGE (TOP VIEW)

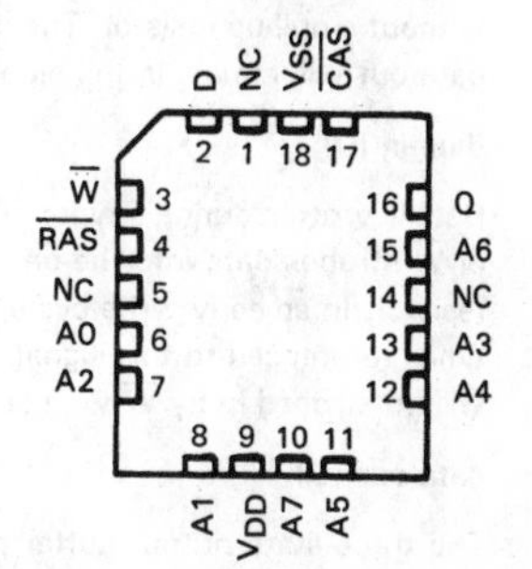

PIN NOMENCLATURE	
A0-A7	Address Inputs
$\overline{\text{CAS}}$	Column Address Strobe
D	Data In
NC	No-Connect
Q	Data Out
$\overline{\text{RAS}}$	Row Address Strobe
$\overline{\text{W}}$	Write Enable
V_{DD}	+5 V Supply
V_{SS}	Ground

description

The TMS 4164 is a high-speed, 65,536-bit, dynamic random-access memory, organized as 65,536 words of one bit each. It employs state-of-the-art SMOS (scaled MOS) N-channel double-level polysilicon gate technology for very high performance combined with low cost and improved reliability.

The TMS 4164 features $\overline{\text{RAS}}$ access times of 120 ns, 150 ns, 200 ns, or 250 ns maximum. Power dissipation is 125 mW typical operating, 17.5 mW typical standby.

Refresh period is extended to 4 milliseconds, and during this period each of the 256 rows must be strobed with $\overline{\text{RAS}}$ in order to retain data. $\overline{\text{CAS}}$ can remain high during the refresh sequence to conserve power.

All inputs and outputs, including clocks, are compatible with Series 74 TTL. All address lines and data-in are latched on chip to simplify system design. Data-out is unlatched to allow greater system flexibility. Pin 1 has no internal connection to allow compatibility with other 64K RAMs that use this pin for an additional function.

POST OFFICE BOX 225012 • DALLAS, TEXAS 75265

The TMS 4164 is offered in a 16-pin dual-in-line ceramic sidebraze package or plastic package and is guaranteed for operation from 0 °C to 70 °C. These packages are designed for insertion in mounting-hole rows on 300 mil (7.62 mm) centers. An 18-pin plastic chip carrier (FP suffix) package is also available.

operation

address (A0 through A7)

Sixteen address bits are required to decode 1 of 65,536 storage cell locations. Eight row-address bits are set up on pins A0 through A7 and latched onto the chip by the row-address strobe ($\overline{RAS}$). Then the eight column-address bits are set up on Pins A0 through A7 and latched onto the chip by the column-address strobe ($\overline{CAS}$). All addresses must be stable on or before the falling edges of $\overline{RAS}$ and $\overline{CAS}$. $\overline{RAS}$ is similar to a chip enable in that it activates the sense amplifiers as well as the row decoder. $\overline{CAS}$ is used as a chip select activating the column decoder and the input and output buffers.

write enable ($\overline{W}$)

The read or write mode is selected through the write enable ($\overline{W}$) input. A logic high on the $\overline{W}$ input selects the read mode and a logic low selects the write mode. The write enable terminal can be driven from standard TTL circuits without a pull-up resistor. The data input is disabled when the read mode is selected. When $\overline{W}$ goes low prior to $\overline{CAS}$, data-out will remain in the high-impedance state for the entire cycle permitting common I/O operation.

data-in (D)

Data is written during a write or read-modify write cycle. Depending on the mode of operation, the falling edge of $\overline{CAS}$ or $\overline{W}$ strobes data into the on-chip data latch. This latch can be driven from standard TTL circuits without a pull-up resistor. In an early-write cycle, $\overline{W}$ is brought low prior to $\overline{CAS}$ and the data is strobed in by $\overline{CAS}$ with setup and hold times referenced to this signal. In a delayed write or read-modify write cycle, $\overline{CAS}$ will already be low, thus the data will be strobed in by $\overline{W}$ with setup and hold times referenced to this signal.

data-out (Q)

The three-state output buffer provides direct TTL compatibility (no pull-up resistor required) with a fan-out of two Series 74 TTL loads. Data-out is the same polarity as data-in. The output is in the high-impedance (floating) state until $\overline{CAS}$ is brought low. In a read cycle the output goes active after the access time interval $t_{a(C)}$ that begins with the negative transition of $\overline{CAS}$ as long as $t_{a(R)}$ is satisfied. The output becomes valid after the access time has elapsed and remains valid while $\overline{CAS}$ is low; $\overline{CAS}$ going high returns it to a high-impedance state. In an early-write cycle, the output is always in the high-impedance state. In a delayed-write or read-modify-write cycle, the output will follow the sequence for the read cycle.

refresh

A refresh operation must be performed at least every four milliseconds to retain data. Since the output buffer is in the high-impedance state unless $\overline{CAS}$ is applied, the $\overline{RAS}$-only refresh sequence avoids any output during refresh. Strobing each of the 256 row addresses (A0 through A7) with $\overline{RAS}$ causes all bits in each row to be refreshed. $\overline{CAS}$ can remain high (inactive) for this refresh sequence to conserve power.

page-mode

Page-mode operation allows effectively faster memory access by keeping the same row address and strobing successive column addresses onto the chip. Thus, the time required to setup and strobe sequential row addresses for the same page is eliminated. To extend beyond the 256 column locations on a single RAM, the row address and $\overline{RAS}$ are applied to multiple 64K RAMs. $\overline{CAS}$ is then decoded to select the proper RAM.

power-up

After power-up, the power supply must remain at its steady-state value for 1 ms. In addition, $\overline{RAS}$ must remain high for 100 μs immediately prior to initialization. Initialization consists of performing eight $\overline{RAS}$ cycles before proper device operation is achieved.

logic symbol†

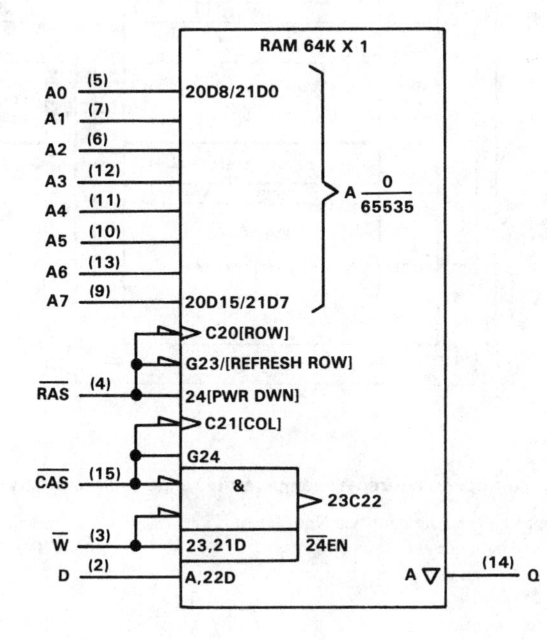

† This symbol is in accordance with IEEE Std 91/ANSI Y32.14 and recent decisions by IEEE and IEC. See explanation on page 289.

functional block diagram

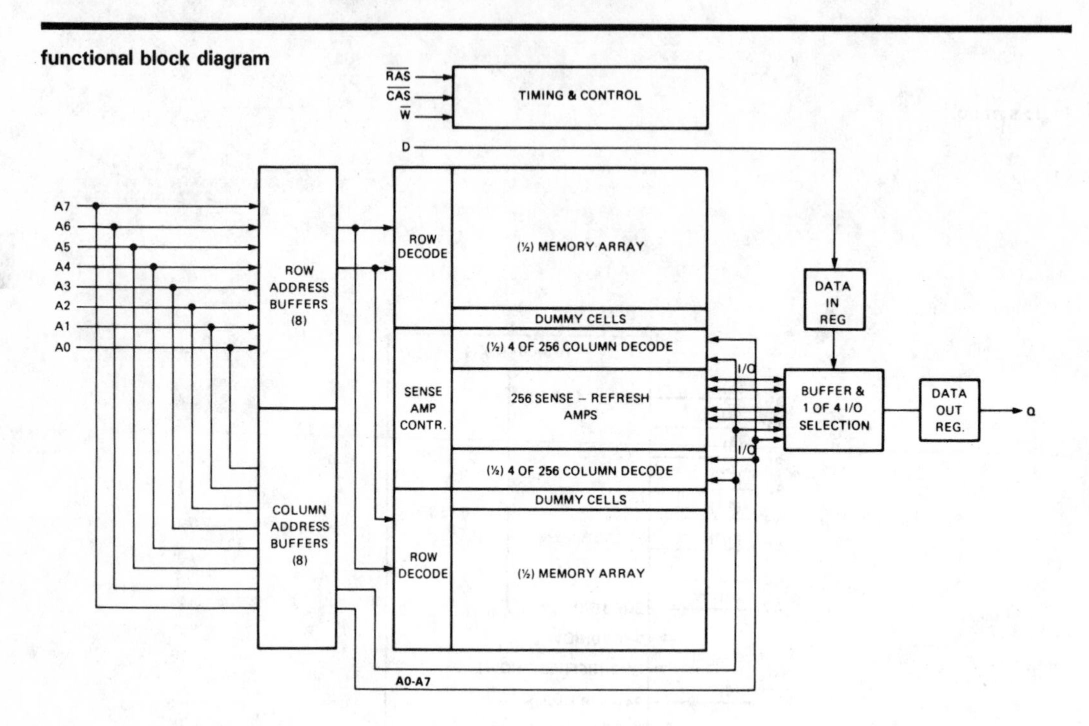

absolute maximum ratings over operating free-air temperature range (unless otherwise noted)*

Voltage on any pin except V_{DD} and data out (see Note 1)	−1.5 to 10 V
Voltage on V_{DD} supply and data out with respect to V_{SS}	−1 to 6 V
Short circuit output current	50 mA
Power dissipation	1 W
Operating free-air temperature range	0°C to 70°C
Storage temperature range	−65°C to 150°C

NOTE 1: All voltage values in this data sheet are with respect to V_{SS}.

* Stresses beyond those listed under "Absolute Maximum Ratings" may cause permanent damage to the device. This is a stress rating only and functional operation of the device at these or any other conditions beyond those indicated in the "Recommended Operating Conditions" section of this specification is not implied. Exposure to absolute-maximum-rated conditions for extended periods may affect device reliability.

recommended operating conditions

PARAMETER	MIN	NOM	MAX	UNIT
Supply voltage, V_{DD}	4.5	5	5.5	V
Supply voltage, V_{SS}		0		V
High-level input voltage, V_{IH}	2.4		V_{DD}+0.3	V
Low-level input voltage, V_{IL} (see Note 2)	−1		0.8	V
Operating free-air temperature, T_A	0		70	°C

NOTE 2: The algebraic convention, where the more negative (less positive) limit is designated as minimum, is used in this data sheet for logic voltage levels only.

TEXAS INSTRUMENTS
INCORPORATED
POST OFFICE BOX 225012 • DALLAS, TEXAS 75265

electrical characteristics over full ranges of recommended operating conditions (unless otherwise noted)

PARAMETER		TEST CONDITIONS	TMS 4164-12			TMS 4164-15			UNIT
			MIN	TYP†	MAX	MIN	TYP†	MAX	
V_{OH}	High-level output voltage	I_{OH} = −5 mA	2.4			2.4			V
V_{OL}	Low-level output voltage	I_{OL} = 4.2 mA			0.4			0.4	V
I_I	Input current (leakage)	V_I=0 V to 5.8 V, V_{DD}=5 V, All other pins = 0 V			±10			±10	μA
I_O	Output current (leakage)	V_O = 0.4 to 5.5 V, V_{DD} = 5 V, $\overline{CAS}$ high			±10			±10	μA
I_{DD1}*	Average operating current during read or write cycle	t_c = minimum cycle		35	45		28	39	mA
I_{DD2}**	Standby current	After 1 memory cycle, $\overline{RAS}$ and $\overline{CAS}$ high		3.5	5		3.5	5	mA
I_{DD3}*	Average refresh current	t_c = minimum cycle, $\overline{RAS}$ low, $\overline{CAS}$ high		25	35		22	30	mA
I_{DD4}	Average page-mode current	$t_{c(P)}$ = minimum cycle, $\overline{RAS}$ low, $\overline{CAS}$ cycling		25	35		22	30	mA

PARAMETER		TEST CONDITIONS	TMS 4164-20			TMS 4164-25			UNIT
			MIN	TYP†	MAX	MIN	TYP†	MAX	
V_{OH}	High-level output voltage	I_{OH} = −5 mA	2.4			2.4			V
V_{OL}	Low-level output voltage	I_{OL} = 4.2 mA			0.4			0.4	V
I_I	Input current (leakage)	V_I=0 V to 5.8 V, V_{DD}=5 V, All other pins = 0 V			±10			±10	μA
I_O	Output current (leakage)	V_O = 0.4 to 5.5 V, V_{DD} = 5 V, $\overline{CAS}$ high			±10			±10	μA
I_{DD1}*	Average operating current during read or write cycle	t_c = minimum cycle		24	34		21	29	mA
I_{DD2}**	Standby current	After 1 memory cycle, $\overline{RAS}$ and $\overline{CAS}$ high		3.5	5		3.5	5	mA
I_{DD3}*	Average refresh current	t_c = minimum cycle, $\overline{RAS}$ low, $\overline{CAS}$ high		19	26		16	22	mA
I_{DD4}	Average page-mode current	$t_{c(P)}$ = minimum cycle, $\overline{RAS}$ low, $\overline{CAS}$ cycling		19	26		16	22	mA

† All typical values are at T_A = 25°C and nominal supply voltages.
* Additional information on last page.
** $V_{IL} > -0.6$ V.

capacitance over recommended supply voltage range and operating free-air temperature range, f = 1 MHz

PARAMETER		TYP†	MAX	UNIT
$C_{i(A)}$	Input capacitance, address inputs	4	7	pF
$C_{i(D)}$	Input capacitance, data input	4	7	pF
$C_{i(RC)}$	Input capacitance strobe inputs	8	10	pF
$C_{i(W)}$	Input capacitance, write enable input	8	10	pF
C_o	Output capacitance	5	8	pF

† All typical values are at T_A = 25 °C and nominal supply voltages.

switching characteristics over recommended supply voltage range and operating free-air temperature range

PARAMTER		TEST CONDITIONS	ALT. SYMBOL	TMS 4164-12		TMS 4164-15		UNIT
				MIN	MAX	MIN	MAX	
$t_{a(C)}$	Access time from $\overline{CAS}$	C_L = 100 pF, Load = 2 Series 74 TTL gates	t_{CAC}		75		100	ns
$t_{a(R)}$	Access time from $\overline{RAS}$	t_{RLCL} = MAX, Load = 2 Series 74 TTL gates	t_{RAC}		120		150	ns
$t_{dis(CH)}$	Output disable time after $\overline{CAS}$ high	C_L = 100 pF, Load = 2 Series 74 TTL gates	t_{OFF}	0	40	0	40	ns

PARAMETER		TEST CONDITIONS	ALT. SYMBOL	TMS 4164-20		TMS 4164-25		UNIT
				MIN	MAX	MIN	MAX	
$t_{a(C)}$	Access time from $\overline{CAS}$	C_L = 100 pF, Load = 2 Series 74 TTL gates	t_{CAC}		135		165	ns
$t_{a(R)}$	Access time from $\overline{RAS}$	t_{RLCL} = MAX, Load = 2 Series 74 TTL gates	t_{RAC}		200		250	ns
$t_{dis(CH)}$	Output disable time after $\overline{CAS}$ high	C_L = 100 pF, Load = 2 Series 74 TTL gates	t_{OFF}	0	50	0	60	ns

TEXAS INSTRUMENTS
INCORPORATED
POST OFFICE BOX 225012 • DALLAS, TEXAS 75265

timing requirements over recommended supply voltage range and operating free-air temperature range

PARAMETER		ALT. SYMBOL	TMS 4164-12		TMS 4164-15		UNIT
			MIN	MAX	MIN	MAX	
$t_{c(P)}$	Page mode cycle time	t_{PC}	140		160		ns
$t_{c(rd)}$	Read cycle time*	t_{RC}	230		260		ns
$t_{c(W)}$	Write cycle time	t_{WC}	230		260		ns
$t_{c(rdW)}$	Read-write/read-modify-write cycle time	t_{RWC}	260		285		ns
$t_{w(CH)}$	Pulse width, $\overline{CAS}$ high (precharge time)**	t_{CP}	50		50		ns
$t_{w(CL)}$	Pulse width, $\overline{CAS}$ low†	t_{CAS}	75	10,000	100	10,000	ns
$t_{w(RH)}$	Pulse width, $\overline{RAS}$ high (precharge time)	t_{RP}	100		100		ns
$t_{w(RL)}$	Pulse width, $\overline{RAS}$ low‡	t_{RAS}	120	10,000	150	10,000	ns
$t_{w(W)}$	Write pulse width	t_{WP}	45		45		ns
t_t	Transition times (rise and fall) for $\overline{RAS}$ and $\overline{CAS}$	t_T	3	50	3	50	ns
$t_{su(CA)}$	Column address setup time	t_{ASC}	0		−5		ns
$t_{su(RA)}$	Row address setup time	t_{ASR}	0		0		ns
$t_{su(D)}$	Data setup time	t_{DS}	0		0		ns
$t_{su(rd)}$	Read command setup time	t_{RCS}	0		0		ns
$t_{su(WCH)}$	Write command setup time before $\overline{CAS}$ high	t_{CWL}	50		60		ns
$t_{su(WRH)}$	Write command setup time before $\overline{RAS}$ high	t_{RWL}	50		60		ns
$t_{h(CLCA)}$	Column address hold time after $\overline{CAS}$ low	t_{CAH}	45		45		ns
$t_{h(RA)}$	Row address hold time	t_{RAH}	15		20		ns
$t_{h(RLCA)}$	Column address hold time after $\overline{RAS}$ low	t_{AR}	90		95		ns
$t_{h(CLD)}$	Data hold time after $\overline{CAS}$ low	t_{DH}	50		60		ns
$t_{h(RLD)}$	Data hold time after $\overline{RAS}$ low	t_{DHR}	95		110		ns
$t_{h(WLD)}$	Data hold time after $\overline{W}$ low	t_{DH}	45		45		ns
$t_{h(CHrd)}$	Read command hold time after $\overline{CAS}$ high	t_{RCH}	0		0		ns
$t_{h(RHrd)}$	Read command hold time after $\overline{RAS}$ high	t_{RRH}	5		5		ns
$t_{h(CLW)}$	Write command hold time after $\overline{CAS}$ low	t_{WCH}	50		60		ns
$t_{h(RLW)}$	Write command hold time after $\overline{RAS}$ low	t_{WCR}	95		110		ns
t_{RLCH}	Delay time, $\overline{RAS}$ low to $\overline{CAS}$ high	t_{CSH}	120		150		ns
t_{CHRL}	Delay time, $\overline{CAS}$ high to $\overline{RAS}$ low	t_{CRP}	0		0		ns
t_{CLRH}	Delay time, $\overline{CAS}$ low to $\overline{RAS}$ high	t_{RSH}	80		100		ns
t_{CLWL}	Delay time, $\overline{CAS}$ low to $\overline{W}$ low (read, modify-write-cycle only)	t_{CWD}	50		60		ns
t_{RLCL}	Delay time, $\overline{RAS}$ low to $\overline{CAS}$ low (maximum value specified only to guarantee access time)	t_{RCD}	15	45	20	50	ns
t_{RLWL}	Delay time, $\overline{RAS}$ low to $\overline{W}$ low (read, modify-write-cycle only)	t_{RWD}	95		110		ns
t_{WLCL}	Delay time, $\overline{W}$ low to $\overline{CAS}$ low (early write cycle)	t_{WCS}	−5		−5		ns
t_{rf}	Refresh time interval	t_{REF}		4		4	ms

NOTE: Timing measurements are made at the 10% and 90% points of input and clock transitions. In addition, V_{IL} max and V_{IH} min must be met at the 10% and 90% points.

* All cycle times assume t_t = 5 ns.

** Page mode only.

† In a read-modify-write cycle, t_{CLWL} and $t_{su(WCH)}$ must be observed. Depending on the user's transition times, this may require additional $\overline{CAS}$ low time ($t_{w(CL)}$). This applies to page mode read-modify-write also.

‡ In a read-modify-write cycle, t_{RLWL} and $t_{su(WRH)}$ must be observed. Depending on the user's transition times, this may require additional $\overline{RAS}$ low time ($t_{w(RL)}$).

timing requirements over recommended supply voltage range and operating free-air temperature range

PARAMETER		ALT. SYMBOL	TMS 4164-20		TMS 4164-25		UNIT
			MIN	MAX	MIN	MAX	
$t_{c(P)}$	Page mode cycle time	t_{PC}	225		275		ns
$t_{c(rd)}$	Read cycle time*	t_{RC}	330		410		ns
$t_{c(W)}$	Write cycle time	t_{WC}	330		410		ns
$t_{c(rdW)}$	Read-write/read-modify-write cycle time	t_{RWC}	345		455		ns
$t_{w(CH)}$	Pulse width, $\overline{CAS}$ high (precharge time)**	t_{CP}	80		100		ns
$t_{w(CL)}$	Pulse width, $\overline{CAS}$ low†	t_{CAS}	135	10,000	165	10,000	ns
$t_{w(RH)}$	Pulse width, $\overline{RAS}$ high (precharge time)	t_{RP}	120		150		ns
$t_{w(RL)}$	Pulse width, $\overline{RAS}$ low‡	t_{RAS}	200	10,000	250	10,000	ns
$t_{w(W)}$	Write pulse width	t_{WP}	55		75		ns
t_t	Transition times (rise and fall) for $\overline{RAS}$ and $\overline{CAS}$	t_T	3	50	3	50	ns
$t_{su(CA)}$	Column address setup time	t_{ASC}	−5		−5		ns
$t_{su(RA)}$	Row address setup time	t_{ASR}	0		0		ns
$t_{su(D)}$	Data setup time	t_{DS}	0		0		ns
$t_{su(rd)}$	Read command setup time	t_{RCS}	0		0		ns
$t_{su(WCH)}$	Write command setup time before $\overline{CAS}$ high	t_{CWL}	80		100		ns
$t_{su(WRH)}$	Write command setup time before $\overline{RAS}$ high	t_{RWL}	80		100		ns
$t_{h(CLCA)}$	Column address hold time after $\overline{CAS}$ low	t_{CAH}	55		75		ns
$t_{h(RA)}$	Row address hold time	t_{RAH}	25		35		ns
$t_{h(RLCA)}$	Column address hold time after $\overline{RAS}$ low	t_{AR}	140		190		ns
$t_{h(CLD)}$	Data hold time after $\overline{CAS}$ low	t_{DH}	80		110		ns
$t_{h(RLD)}$	Data hold time after $\overline{RAS}$ low	t_{DHR}	145		195		ns
$t_{h(WLD)}$	Data hold time after $\overline{W}$ low	t_{DH}	55		75		ns
$t_{h(CHrd)}$	Read command hold time after $\overline{CAS}$ high	t_{RCH}	0		0		ns
$t_{h(RHrd)}$	Read command hold time after $\overline{RAS}$ high	t_{RRH}	5		5		ns
$t_{h(CLW)}$	Write command hold time after $\overline{CAS}$ low	t_{WCH}	80		110		ns
$t_{h(RLW)}$	Write command hold time after $\overline{RAS}$ low	t_{WCR}	145		195		ns
t_{RLCH}	Delay time, $\overline{RAS}$ low to $\overline{CAS}$ high	t_{CSH}	200		250		ns
t_{CHRL}	Delay time, $\overline{CAS}$ high to $\overline{RAS}$ low	t_{CRP}	0		0		ns
t_{CLRH}	Delay time, $\overline{CAS}$ low to $\overline{RAS}$ high	t_{RSH}	135		165		ns
t_{CLWL}	Delay time, $\overline{CAS}$ low to $\overline{W}$ low (read, modify-write-cycle only)	t_{CWD}	65		105		ns
t_{RLCL}	Delay time, $\overline{RAS}$ low to $\overline{CAS}$ low (maximum value specified only to guarantee access time)	t_{RCD}	25	65	35	85	ns
t_{RLWL}	Delay time, $\overline{RAS}$ low to $\overline{W}$ low (read, modify-write-cycle only)	t_{RWD}	130		190		ns
t_{WLCL}	Delay time, $\overline{W}$ low to $\overline{CAS}$ low (early write cycle)	t_{WCS}	−5		−5		ns
t_{rf}	Refresh time interval	t_{REF}		4		4	ms

NOTE: Timing measurements are made at the 10% and 90% points of input and clock transitions. In addition, V_{IL} max and V_{IH} min must be met at the 10% and 90% points.

* All cycle times assume $t_t = 5$ ns.

** Page mode only.

† In a read-modify-write cycle, t_{CLWL} and $t_{su(WCH)}$ must be observed. Depending on the user's transition times, this may require additional $\overline{CAS}$ low time ($t_{w(CL)}$). This applies to page mode read-modify-write also.

‡ In a read-modify-write cycle, t_{RLWL} and $t_{su(WRH)}$ must be observed. Depending on the user's transition times, this may require additional $\overline{RAS}$ low time ($t_{w(RL)}$).

read cycle timing

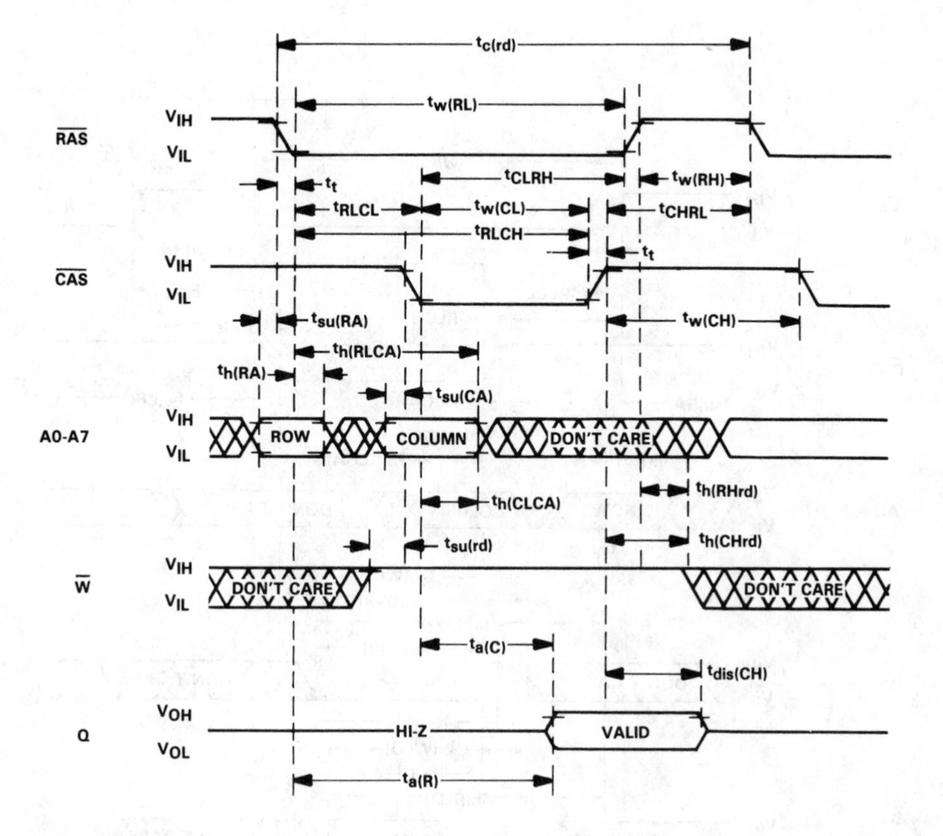

3

early write cycle timing

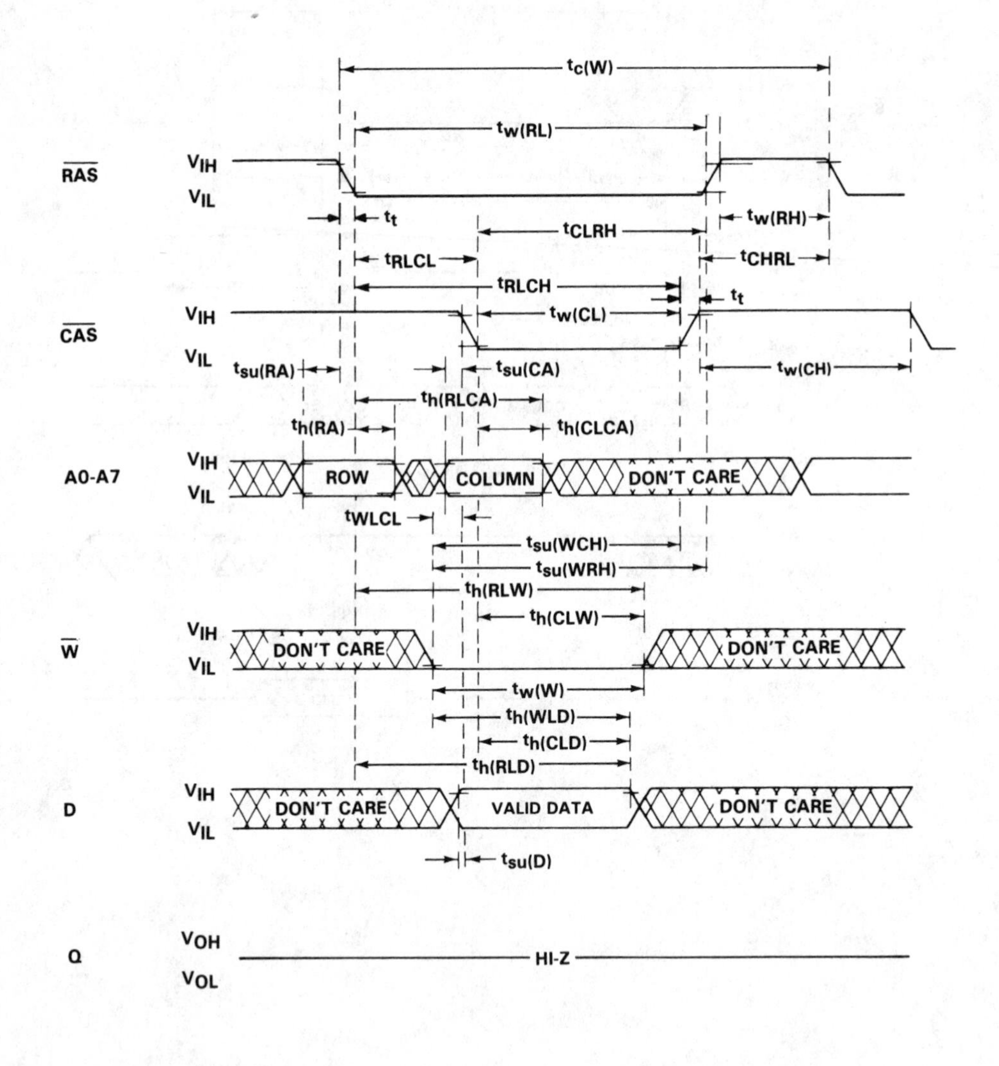

write cycle timing

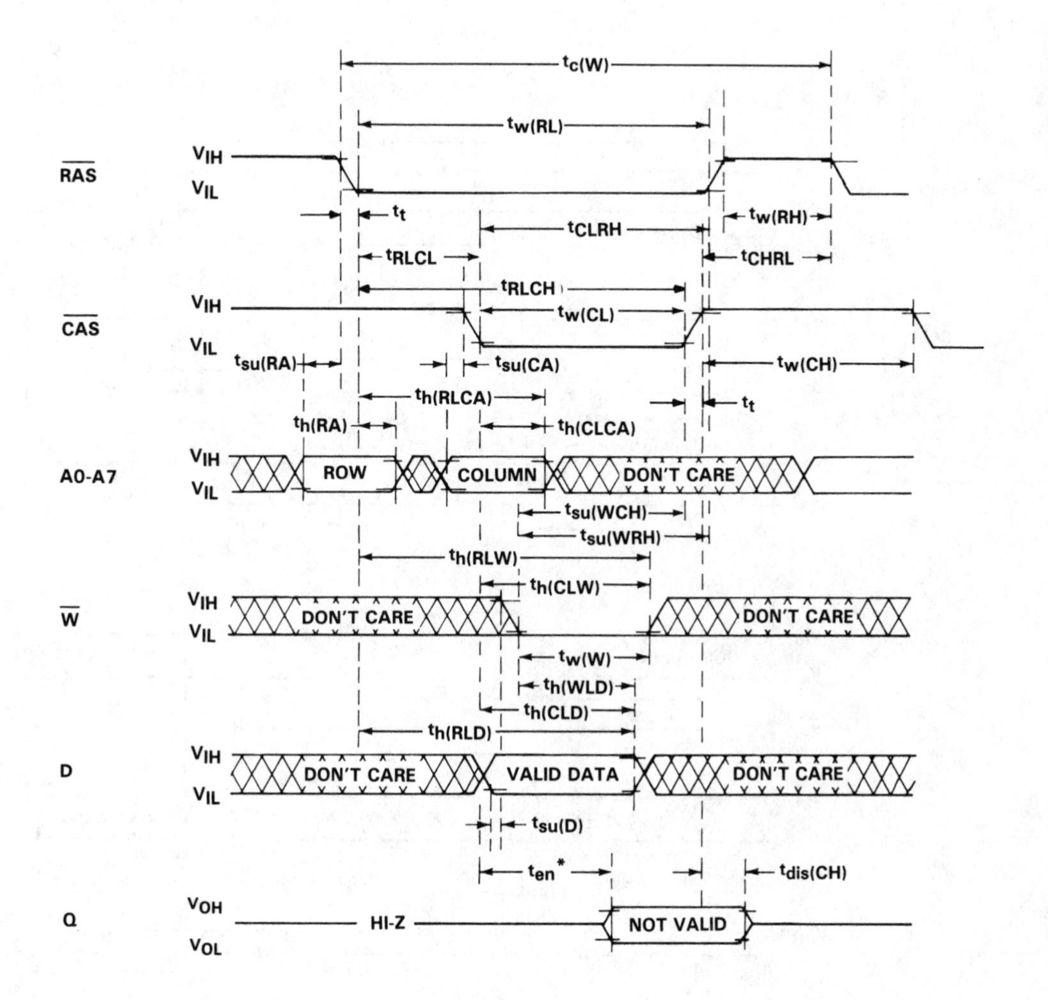

* The enable time (t_{en}) for a write cycle is equal in duration to the access time from $\overline{CAS}$ ($t_{a(C)}$) in a read cycle; but the active levels at the output are invalid.

read-write/read-modify-write cycle timing

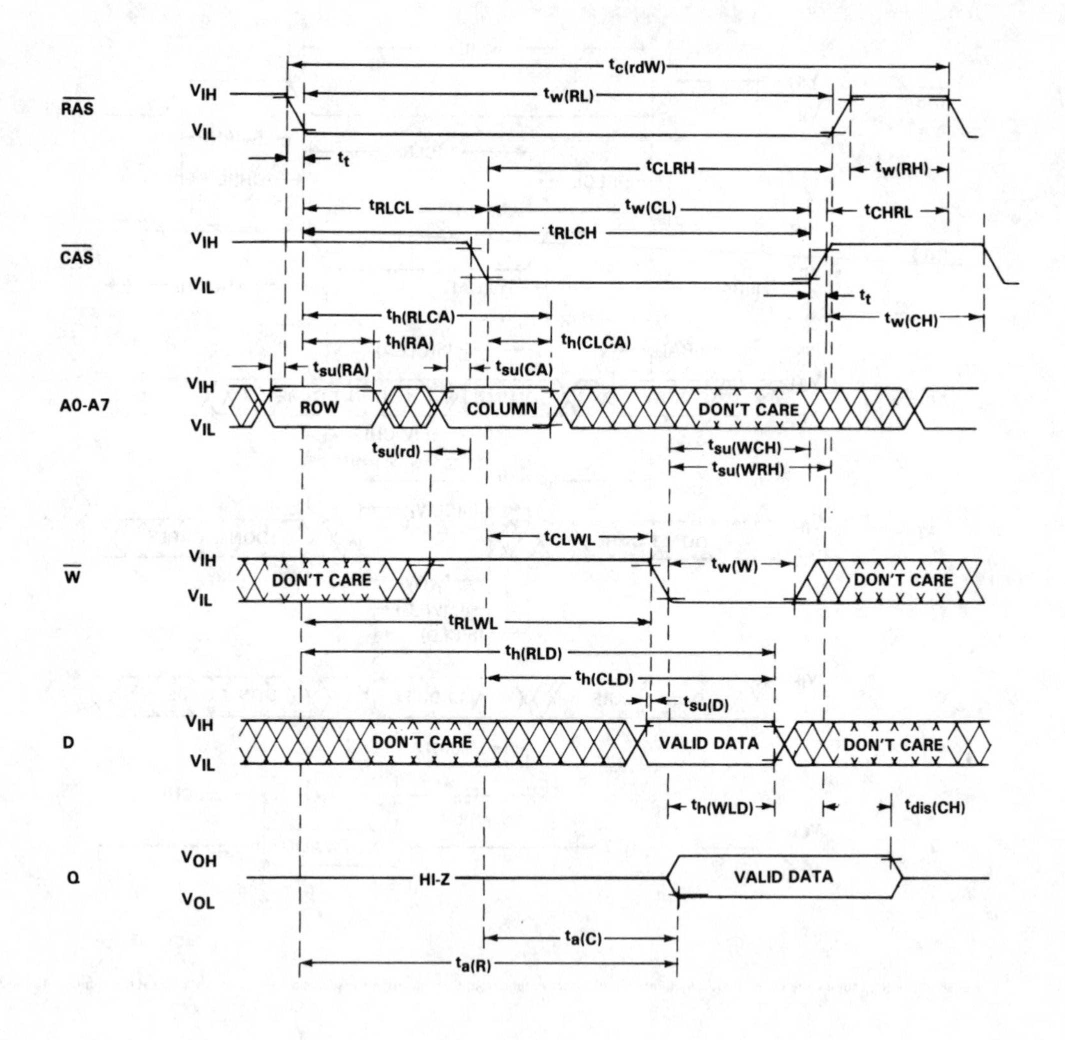

TEXAS INSTRUMENTS
INCORPORATED
POST OFFICE BOX 225012 • DALLAS, TEXAS 75265

page-mode read cycle timing

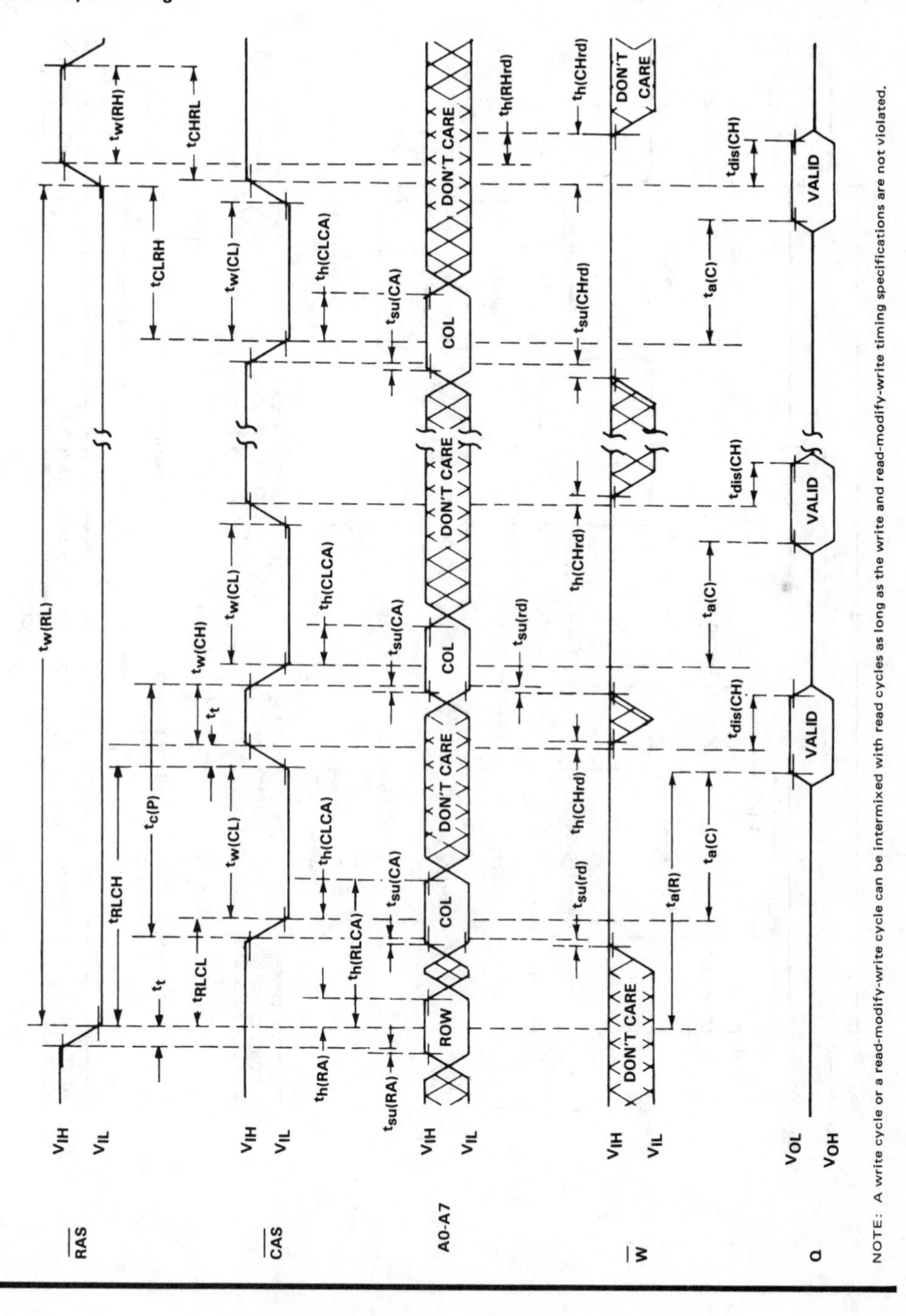

page-mode write cycle timing

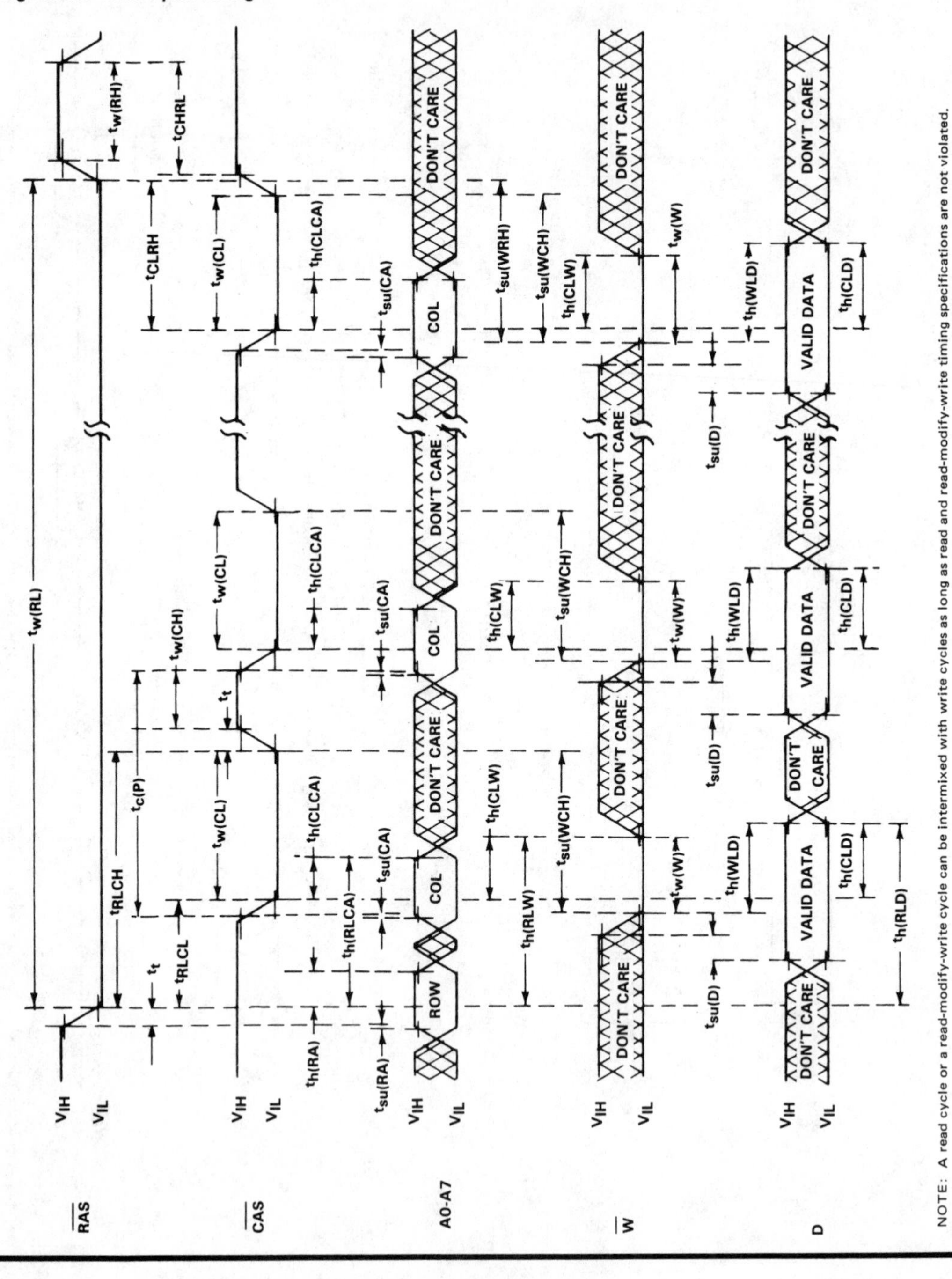

NOTE: A read cycle or a read-modify-write cycle can be intermixed with write cycles as long as read and read-modify-write timing specifications are not violated.

TEXAS INSTRUMENTS
INCORPORATED
POST OFFICE BOX 225012 • DALLAS, TEXAS 75265

page-mode read-modify-write cycle timing

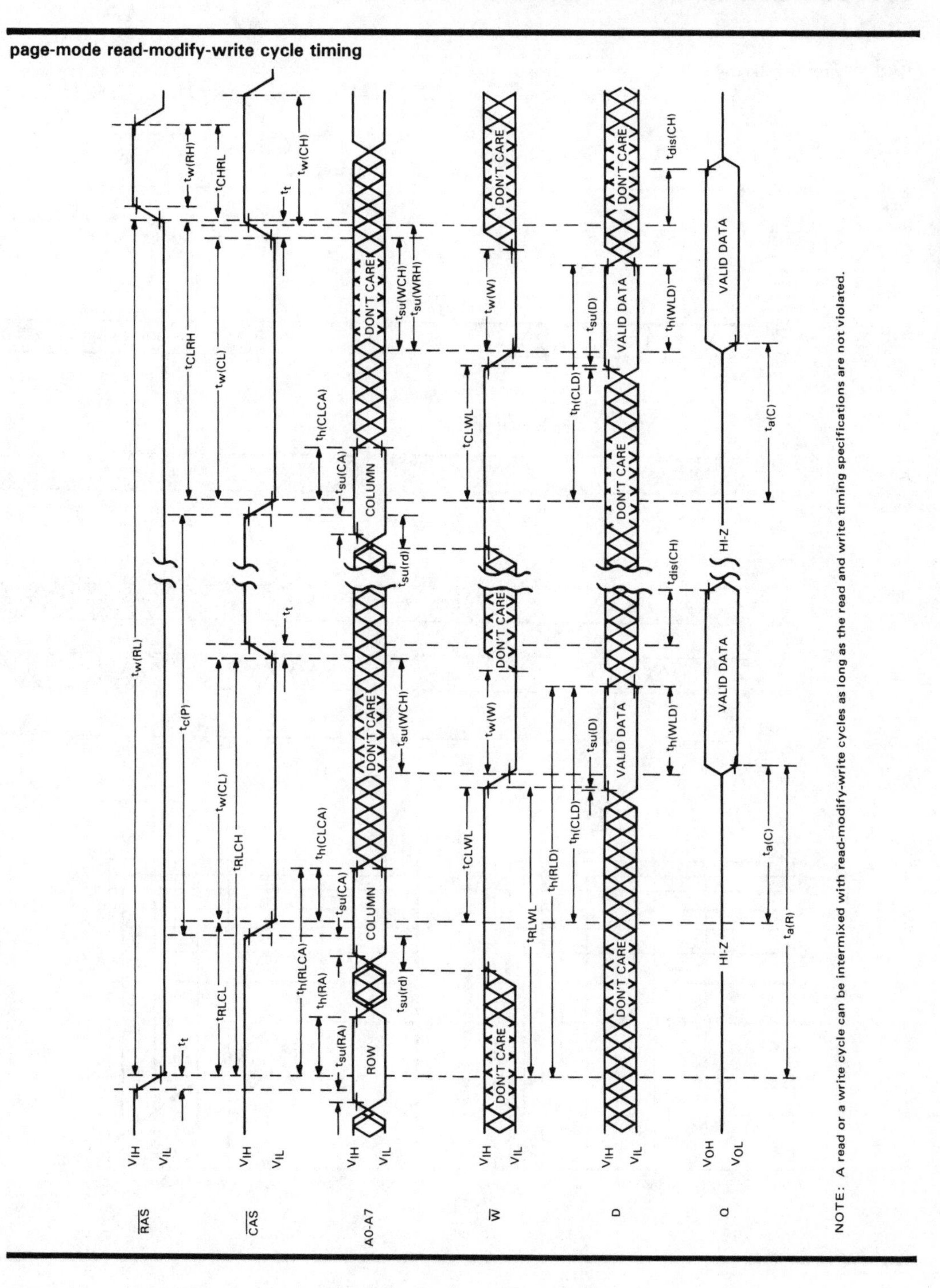

NOTE: A read or a write cycle can be intermixed with read-modify-write cycles as long as the read and write timing specifications are not violated.

TEXAS INSTRUMENTS
INCORPORATED
POST OFFICE BOX 225012 • DALLAS, TEXAS 75265

RAS-only refresh timing

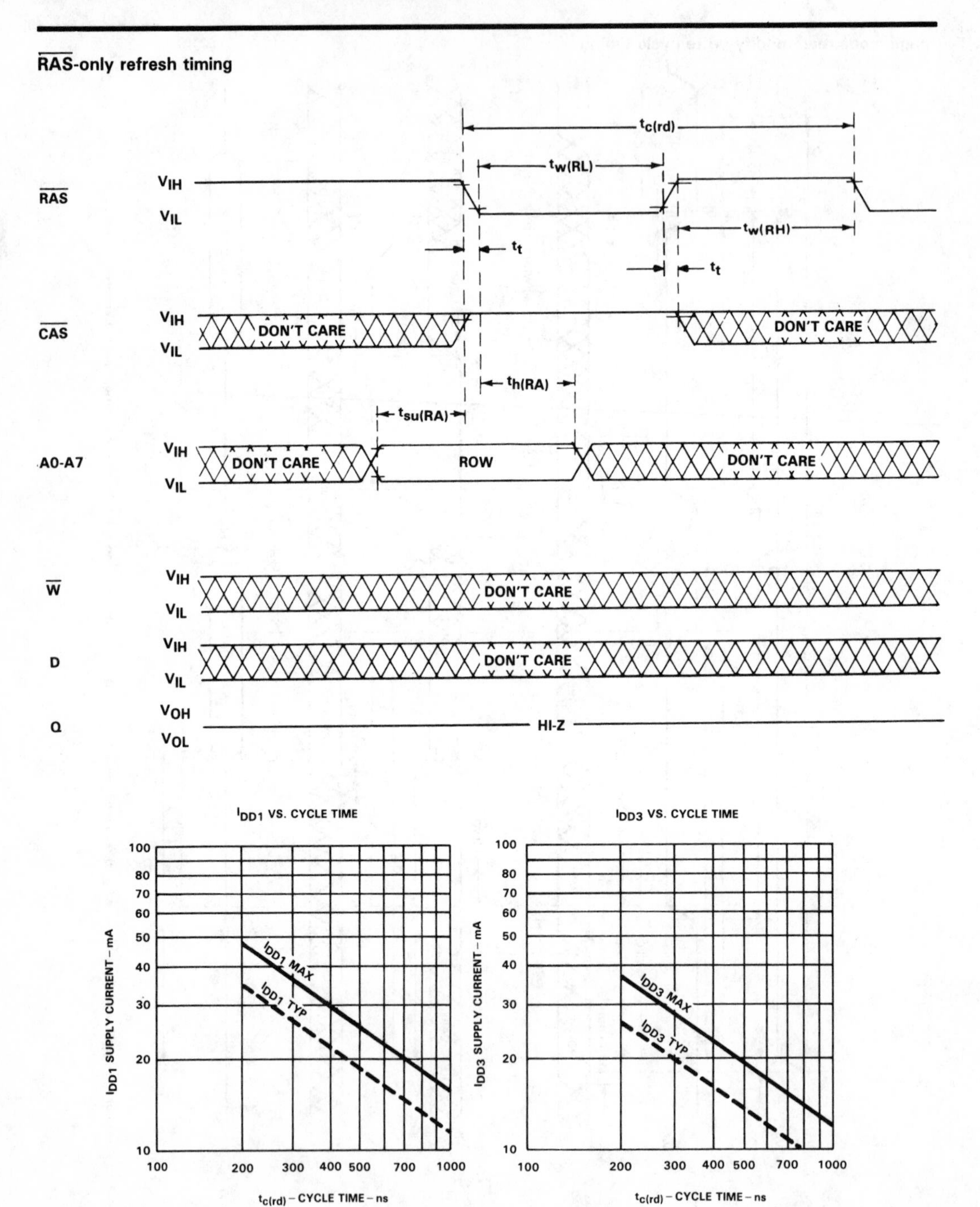

TMS 4416 NL
16,384-WORD BY 4-BIT DYNAMIC RAM

AUGUST 1980—REVISED MAY 1982

- 16,384 X 4 Organization
- Single +5 V Supply (10% Tolerance)
- Performance Ranges:

	ACCESS TIME ROW ADDRESS (MAX)	ACCESS TIME COLUMN ADDRESS (MAX)	READ OR WRITE CYCLE (MIN)	READ, MODIFY, WRITE CYCLE (MIN)
TMS 4416-15	150 ns	80 ns	260 ns	360 ns
TMS 4416-20	200 ns	120 ns	330 ns	460 ns

- Long Refresh Period . . . 4 milliseconds
- Low Refresh Overhead Time . . . As Low As 1.7% of Total Refresh Period
- All Inputs, Outputs, Clocks Fully TTL Compatible
- 3-State Unlatched Outputs
- Early Write or $\overline{G}$ to Control Output Buffer Impedance
- Page-mode Operation for Faster Access
- Low Power Dissipation
  - Operating . . . 130 mW (typ)
  - Standby . . . 17.5 mW (typ.)
- New SMOS (Scaled-MOS) N-Channel Technology

18-PIN PLASTIC DUAL-IN-LINE PACKAGE (TOP VIEW)

Signal	Pin	Pin	Signal
$\overline{G}$	1	18	V_{SS}
DQ1	2	17	DQ4
DQ2	3	16	$\overline{CAS}$
$\overline{W}$	4	15	DQ3
$\overline{RAS}$	5	14	A0
A6	6	13	A1
A5	7	12	A2
A4	8	11	A3
V_{DD}	9	10	A7

PIN NOMENCLATURE	
A0-A7	Address Inputs
$\overline{CAS}$	Column Address Strobe
DQ1-DQ4	Data In/Data Out
$\overline{G}$	Output Enable
$\overline{RAS}$	Row Address Strobe
$\overline{W}$	Write Enable
V_{DD}	+5 V Supply
V_{SS}	Ground

description

The TMS 4416 NL is a high speed, 65,536-bit, dynamic, random-access memory, organized as 16,384 words of 4 bits each. It employs state-of-the-art SMOS (scaled MOS) N-channel double-level polysilicon gate technology for very high performance combined with low cost and improved reliability.

The TMS 4416 NL features $\overline{RAS}$ access times to 150 ns maximum. Power dissipation is 125 mW typical operating, 17.5 mW typical standby.

New SMOS technology permits operation from a single +5 V supply, reducing system power supply and decoupling requirements, and easing board layout. I_{DD} peaks have been reduced to 60 mA typical, and a −1 V input voltage undershoot can be tolerated, minimizing system noise considerations. Input clamp diodes are used to ease system design.

Refresh period is extended to 4 milliseconds, and during this period each of the 256 rows must be strobed with $\overline{RAS}$ in order to retain data. $\overline{CAS}$ can remain high during the refresh sequence to conserve power.

All inputs and outputs, including clocks, are compatible with Series 74 TTL. All address lines and data-in are latched on chip to simplify system design. Data-out is unlatched to allow greater system flexibility.

The TMS 4416 NL is offered in an 18-pin dual-in-line plastic package and is guaranteed for operation from 0°C to 70°C. Packages are designed for insertion in mounting-hole rows on 300 mil (7.62mm) centers.

TEXAS INSTRUMENTS
INCORPORATED
POST OFFICE BOX 225012 • DALLAS, TEXAS 75265

operation

address (A0 through A7)

Fourteen address bits are required to decode 1 of 16,384 storage locations. Eight row-address bits are set up on pins A0 through A7 and latched onto the chip by the row-address strobe ($\overline{RAS}$). Then the six column-address bits are set up on pins A1 through A6 and latched onto the chip by the column-address strobe ($\overline{CAS}$). All addresses must be stable on or before the falling edges of $\overline{RAS}$ and $\overline{CAS}$. $\overline{RAS}$ is similar to a chip enable in that it activates the sense amplifiers as well as the row decoder. $\overline{CAS}$ is used as a chip select activating the column decoder and the input and output buffers.

write enable ($\overline{W}$)

The read or write mode is selected through the write enable ($\overline{W}$) input. A logic high on the $\overline{W}$ input selects the read mode and a logic low selects the write mode. The write enable terminal can be driven from standard TTL circuits without a pull-up resistor. The data input is disabled when the read mode is selected. When $\overline{W}$ goes low prior to $\overline{CAS}$, data-out will remain in the high-impedance state allowing a write cycle with $\overline{G}$ grounded.

data-in (DQ1 through DQ4)

Data is written during a write or read-modify write cycle. Depending on the mode of operation, the falling edge of $\overline{CAS}$ or $\overline{W}$ strobes data into the on-chip data latches. These latches can be driven from standard TTL circuits without a pull-up resistor. In an early-write cycle, $\overline{W}$ is brought low prior to $\overline{CAS}$ and the data is strobed in by $\overline{CAS}$ with setup and hold times referenced to this signal. In a delayed write or read-modify-write cycle, $\overline{CAS}$ will already be low, thus the data will be strobed in by $\overline{W}$ with setup and hold times referenced to this signal. In delayed or read-modify-write, $\overline{G}$ must be high to bring the output buffers to high impedance prior to impressing data on the I/O lines.

data-out (DQ1 through DQ4)

The three-state output buffer provides direct TTL compatibility (no pull-up resistor required) with a fan-out of two Series 74 TTL loads. Data-out is the same polarity as data-in. The output is in the high-impedance (floating) state until $\overline{CAS}$ is brought low. In a read cycle the output goes active after the access time interval $t_{a(C)}$ that begins with the negative transition of $\overline{CAS}$ as long as $t_{a(R)}$ and $t_{a(E)}$ are satisfied. The output becomes valid after the access time has elapsed and remains valid while $\overline{CAS}$ or $\overline{G}$ are low. $\overline{CAS}$ or $\overline{G}$ going high returns it to a high impedance state. In an early-write cycle, the output is always in the high-impedance state. In a delayed-write or read-modify-write cycle, the output must be put in the high impedance state prior to applying data to the DQ input. This is accomplished by bringing $\overline{G}$ high prior to applying data, thus satisfying t_{GHD}.

output enable ($\overline{G}$)

The $\overline{G}$ controls the impedance of the output buffers. When $\overline{G}$ is high, the buffers will remain in the high impedance state. Bringing $\overline{G}$ low during a normal cycle will activate the output buffers putting them in the low impedance state. It is necessary for both $\overline{RAS}$ and $\overline{CAS}$ to be brought low for the output buffers to go into the low impedance state. Once in the low impedance state, they will remain in the low impedance state until $\overline{G}$ or $\overline{CAS}$ is brought high.

refresh

A refresh operation must be performed at least every four milliseconds to retain data. Since the output buffer is in the high-impedance state unless $\overline{CAS}$ is applied, the $\overline{RAS}$-only refresh sequence avoids any output during refresh. Strobing each of the 256 row addresses (A0 through A7) with $\overline{RAS}$ causes all bits in each row to be refreshed. $\overline{CAS}$ can remain high (inactive) for this refresh sequence to conserve power.

page mode

Page mode operation allows effectively faster memory access by keeping the same row address and strobing successive column addresses onto the chip. Thus, the time required to setup and strobe sequential row addresses for the same page is eliminated. To extend beyond the 64 column locations on a single RAM, the row address and $\overline{RAS}$ are applied to multiple 16K X 4 RAMs. $\overline{CAS}$ is then decoded to select the proper RAM.

power-up

After power-up, the power supply must remain at its steady-state value for 1 ms. In addition, the $\overline{RAS}$ input must remain high for 100 μs immediately prior to initialization. Initialization consists of performing eight $\overline{RAS}$ cycles before proper device operation is achieved.

logic symbol†

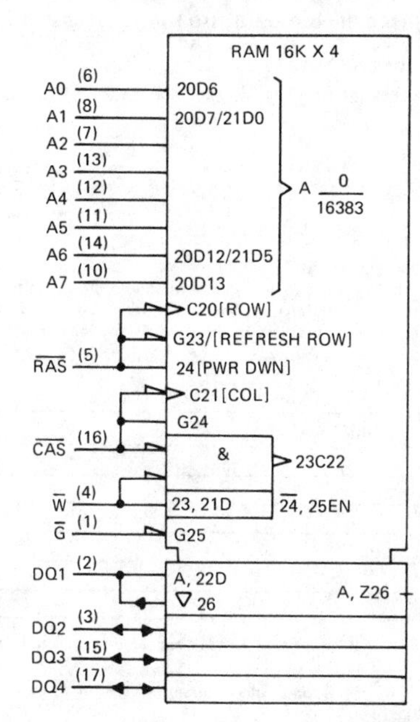

† This symbol is in accordance with IEEE Std 91/ANSI Y32.14 and recent decisions by IEEE and IEC. See explanation on page 289.

functional block diagram

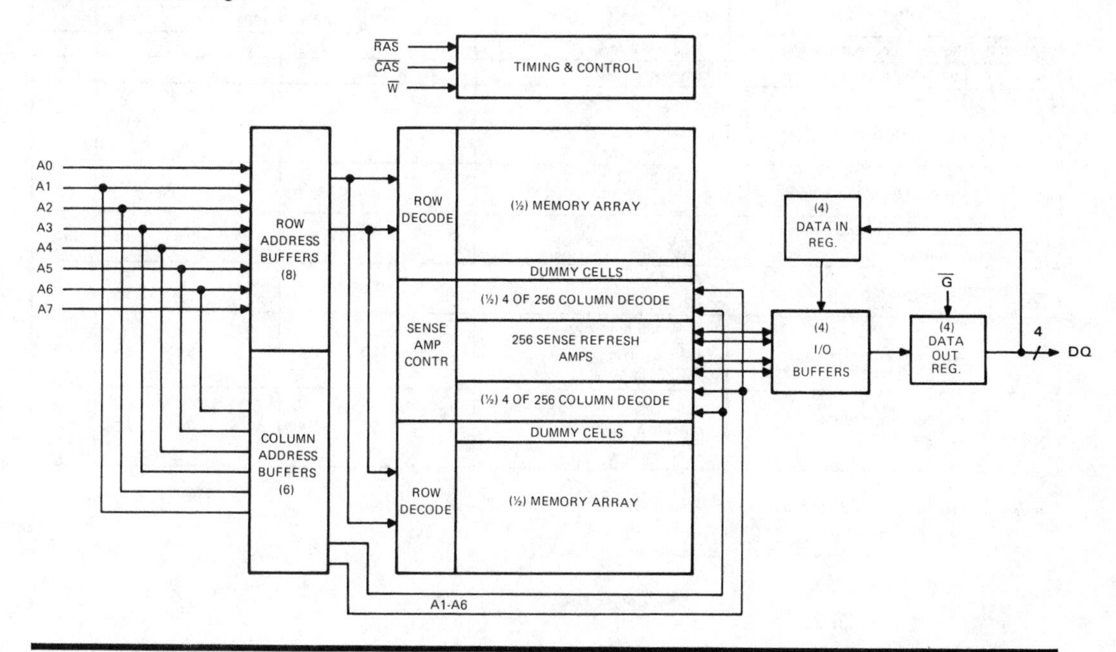

absolute maximum ratings over operating free-air temperature range (unless otherwise noted)*

- Voltage on any pin except V_{DD} and data out (see Note 1) . . . . . −1.5 to 10 V
- Voltage on V_{DD} supply and data out with respect to V_{SS} . . . . . −1 to 6 V
- Short circuit output current . . . . . 50 mA
- Power dissipation . . . . . 1 W
- Operating free-air temperature range . . . . . 0°C to 70°C
- Storage temperature range . . . . . −65°C to 150°C

NOTE 1: All voltage values in this data sheet are with respect to V_{SS}.

*Stresses beyond those listed under "Absolute Maximum Ratings" may cause permanent damage to the device. This is a stress rating only and functional operation of the device at these or any other conditions beyond those indicated in the "Recommended Operating Conditions" section of this specification is not implied. Exposure to absolute-maximum-rated conditions for extended periods may affect device reliability.

recommended operating conditions

PARAMETER	MIN	NOM	MAX	UNIT
Supply voltage, V_{DD}	4.5	5	5.5	V
Supply voltage, V_{SS}		0		V
High-level input voltage, V_{IH}	2.7		V_{DD}+0.3	V
Low-level input voltage, V_{IL} (see Note 2)	V_{IK}		0.8	V
Input clamp voltage, V_{IK} (I_I = −15 mA) (see Note 3)	−1.2			V
Operating free-air temperature, T_A	0		70	°C

NOTES: 2. The algebraic convention, where the more negative (less positive) limit is designated as minimum, is used in this data sheet for logic voltage levels only.

3. V_{IK} is the guaranteed minimum DC clamp voltage with a forced input current of −15 mA (See Figure 1)

electrical characteristics over full ranges of recommended operating conditions (unless otherwise noted)

	PARAMETER	TEST CONDITIONS	TMS 4416-15			TMS 4416-20			UNIT
			MIN	TYP†	MAX	MIN	TYP†	MAX	
V_{OH}	High-level output voltage	I_{OH} = −2 mA	2.4			2.4			V
V_{OL}	Low-level output voltage	I_{OL} = 4.2 mA			0.4			0.4	V
I_I	Input current (leakage)	V_I = 0 V to 5.8 V, V_{DD} = 5 V, All other pins = 0 V			±10			±10	μA
I_O	Output current (leakage)	V_O = 0.4 to 5.5 V, V_{DD} = 5 V, $\overline{CAS}$ high			±10			±10	μA
I_{DD1}	Average operaing current during read or write cycle	At t_c = minimum cycle		30	39		26	33	mA
I_{DD2}*	Standby current	After 1 memory cycle, $\overline{RAS}$ and $\overline{CAS}$ high		3.5	5		3.5	5	mA
I_{DD3}	Average refresh current	t_c = minimum cycle, $\overline{RAS}$ cycling, $\overline{CAS}$ high		25	34		21	28	mA
I_{DD4}	Average page-mode current	$t_{c(P)}$ = minimum cycle, $\overline{RAS}$ low, $\overline{CAS}$ cycling		25	34		21	28	mA

† All typical values are at T_A = 25°C and nominal supply voltages.

* $V_{IL} \geqslant -0.7$ V on all inputs.

TEXAS INSTRUMENTS
INCORPORATED
POST OFFICE BOX 225012 • DALLAS, TEXAS 75265

capacitance over recommended supply voltage range and operating free-air temperature range, f = 1 MHz

PARAMETER		TYP†	MAX	UNIT
$C_{i(A)}$	Input capacitance, address inputs	5	7	pF
$C_{i(RC)}$	Input capacitance, strobe inputs	8	10	pF
$C_{i(W)}$	Input capacitance, write enable input	8	10	pF
$C_{i/o}$	Input/output capacitance, data ports	8	10	pF

†All typical values are at $T_A = 25°C$ and nominal supply voltages.

switching characteristics over recommended supply voltage range and operating free-air temperature range

PARAMETER		TEST CONDITIONS	ALT. SYMBOL	TMS 4416-15		TMS 4416-20		UNIT
				MIN	MAX	MIN	MAX	
$t_{a(C)}$	Access time from $\overline{CAS}$	C_L = 100 pF, Load = 2 Series 74 TTL gates	t_{CAC}		80		120	ns
$t_{a(R)}$	Access time from $\overline{RAS}$	t_{RLCL} = MAX, C_L = 100 pF Load = 2 Series 74 TTL gates	t_{RAC}		150		200	ns
$t_{a(G)}$	Access time after $\overline{G}$ low	C_L = 100 pF, Load = 2 Series 74 TTL gates			40		50	ns
$t_{dis(CH)}$	Output disable time after $\overline{CAS}$ high	C_L = 100 pF, Load = 2 Series 74 TTL gates	t_{OFF}	0	30	0	40	ns
$t_{dis(G)}$	Output disable time after $\overline{G}$ high	C_L = 100 pF, Load = 2 Series 74 TTL gates		0	30	0	40	ns

timing requirements over recommended supply voltage range and operating free-air temperature range

PARAMETER		ALT. SYMBOL	TMS4416-15		TMS4416-20		UNIT
			MIN	MAX	MIN	MAX	
$t_{c(P)}$	Page mode cycle time	t_{PC}	140		210		ns
$t_{c(rd)}$	Read cycle time*	t_{RC}	260		330		ns
$t_{c(W)}$	Write cycle time	t_{WC}	260		330		ns
$t_{c(rdw)}$	Read-write/read-modify-write cycle time	t_{RWC}	360		460		ns
$t_{w(CH)}$	Pulse width, $\overline{\text{CAS}}$ high (precharge time)**	t_{CP}	50		80		ns
$t_{w(CL)}$	Pulse width, $\overline{\text{CAS}}$ low	t_{CAS}	80	10,000	120	10,000	ns
$t_{w(RH)}$	Pulse width $\overline{\text{RAS}}$ high (precharge time)	t_{RP}	100		120		ns
$t_{w(RL)}$	Pulse width, $\overline{\text{RAS}}$ low	t_{RAS}	150	10,000	200	10,000	ns
$t_{w(W)}$	Write pulse width	t_{WP}	40		50		ns
t_t	Transition times (rise and fall) for $\overline{\text{RAS}}$ and $\overline{\text{CAS}}$	t_T	3	50	3	50	ns
$t_{su(CA)}$	Column address setup time	t_{ASC}	0		0		ns
$t_{su(RA)}$	Row address setup time	t_{ASR}	0		0		ns
$t_{su(D)}$	Data setup time	t_{DS}	0		0		ns
$t_{su(rd)}$	Read command setup time	t_{RCS}	0		0		ns
$t_{su(WCH)}$	Write command setup time before $\overline{\text{CAS}}$ high	t_{CWL}	60		80		ns
$t_{su(WRH)}$	Write command setup time before $\overline{\text{RAS}}$ high	t_{RWL}	60		80		ns
$t_{h(CLCA)}$	Column address hold time after $\overline{\text{CAS}}$ low	t_{CAH}	40		50		ns
$t_{h(RA)}$	Row address hold time	t_{RAH}	20		25		ns
$t_{h(RLCA)}$	Column address hold time after $\overline{\text{RAS}}$ low	t_{AR}	110		130		ns
$t_{h(CLD)}$	Data hold time after $\overline{\text{CAS}}$ low	t_{DH}	60		80		ns
$t_{h(RLD)}$	Data hold time after $\overline{\text{RAS}}$ low	t_{DHR}	130		160		ns
$t_{h(WLD)}$	Data hold time after $\overline{\text{W}}$ low	t_{DH}	40		50		ns
$t_{h(RHrd)}$	Read command hold time after $\overline{\text{RAS}}$ high	t_{RRH}	10		10		ns
$t_{h(CHrd)}$	Read command hold time after $\overline{\text{CAS}}$ high	t_{RCH}	0		0		ns
$t_{h(CLW)}$	Write command hold time after $\overline{\text{CAS}}$ low	t_{WCH}	60		80		ns
$t_{h(RLW)}$	Write command hold time after $\overline{\text{RAS}}$ low	t_{WCR}	130		160		ns
t_{RLCH}	Delay time $\overline{\text{RAS}}$ low to $\overline{\text{CAS}}$ high	t_{CSH}	150		200		ns
t_{CHRL}	Delay time, $\overline{\text{CAS}}$ high to $\overline{\text{RAS}}$ low	t_{CRP}	0		0		ns
t_{CLRH}	Delay time, $\overline{\text{CAS}}$ low to $\overline{\text{RAS}}$ high	t_{RSH}	80		120		ns
t_{CLWL}	Delay time, $\overline{\text{CAS}}$ low to $\overline{\text{W}}$ low (read, modify-write-cycle only)***	t_{CWD}	120		150		ns
t_{RLCL}	Delay time, $\overline{\text{RAS}}$ low to $\overline{\text{CAS}}$ low (maximum value specified only to guarantee access time)	t_{RCD}	20	70	25	80	ns
t_{RLWL}	Delay time, $\overline{\text{RAS}}$ low to $\overline{\text{W}}$ low (read, modify-write-cycle only)***	t_{RWD}	190		250		ns
t_{WLCL}	Delay time, $\overline{\text{W}}$ low to $\overline{\text{CAS}}$ low (early write cycle)	t_{WCS}	−5		−5		ns
t_{GHD}	Delay time, $\overline{\text{G}}$ high before data applied at DQ		30		40		ns
t_{rf}	Refresh time interval	t_{REF}		4		4	ms

* Note: All cycle times assume t_t = 5 ns.

** Page mode only

*** Necessary to ensure $\overline{\text{G}}$ has disabled the output buffers prior to applying data to the device.

3

PARAMETER MEASUREMENT INFORMATION

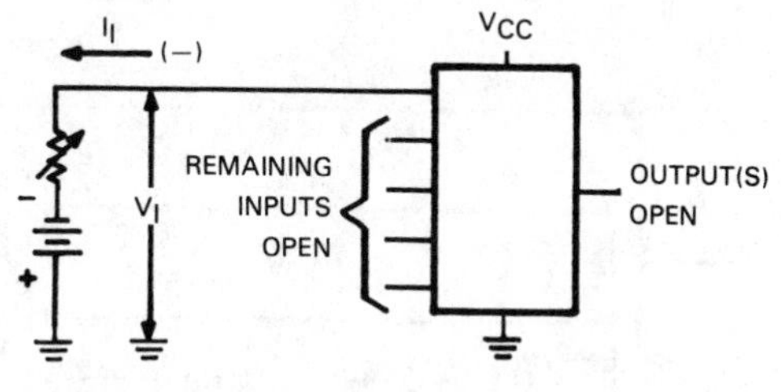

NOTE: Each input is tested separately.

FIGURE 1 — INPUT CLAMP VOLTAGE TEST CIRCUIT

read cycle timing

early write cycle timing

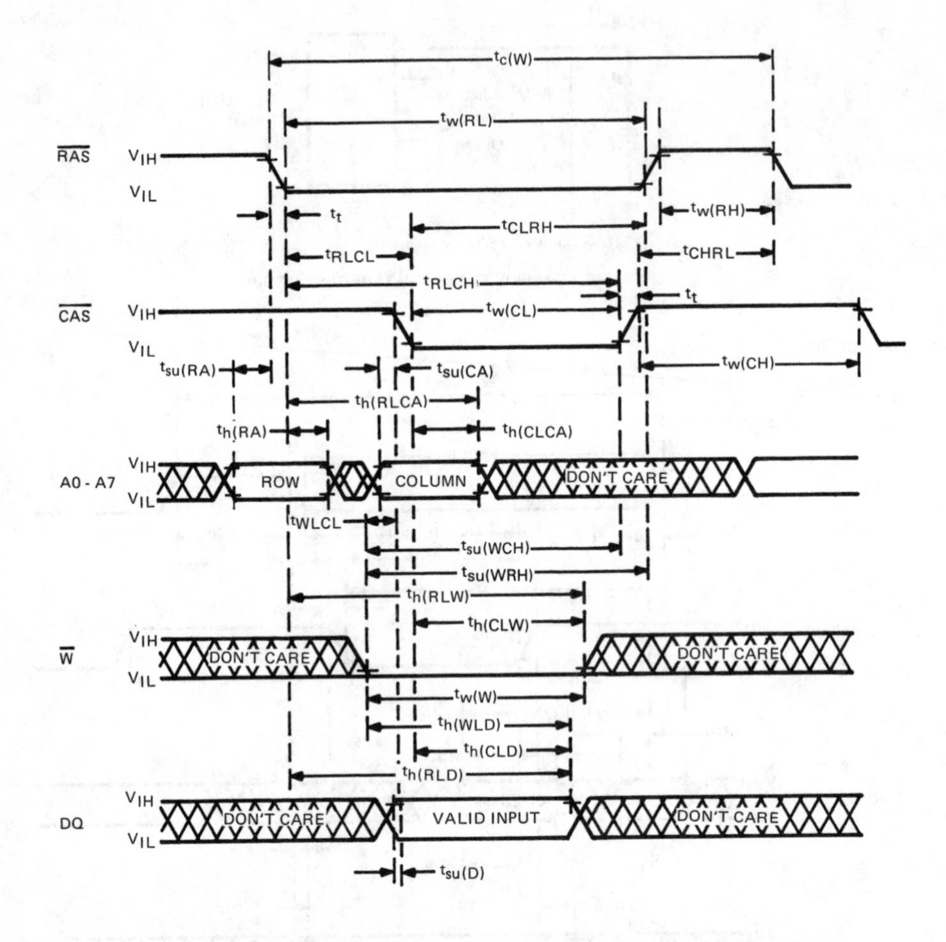

TEXAS INSTRUMENTS
INCORPORATED
POST OFFICE BOX 225012 • DALLAS, TEXAS 75265

write cycle timing

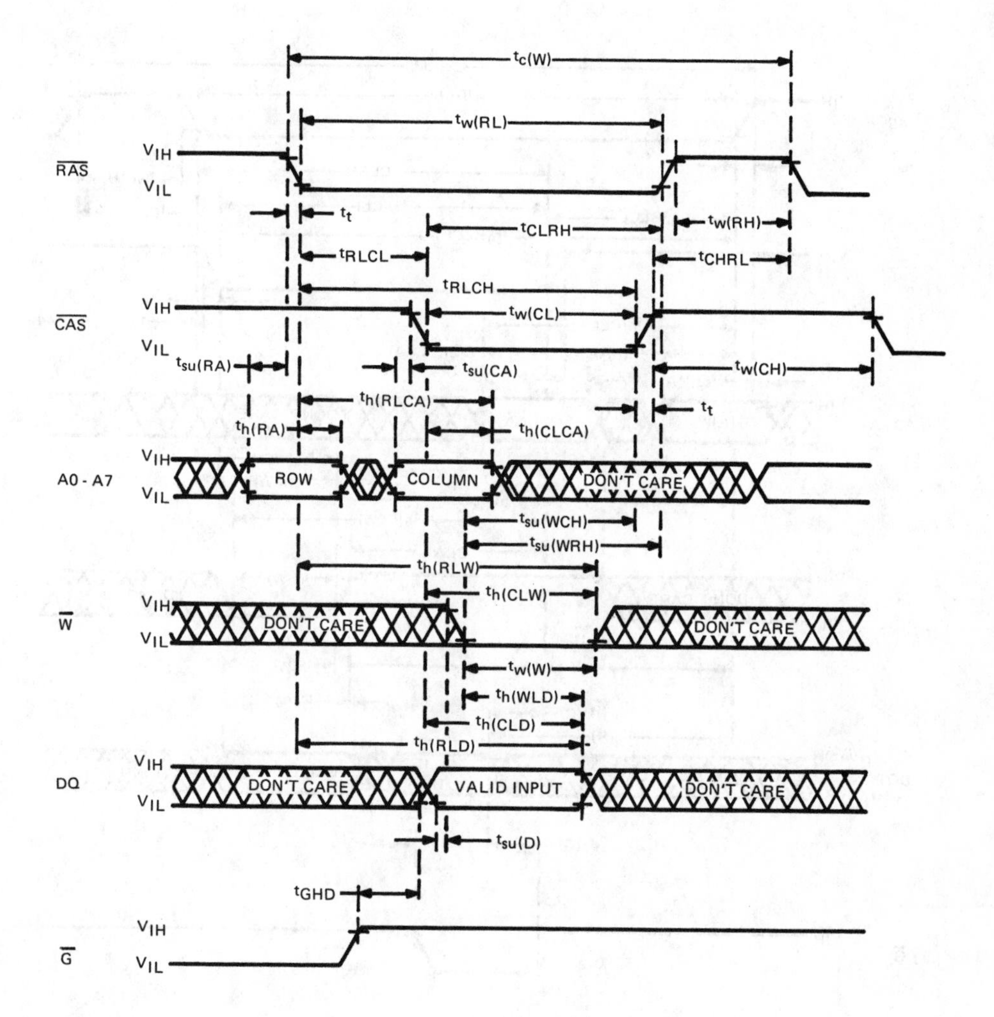

read-write/read-modify-write cycle timing

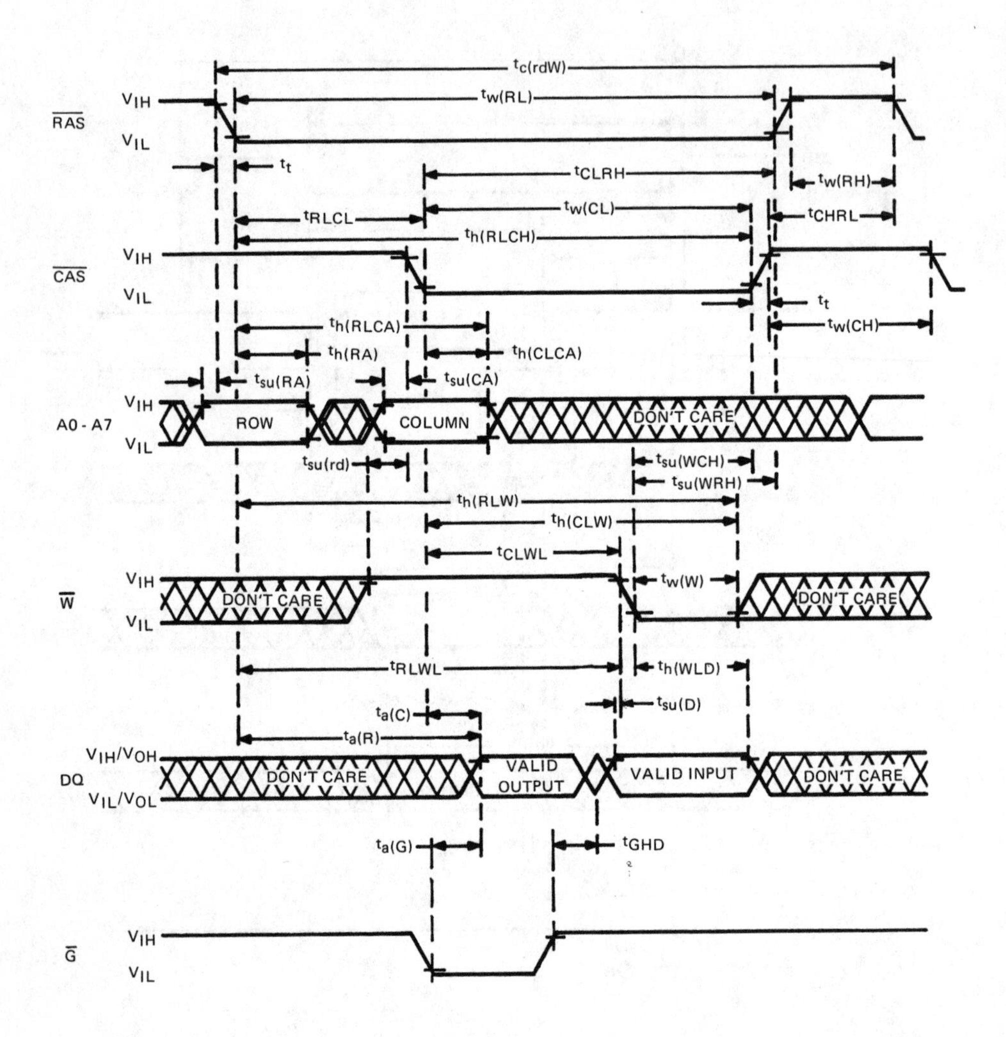

TEXAS INSTRUMENTS
INCORPORATED
POST OFFICE BOX 225012 • DALLAS, TEXAS 75265

page-mode read cycle timing

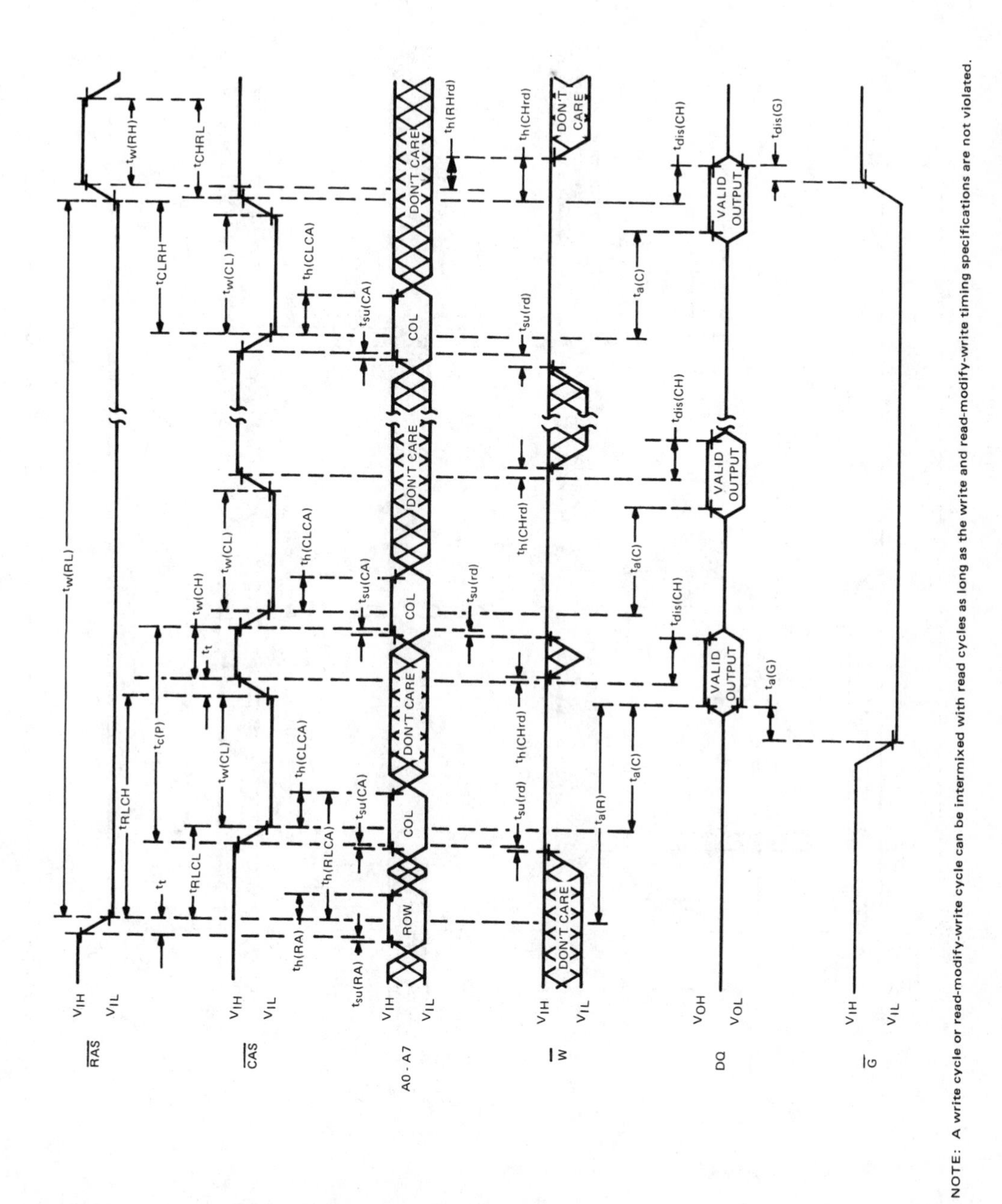

NOTE: A write cycle or read-modify-write cycle can be intermixed with read cycles as long as the write and read-modify-write timing specifications are not violated.

page-mode write cycle timing

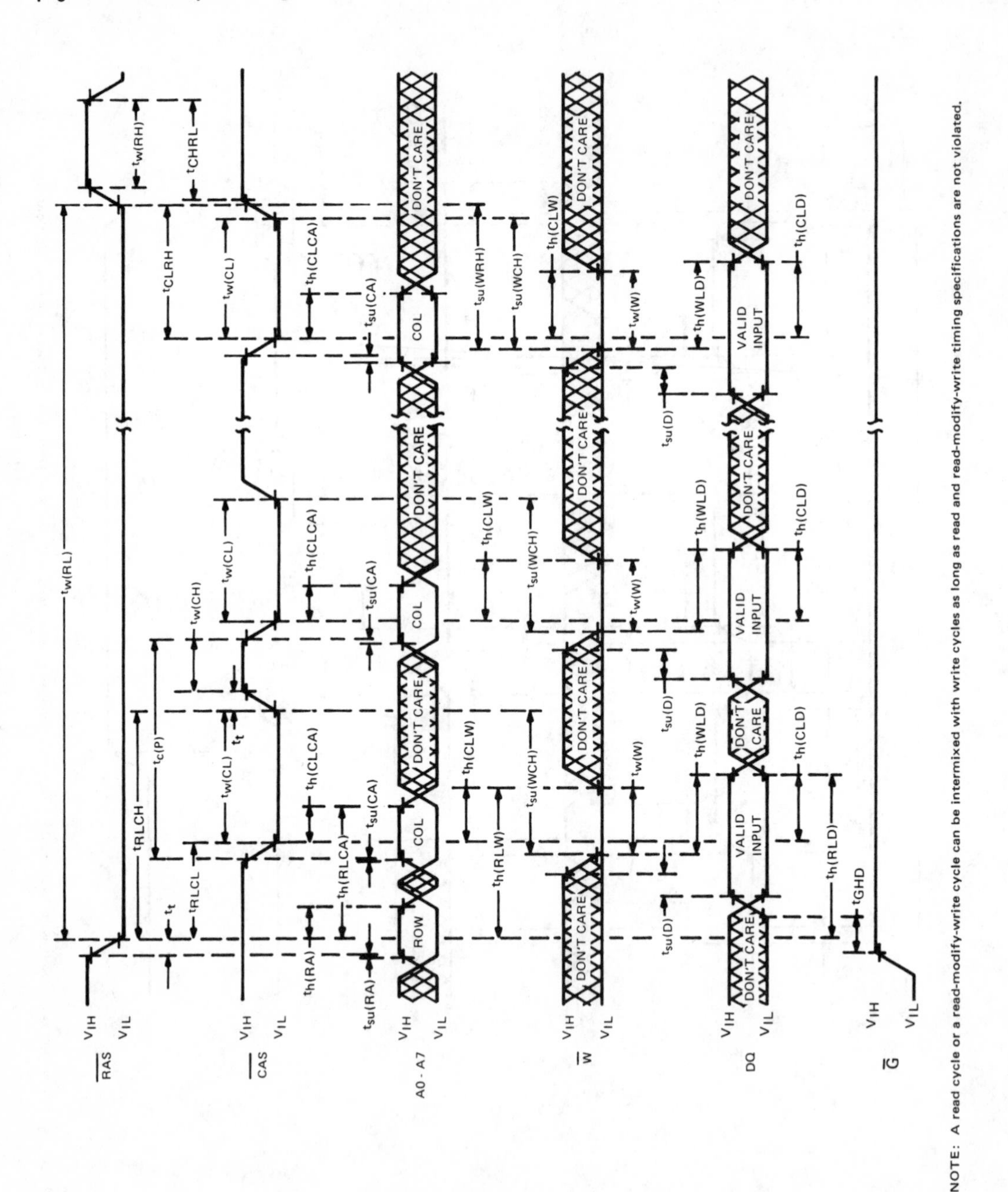

NOTE: A read cycle or a read-modify-write cycle can be intermixed with write cycles as long as read and read-modify-write timing specifications are not violated.

TEXAS INSTRUMENTS
INCORPORATED
POST OFFICE BOX 225012 • DALLAS, TEXAS 75265

page mode read-modify-write timing

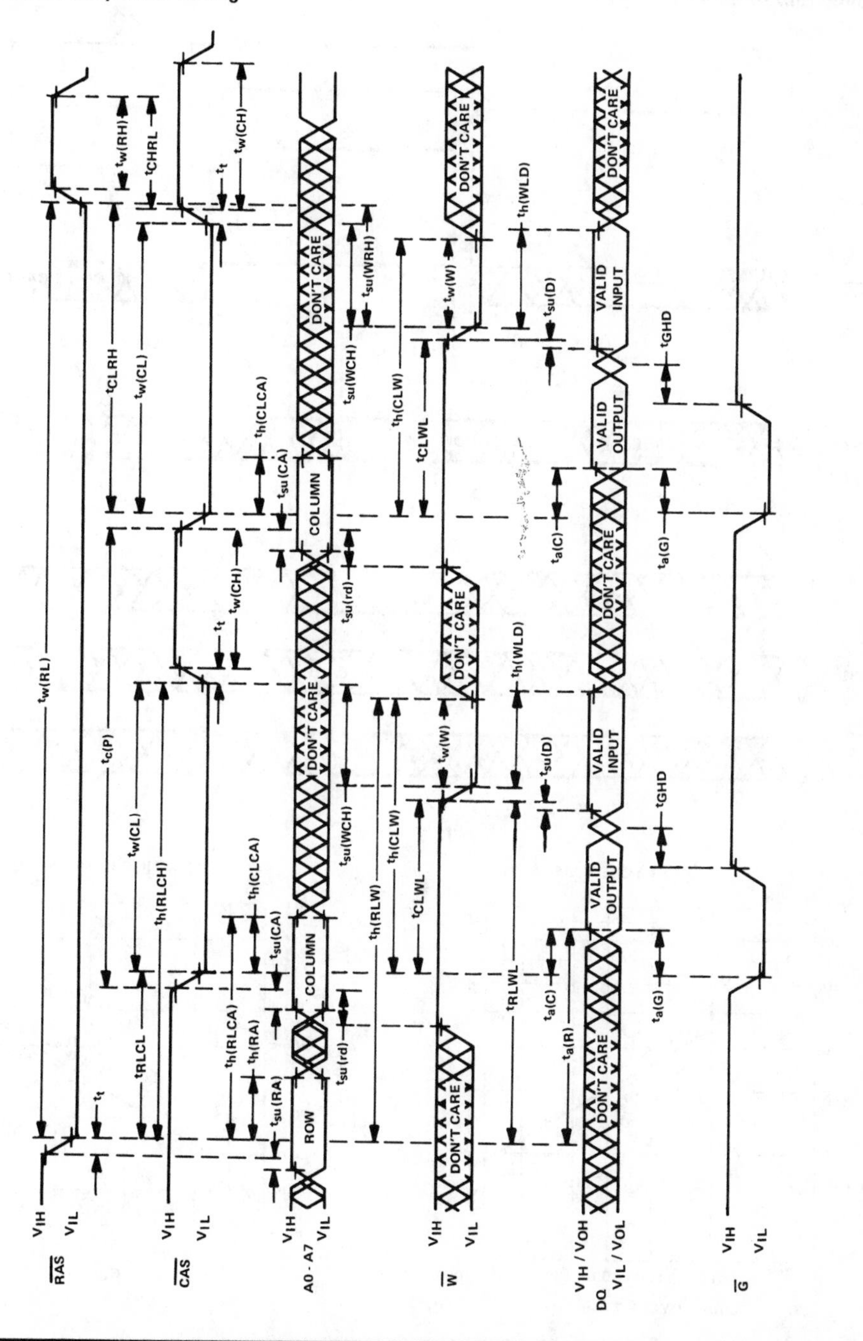

NOTE: A read cycle or a write cycle can be intermixed with read-modify-write cycles as long as read and write timing specifications are not violated.

RAS-only refresh timing

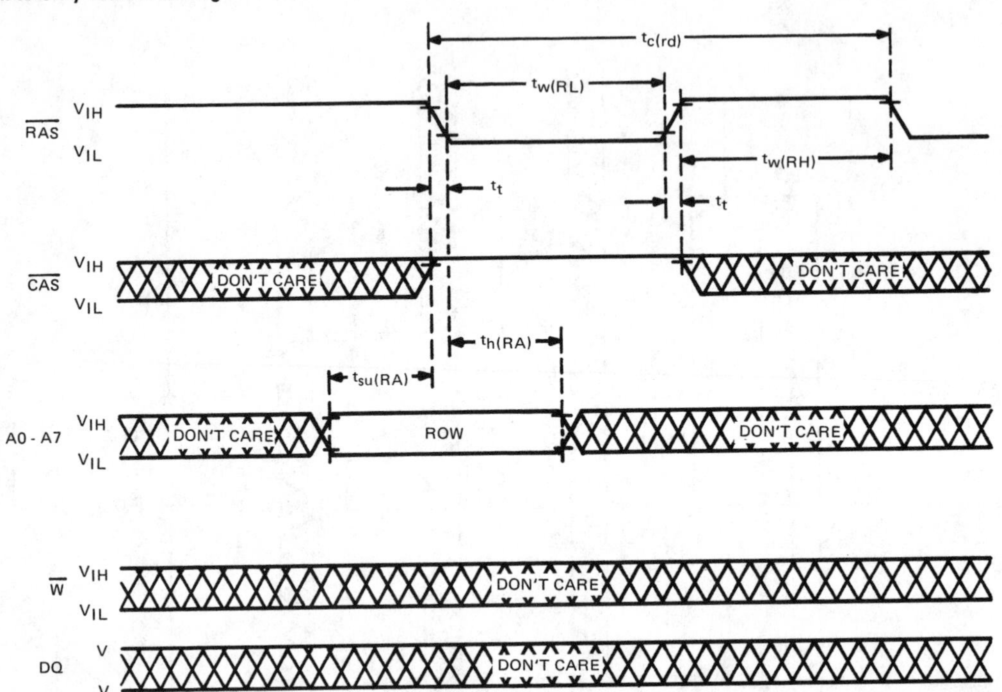

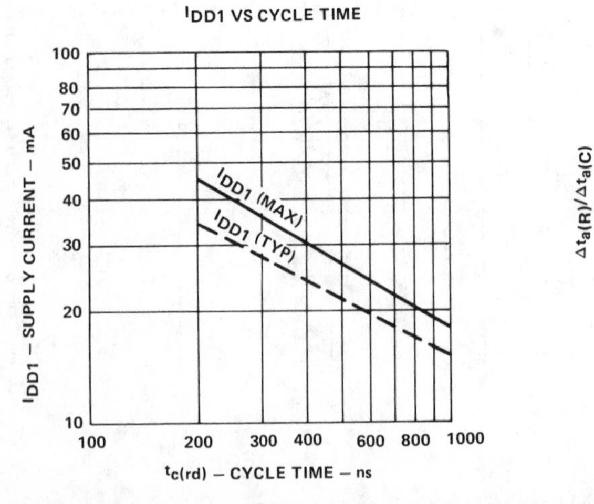

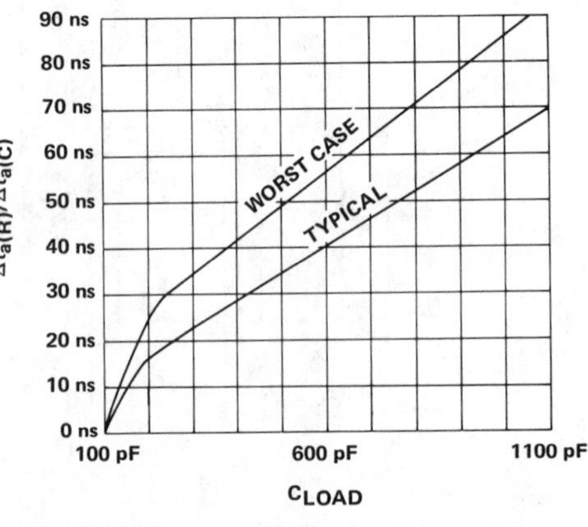

TEXAS INSTRUMENTS
INCORPORATED
POST OFFICE BOX 225012 • DALLAS, TEXAS 75265

TMS 4500A NL
DYNAMIC RAM CONTROLLER

JANUARY 1982

- Controls Operation of 8K/16K/32K/64K Dynamic RAMs
- Creates Static RAM Appearance
- One Package Contains Address Multiplexer, Refresh Control, and Timing Control
- Directly Addresses and Drives Up to 256K Bytes of Memory Without External Drivers
- Operates from Microprocessor Clock
  - No Crystals, Delay Lines, or RC Networks
  - Eliminates Arbitration Delays
- Refresh May Be Internally or Externally Initiated
- Versatile
  - Strap-Selected Refresh Rate
  - Synchronous, Predictable Refresh
  - Selection of Distributed, Transparent, and Cycle-Steal Refresh Modes
  - Interfaces Easily to Popular Microprocessors
- Strap-Selected Wait State Generation for Microprocessor/Memory Speed Matching
- Ability to Synchronize or Interleave Controller with the Microprocessor System (Including Multiple Controllers)
- Three-State Outputs Allow Multiport Memory Configuration

TMS 4500A
40-PIN 600-MIL PLASTIC
DUAL-IN-LINE PACKAGE
(TOP VIEW)

Signal	Pin	Pin	Signal
CLK	1	40	V_{CC}
RDY	2	39	$\overline{REFREQ}$
REN1	3	38	TWST
$\overline{CS}$	4	37	FS0
ALE	5	36	FS1
$\overline{RAS0}$	6	35	RA7
$\overline{RAS1}$	7	34	CA7
$\overline{ACR}$	8	33	MA7
$\overline{ACW}$	9	32	MA6
$\overline{CAS}$	10	31	CA6
RA0	11	30	RA6
CA0	12	29	RA5
MA0	13	28	CA5
MA1	14	27	MA5
CA1	15	26	RA4
RA1	16	25	CA4
RA2	17	24	MA4
CA2	18	23	RA3
MA2	19	22	CA3
GND	20	21	MA3

3

description

The TMS 4500A is a monolithic DRAM system controller designed to provide address multiplexing, timing, control and refresh/access arbitration functions to simplify the interface of dynamic RAMs to microprocessor systems.

The controller contains a 16-bit multiplexer that generates the address lines for the memory device from the 16 system address bits and provides the strobe signals required by the memory to decode the address. An 8-bit refresh counter generates the 256-row addresses required for refresh.

A refresh timer is provided that generates the necessary timing to refresh the dynamic memories and assure data retention.

The TMS 4500A also contains refresh/access arbitration circuitry to resolve conflicts between memory access requests and memory refresh cycles. The TMS4500A is offered in a 40-pin, 600-mil dual-in-line plastic package and is guaranteed for operation from 0°C to 70°C.

TEXAS INSTRUMENTS
INCORPORATED
POST OFFICE BOX 225012 • DALLAS, TEXAS 75265

BLOCK DIAGRAM

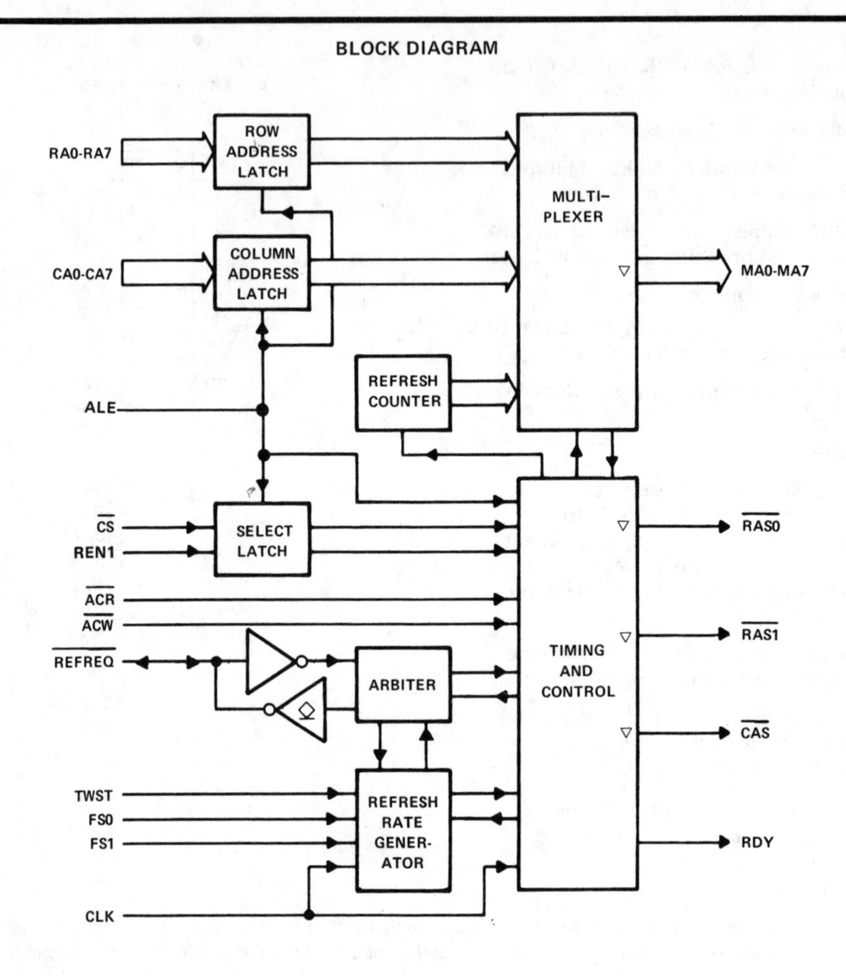

pin descriptions

RA0 - RA7	Input	Row Address – These address inputs are used to generate the row address for the multiplexer.
CA0 - CA7	Input	Column Address – These address inputs are used to generate the column address for the multiplexer.
MA0 - MA7	Output	Memory Address – These three-state outputs are designed to drive the addresses of the dynamic RAM array.
ALE	Input	Address Latch Enable – This input is used to latch the 16 address inputs, CS and REN1. This also initiates an access cycle if chip select is valid. The rising edge (low level to high level) of ALE returns RAS to the high level.

TEXAS INSTRUMENTS
INCORPORATED
POST OFFICE BOX 225012 • DALLAS, TEXAS 75265

pin descriptions (continued)

$\overline{CS}$	Input	Chip Select – A low on this input enables an access cycle. The trailing edge of ALE latches the chip select input.
REN1	Input	RAS Enable 1 – This input is used to select one of two banks of RAM via the $\overline{RAS}$ 0 and $\overline{RAS}$ 1 outputs when chip select is present.
$\overline{ACR}$, $\overline{ACW}$	Input	Access Control, Read; Access Control, Write – A low on either of these inputs causes the column address to appear on MA0 - MA7 and the column address strobe. The rising edge of $\overline{ACR}$ or $\overline{ACW}$ terminates the cycle by ending $\overline{RAS}$ and $\overline{CAS}$ strobes. When $\overline{ACR}$ and $\overline{ACW}$ are both low, MA0 - MA7, $\overline{RAS0}$, $\overline{RAS1}$, and $\overline{CAS}$ go into a high-impedance (floating) state.
CLK	Input	System Clock – This input provides the master timing to generate refresh cycle timings and refresh rate. Refresh rate is determined by the TWST, FS1, FS0 inputs.
$\overline{REFREQ}$	Input/Output	Refresh Request – (This input should be driven by an open-collector output.) On input, a low-going edge initiates a refresh cycle and will cause the internal refresh timer to be reset on the next falling edge of the CLK. As an output, a low-going edge signals an internal refresh request and that the refresh timer will be reset on the next low-going edge of CLK. $\overline{REFREQ}$ will remain low until the refresh cycle is in progress and the current refresh address is present on MA0-MA7.
$\overline{RAS0}$, $\overline{RAS1}$	Output	Row Address Strobe – These three-state outputs are used to latch the row address into the bank of DRAMs selected by REN1. On refresh both signals are driven.
$\overline{CAS}$	Output	Column Address Strobe – This three-state output is used to latch the column address into the DRAM array.
RDY	Output	Ready – This totem-pole output synchronizes memories that are too slow to guarantee microprocessor access time requirements. This output is also used to inhibit access cycles during refresh when in cycle-steal mode.
TWST	Input	Timing/Wait Strap – A high on this input indicates a wait state should be added to each memory cycle. In addition it is used in conjunction with FS0 and FS1 to determine refresh rate and timing.
FS0, FS1	Inputs	Frequency Select 0; Frequency Select 1 – These are strap inputs to select Mode and Frequency of operation as shown in Table 1.

TABLE 1 – STRAP CONFIGURATION

STRAP INPUT MODES			WAIT STATES FOR MEMORY ACCESS	REFRESH RATE	MINIMUM CLK FREQ. (MHz)	REFRESH FREQ. (kHz)	CLOCK CYCLES FOR EACH REFRESH
TWST	FS1	FS0					
L	L	L†	0	EXTERNAL	–	REFREQ	4
L	L	H	0	CLK ÷ 31	1.984	64 - 95‡	3
L	H	L	0	CLK ÷ 46	2.944	64 - 85‡	3
L	H	H	0	CLK ÷ 61	3.904	64 - 82§	4
H	L	L	1	CLK ÷ 46	2.944	64 - 85‡	3
H	L	H	1	CLK ÷ 61	3.904	64 - 80‡	4
H	H	L	1	CLK ÷ 76	4.864	64 - 77‡	4
H	H	H	1	CLK ÷ 91	5.824	64 - 88¶	4

† This strap configuration resets the Refresh Timer circuitry.
‡ Upper figure in refresh frequency is the frequency that is produced if the minimum CLK frequency of the next select state is used.
§ Refresh frequency if CLK frequency is 5 MHz.
¶ Refresh frequency if CLK frequency is 8 MHz.

functional description

The TMS 4500A consists of six basic blocks; address and select latches, refresh rate generator, refresh counter, the multiplexer, the arbiter, and the timing and control block.

address and select latches

The address and select latches allow the DRAM controller to be used in systems that multiplex address and data on the same lines without external latches. The row address latches are transparent, meaning that while ALE is high, the output at MA0 - MA7 follows the inputs RA0 - RA7.

refresh rate generator

The refresh rate generator is a counter that indicates to the arbiter that it is time for a refresh cycle. The counter divides the clock frequency according to the configuration straps as shown in Table 1. The counter is reset when a refresh cycle is requested or when TWST, FS1 and FS0 are low. The configuration straps allow the matching of memories to the system access time.

Upon Power-Up it is necessary to provide a reset signal by driving all three straps to the controller low to initialize internal counters. A system's low-active, power-on reset ($\overline{\text{RESET}}$) can be used to accomplish this by connecting it to those straps that are desired high during operation.

refresh counter

The refresh counter contains the address of the row to be refreshed. The counter is decremented after each refresh cycle.

multiplexer

The multiplexer provides the DRAM array with row, column, and refresh addresses at the proper times. Its inputs are the address latches and the refresh counter. The outputs provide up to 16 multiplexed addresses on eight lines.

TEXAS INSTRUMENTS
INCORPORATED
POST OFFICE BOX 225012 • DALLAS, TEXAS 75265

arbiter

The arbiter provides two operational cycles: access and refresh. The arbiter resolves conflicts between cycle requests and cycles in execution, and schedules the inhibited cycle when used in cycle-steal mode.

timing and control block

The timing and control block executes the operational cycle at the request of the arbiter. It provides the DRAM array with $\overline{RAS}$ and $\overline{CAS}$ signals. It provides the CPU with a RDY signal. It controls the multiplexer during all cycles. It resets the refresh rate generator and decrements the refresh counter during refresh cycles.

absolute maximum ratings over operating ambient† temperature range (unless otherwise noted)*

Supply voltage range, V_{CC} (see Note 1) . . . . −1.5 to 7 V
Input voltage range (any input) (see Note 1) . . . . −1.5 to 7 V
Continuous power dissipation . . . . 1.2 W
Operating ambient temperature range . . . . 0°C to 70°C
Storage temperature range . . . . −65°C to 150°C

recommended operating conditions

PARAMETER	MIN	NOM	MAX	UNIT
Supply voltage, V_{CC}	4.5	5	5.5	V
High-level input voltage, V_{IH} (except $\overline{REFREQ}$)	2		6	V
High-level input voltage, V_{IH} ($\overline{REFREQ}$)	2.4		6	V
Low-level input voltage, V_{IL} (except $\overline{REFREQ}$)	−1‡		0.8	V
Low-level input voltage, V_{IL} ($\overline{REFREQ}$)	−1‡		1.2	V
Operating ambient† temperature, T_A	0		70	°C

† The ambient temperature conditions assume air moving perpendicular to the longitudinal axis and parallel to the seating plane of the device at a velocity of 400 ft/min (122 m/min) with the device under test soldered to a 4 x 6 x 0.062-inch (102 x 152 x 1.6-mm) double-sided 2-ounce copper-clad circuit board (plating thickness 0.07 mm).

‡ The algebraic convention, where the more negative limit is designated as minimum, is used in this data sheet for logic voltage levels only.

* Stresses beyond those listed under "Absolute Maximum Ratings" may cause permanent damage to the device. This is stress rating only and functional operation of the device at these or any other conditions beyond those indicated in the "Recommended Operating Conditions" section of this specification is not implied. Exposure to absolute-maximum-rated conditions for extended periods may affect device reliability.

Note 1: Voltage values are with respect to the ground terminal.

electrical characteristics over recommended operating ambient temperature† range (unless otherwise noted)

PARAMETER			TEST CONDITIONS†		MIN	TYP‡	MAX	UNIT
V_{OH}	High-level output voltage	MA0-MA7, RDY	I_{OH} = −1 mA	V_{CC} = 4.5 V	2.4			V
		$\overline{\text{RAS0}}$, $\overline{\text{RAS1}}$, $\overline{\text{CAS}}$			2.7			
		$\overline{\text{REFREQ}}$	I_{OH} = 100 µA	V_{CC} = 4.5 V	2.4			
V_{OL}	Low-level output voltage		I_{OL} = 4 mA	V_{CC} = 4.5 V			0.4	V
I_{IH}	High-level input current	$\overline{\text{REFREQ}}$	V_I = 5.5 V				100	µA
		All others					10	
I_{IL}	Low-level input current	$\overline{\text{REFREQ}}$	V_I = 0 V				−1.25	mA
		All others					−10	uA
I_{OZ}	Off-state output current		V_O = 0 to 4.5 V	V_{CC} = 5.5 V			±50	µA
I_{CC}	Operating supply current		T_A = 0°C			100	140	mA
C_i	Input capacitance		V_I = 0 V,	f = 1 MHz		5		pF
C_o	Output capacitance		V_O = 0 V,	f = 1 MHz		6		pF

† The ambient temperature conditions assume air moving perpendicular to the longitudinal axis and parallel to the seating plane of the device at a velocity of 400 ft/min (122 m/min) with the device under test soldered to a 4 x 6 x 0.062-inch (102 x 152 x 1.6-mm) double-sided 2-ounce copper-clad circuit board (plating thickness 0.07 mm).

‡ All typical values are at V_{CC} = 5 V, T_A = 25 °C except where otherwise noted.

timing requirements over recommended supply voltage range and operating ambient† temperature range

PARAMETER				MIN	MAX	UNIT
$t_{c(C)}$	CLK cycle time			100		ns
$t_{w(CH)}$	CLK high pulse width			20		ns
$t_{w(CL)}$	CLK low pulse width			35		ns
t_t	Transition time, all inputs				50	ns
$t_{AEL\text{-}CL}$	Time delay, ALE low to CLK starting low (see Note 1)			10		ns
$t_{CL\text{-}AEL}$	Time delay, CLK low to ALE starting low (see Note 1)			10		ns
$t_{CL\text{-}AEH}$	Time delay, CLK low to ALE starting high (see Note 2)			10		ns
$t_{w(AEH)}$	Pulse width ALE high			35		ns
$t_{AV\text{-}AEL}$	Time delay, address, REN1, $\overline{\text{CS}}$ valid to ALE low			10		ns
$t_{AEL\text{-}AX}$	Time delay, ALE low to address not valid			10		ns
$t_{AEL\text{-}ACL}$	Time delay, ALE low to $\overline{\text{ACX}}$ low (see Notes 3 and 7)	$t_{h(RA)} \geq 30$ ns	C_L = 80 pF	$t_{h(RA)} + 20$		ns
			C_L = 160 pF	$t_{h(RA)} + 30$		
		$t_{h(RA)} < 30$ ns		see Note 4		
$t_{ACH\text{-}CL}$	Time delay, $\overline{\text{ACX}}$ high to CLK low (see Notes 5 and 7)			20		ns
$t_{ACL\text{-}CH}$	Time delay, $\overline{\text{ACX}}$ low to CLK starting high (to remove RDY)			30		ns
$t_{RQL\text{-}CL}$	Time delay, $\overline{\text{REFREQ}}$ low to CLK starting low (see Note 6)			20		ns
$t_{w(RQL)}$	Pulse width, $\overline{\text{REFREQ}}$ low			20		ns

NOTES:
1. Coincidence of the trailing edge of CLK and the trailing edge of ALE should be avoided, as the refresh/access arbitration occurs on the trailing CLK edge. A trailing edge of CLK should occur during the interval from $\overline{\text{ACX}}$ high to ALE low.
2. If ALE rises before $\overline{\text{ACX}}$ and a refresh request is present, the falling edge of CLK after $t_{CL\text{-}AEH}$ will output the refresh address to MA0-MA7 and initiate a refresh cycle.
3. $t_{h(RA)}$ is the dynamic memory Row Address hold time. Capacitive loading is on $\overline{\text{RAS}}$ output.
4. Internal interlocking provides 30 ns minimum Row Address hold time. $\overline{\text{ACX}}$ may occur prior to or coincident with ALE going low.
5. Minimum of 20 ns is specified to ensure arbitration will occur on falling CLK edge. $t_{ACH\text{-}CL}$ also affects precharge time such that the minimum $t_{ACH\text{-}CL}$ should be equal or greater than: $t_{w(RH)} - t_{w(CL)} + 30$ ns (for cycle where $\overline{\text{ACX}}$ high occurs prior to ALE high) where $t_{w(RH)}$ is the DRAM $\overline{\text{RAS}}$ precharge time.
6. This parameter is necessary only if refresh arbitration is to occur on this low-going CLK edge (in systems where refresh is synchronized to external events).
7. These specifications relate to system timings and do not directly reflect device performance.

† The ambient temperature conditions assume air moving perpendicular to the longitudinal axis and parallel to the seating plane of the device at a velocity of 400 ft/min (122 m/min) with the device under test soldered to a 4 x 6 x 0.062-inch (102 x 152 x 1.6-mm) double-sided 2-ounce copper-clad circuit board (plating thickness 0.07 mm).

switching characteristics over recommended supply voltage range and operating ambient† temperature range

PARAMETER		TEST CONDITIONS	MIN	MAX	UNIT
$t_{AEL\text{-}REL}$	Time delay, ALE low to $\overline{\text{RAS}}$ starting low			40	ns
$t_{t(REL)}$	$\overline{\text{RAS}}$ fall time	$\overline{\text{RAS}}$ load = 40 pF		TBD	ns
		$\overline{\text{RAS}}$ load = 160 pF		25	
$t_{RAV\text{-}MAV}$	Time delay, row address valid to memory address valid	Address load = 40 pF		TBD	ns
		Address load = 160 pF		55	
$t_{AEH\text{-}MAV}$	Time delay, ALE high to valid memory address	Address load = 40 pF		TBD	ns
		Address load = 160 pF		70	
$t_{AEL\text{-}RYL}$	Time delay, ALE to RDY starting low (TWST = 1 or refresh in progress)			30	ns
$t_{AEL\text{-}CEL}$	Time delay, ALE low to $\overline{\text{CAS}}$ starting low	Address load = 40 pF		TBD	ns
		Address load = 160 pF		190	
$t_{AEH\text{-}REH}$	Time delay, ALE high to $\overline{\text{RAS}}$ starting high			40	ns
$t_{t(MAV)}$	Address transition time	Address load = 40 pF		TBD	ns
		Address load = 160 pF		25	
$t_{ACL\text{-}MAX}$	Row address hold from $\overline{\text{ACX}}$ low		20		ns
$t_{MAV\text{-}CEL}$	Time delay, memory address valid to $\overline{\text{CAS}}$ starting low		0		ns
$t_{t(CEL)}$	$\overline{\text{CAS}}$ fall time	$\overline{\text{CAS}}$ load = 80 pF		TBD	ns
		$\overline{\text{CAS}}$ load = 320 pF		25	
$t_{ACL\text{-}CEX}$	Time delay, $\overline{\text{ACX}}$ low to $\overline{\text{CAS}}$ starting low	Address load = 80 pF		TBD	ns
		Address load = 320 pF	30	125	
$t_{ACH\text{-}REH}$	Time delay, $\overline{\text{ACX}}$ to $\overline{\text{RAS}}$ starting high	$\overline{\text{RAS}}$ load = 160 pF		50	ns
$t_{t(REH)}$	$\overline{\text{RAS}}$ rise time	$\overline{\text{RAS}}$ load = 40 pF		TBD	ns
		$\overline{\text{RAS}}$ load = 160 pF		25	
$t_{ACH\text{-}CEH}$	Time delay, $\overline{\text{ACX}}$ high to $\overline{\text{CAS}}$ starting high		10	40	ns
$t_{t(CEH)}$	$\overline{\text{CAS}}$ rise time	$\overline{\text{CAS}}$ load = 80 pF		TBD	ns
		$\overline{\text{CAS}}$ load = 320 pF		35	
$t_{ACH\text{-}MAX}$	Column address hold from $\overline{\text{ACX}}$ high		20		ns
$t_{CH\text{-}RYH}$	Time delay, CLK high to RDY starting high (after $\overline{\text{ACX}}$ low)			45	ns

TBD = to be determined.

(continued next page)

switching characteristics over recommended supply voltage range and operating ambient† temperature range

PARAMETER		TEST CONDITIONS	MIN	MAX	UNIT
$t_{RFL-RFL}$	Time delay, $\overline{REFREQ}$ external till supported by $\overline{REFREQ}$ internal			20	ns
t_{CH-RFL}	Time delay, CLK high till $\overline{REFREQ}$ internal starting low			35	ns
t_{CL-MAV}	Time delay, CLK low till refresh address valid	Address load = 40 pF		TBD	ns
		Address load = 160 pF		100	
t_{CH-RRL}	Time delay, CLK high till refresh $\overline{RAS}$ starting low		15	60	ns
$t_{MAV-RRL}$	Time delay, refresh address valid till refresh $\overline{RAS}$ low		5		ns
t_{CL-RFH}	Time delay, CLK low to $\overline{REFREQ}$ starting high (3 cycle refresh)			55	ns
t_{CH-RFH}	Time delay, CLK high to $\overline{REFREQ}$ starting high (4 cycle refresh)			55	ns
t_{CH-RRH}	Time delay, CLK high to refresh $\overline{RAS}$ starting high		10	45	ns
t_{CH-MAX}	Time delay, refresh address hold after CLK high		20		ns
t_{CH-REL}	Time delay, CLK high till access $\overline{RAS}$ starting low			70	ns
t_{CL-CEL}	Time delay, CLK low to access $\overline{CAS}$ starting low (see Note 8)			170	ns
t_{CL-MAX}	Row address hold after CLK low		30		ns
$t_{w(ACL)}$	$\overline{ACX}$ low width		30		ns
$t_{REL-MAX}$	Row address hold from $\overline{RAS}$ low		30		ns
$t_{t(RYL)}$	RDY fall time	40 pF load		15	ns
$t_{t(RYH)}$	RDY rise time	40 pF load		25	ns
t_{dis}	Output disable time (3-state outputs)		55	125	ns
$t_{AEH-MAX}$	Column address hold from ALE high		15		ns
t_{en}	Output enable time (3-state outputs)		0	80	ns
$t_{CAV-CEL}$	Column address setup to $\overline{CAS}$ after refresh		0		ns
t_{CH-CEL}	Time delay, CLK high to access $\overline{CAS}$ starting low (see Note 8)			170	ns

NOTE 8: On the access grant cycle following refresh, the occurrence of $\overline{CAS}$ low depends on the relative occurrence of ALE low and $\overline{ACX}$ low. If $\overline{ACX}$ occurs prior to or coincident with ALE then $\overline{CAS}$ is timed from the CLK high transition. If $\overline{ACX}$ occurs 20 ns after ALE then $\overline{CAS}$ is timed from the CLK low transition.

† The ambient temperature conditions assume air moving perpendicular to the longitudinal axis and parallel to the seating plane of the device at a velocity of 400 ft/min (122 m/min) with the device under test soldered to a 4 x 6 x 0.062-inch (102 x 152 x 1.6-mm) double-sided 2-ounce copper-clad circuit board (plating thickness 0.07 mm).

TEXAS INSTRUMENTS
INCORPORATED
POST OFFICE BOX 225012 • DALLAS, TEXAS 75265

3

access cycle timing

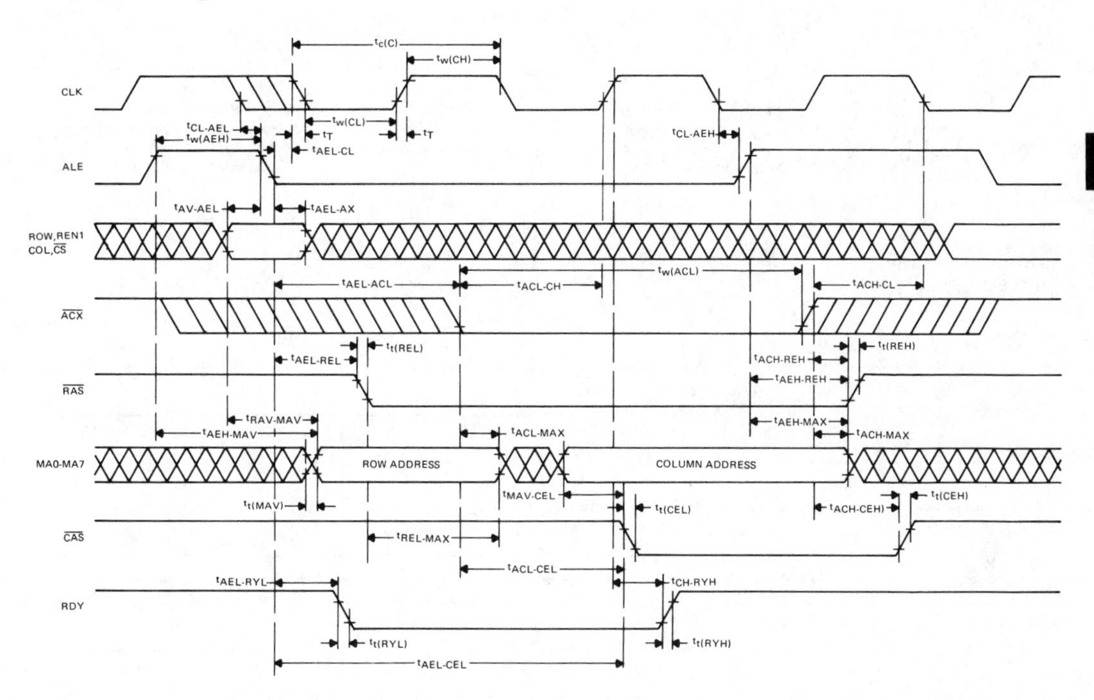

refresh request timing

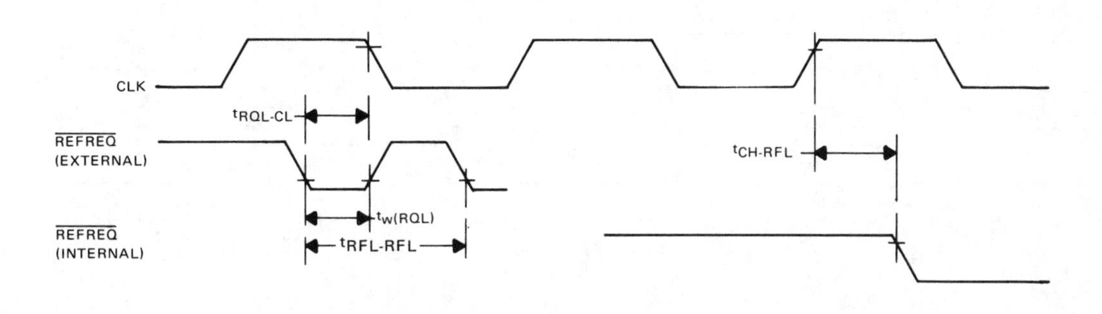

output three-state timing

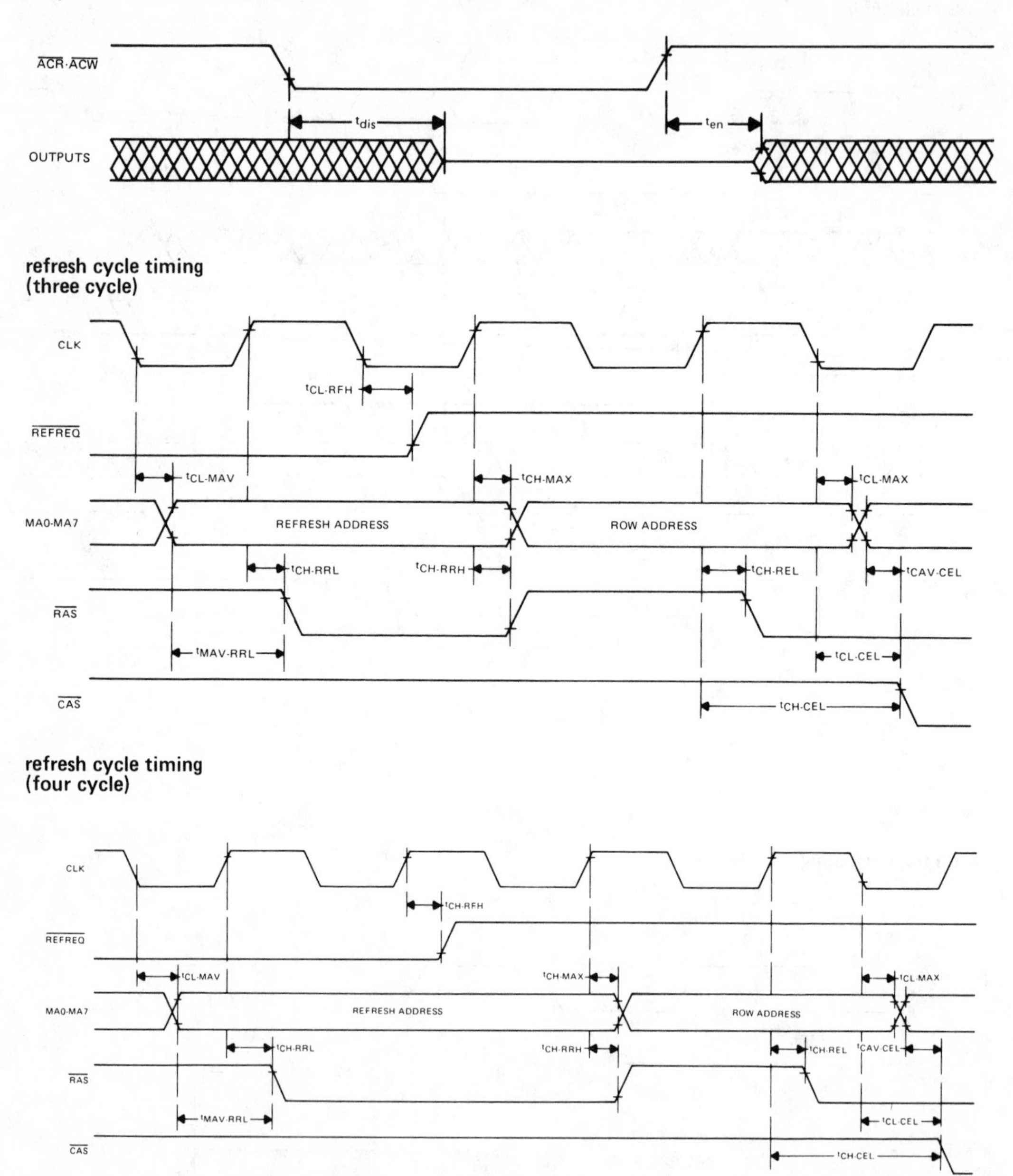

TEXAS INSTRUMENTS
INCORPORATED
POST OFFICE BOX 225012 • DALLAS, TEXAS 75265

typical access/refresh/access cycle
(three cycle, TWST = 0)

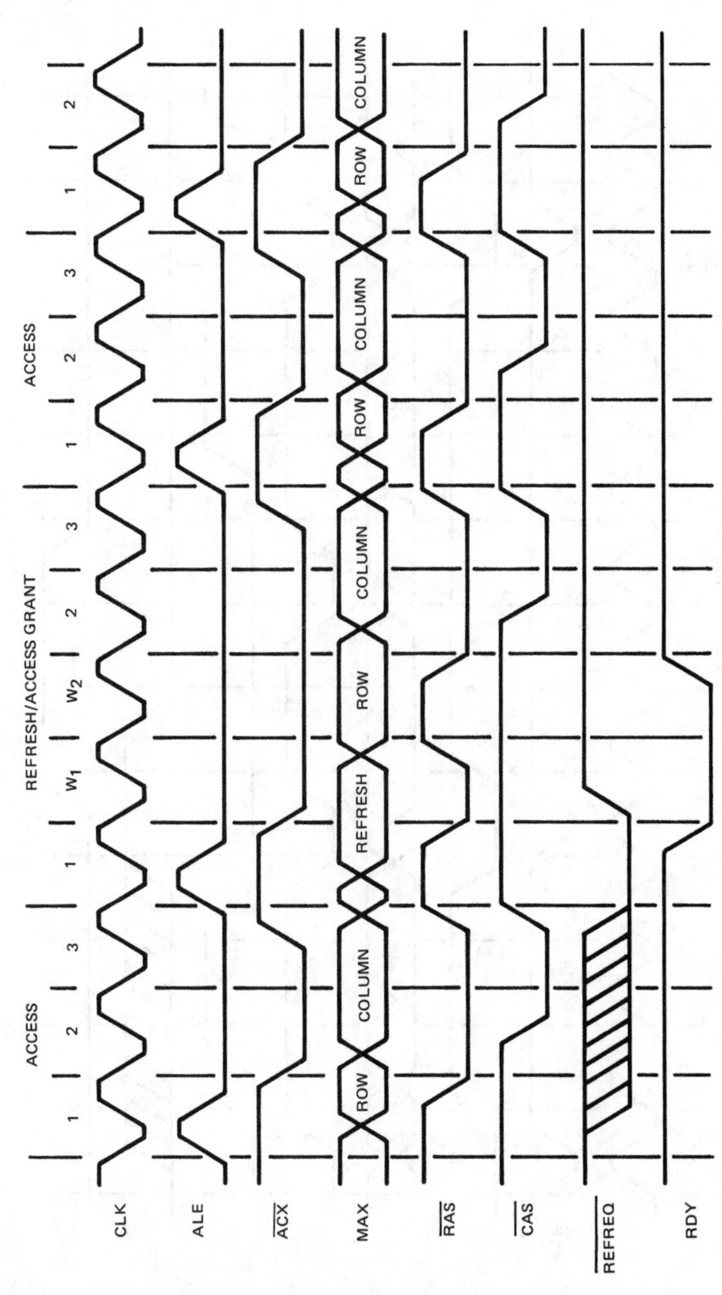

TEXAS INSTRUMENTS
INCORPORATED
POST OFFICE BOX 225012 • DALLAS, TEXAS 75265

typical access/refresh/access cycle
(four cycle, TWST = 0)

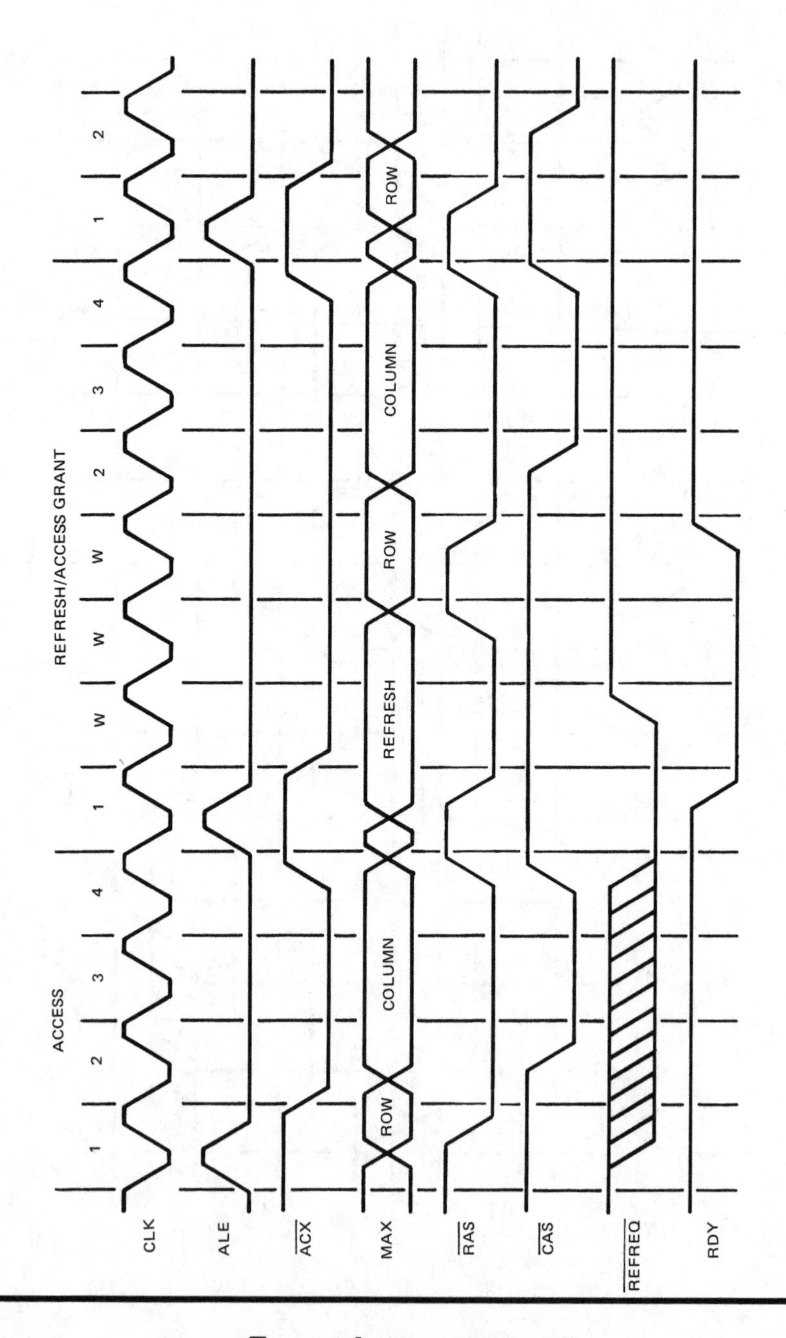

TEXAS INSTRUMENTS
INCORPORATED
POST OFFICE BOX 225012 • DALLAS, TEXAS 75265

typical access/refresh/access cycle
(three cycle, TWST = 1)

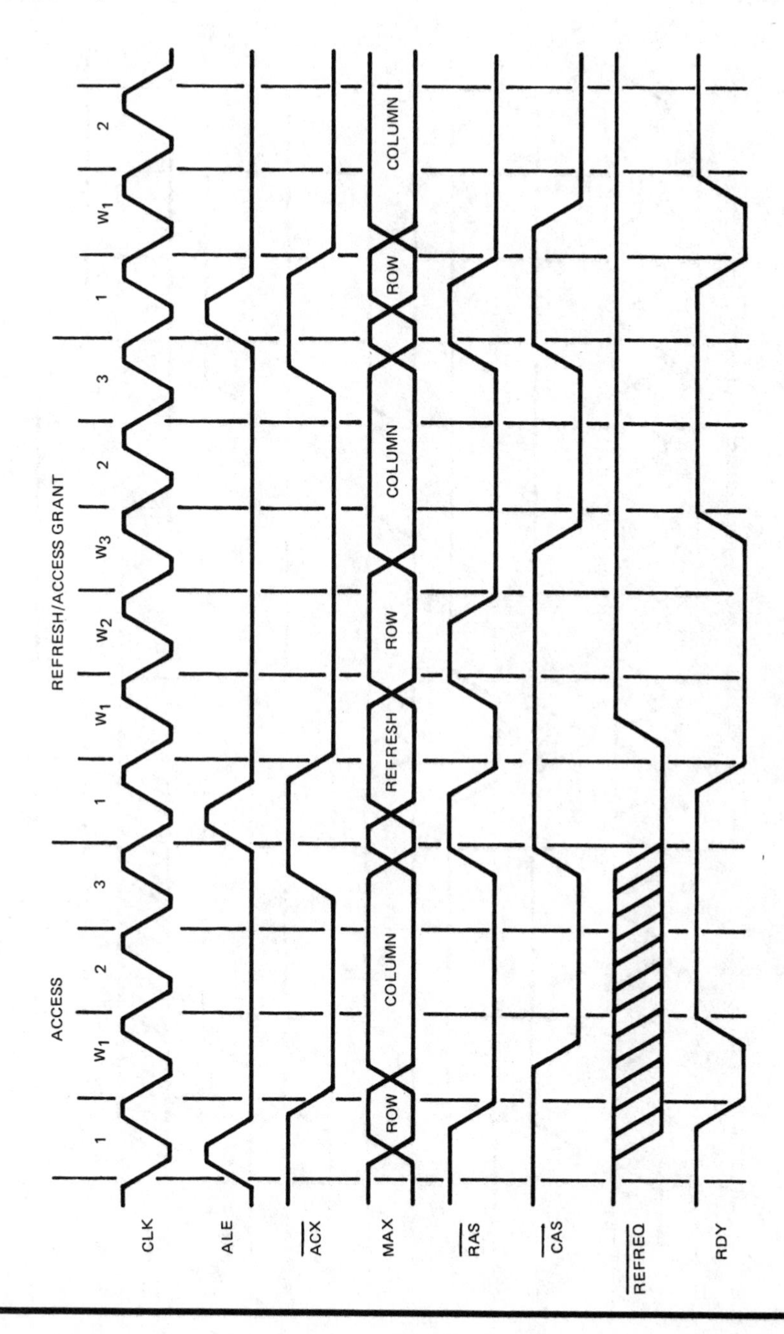

TEXAS INSTRUMENTS
INCORPORATED
POST OFFICE BOX 225012 • DALLAS, TEXAS 75265

typical access/refresh/access cycle
(four cycle, TWST = 1)

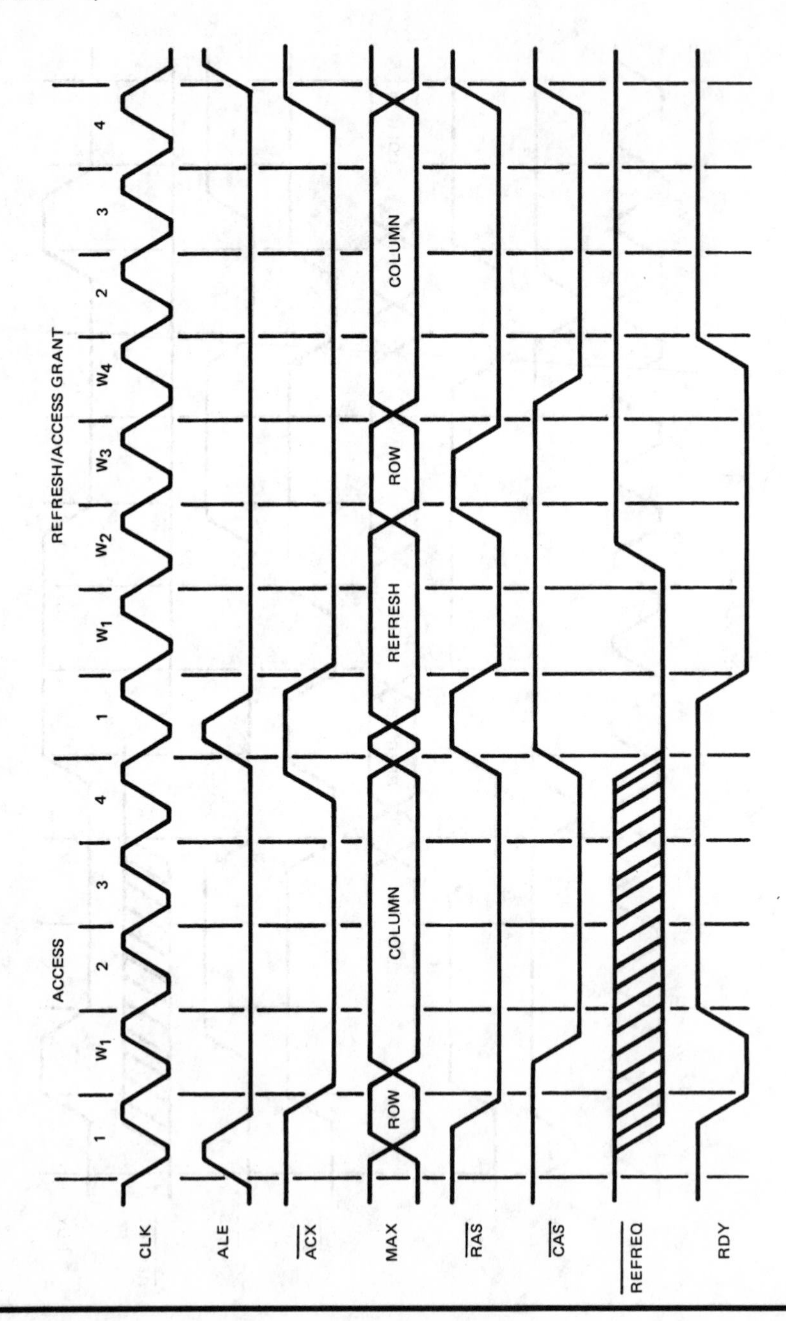

TEXAS INSTRUMENTS
INCORPORATED
POST OFFICE BOX 225012 • DALLAS, TEXAS 75265

Static RAM
and
Memory Support
Data Sheets

TMS 2114 NL; TMS 2114L NL
1024-WORD BY 4-BIT STATIC RAMS

DECEMBER 1979 – REVISED MAY 1982

- Previously Called TMS 4045/TMS 40L45
- 1024 X 4 Organization
- Single +5 V Supply
- High Density 300-mil (7.62 mm) 18-Pin Package
- Fully Static Operation (No Clocks, No Refresh, No Timing Strobe)
- 4 Performance Ranges:

	ACCESS TIME (MAX)	READ OR WRITE CYCLE (MIN)
TMS 2114-15, TMS 2114L-15	150 ns	150 ns
TMS 2114-20, TMS 2114L-20	200 ns	200 ns
TMS 2114-25, TMS 2114L-25	250 ns	250 ns
TMS 2114-45, TMS 2114L-45	450 ns	450 ns

- 400-mV Guaranteed DC Noise Immunity with Standard TTL Loads – No Pull-Up Resistors Required
- Common I/O Capability
- 3-State Outputs and Chip Select Control for OR-Tie Capability
- Fan-Out to 2 Series 74, 1 Series 74S, or 8 Series 74LS TTL Loads
- Low Power Dissipation

	MAX (OPERATING)
TMS 2114	550 mW
TMS 2114L	330 mW

TMS 2114/TMS 2114L
18-PIN PLASTIC
DUAL-IN-LINE PACKAGE
(TOP VIEW)

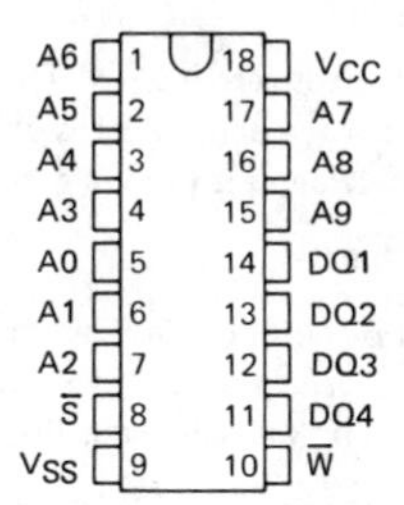

PIN NAMES

A0-A9	Addresses
DQ	Data In/Data Out
S̄	Chip Select
V_{CC}	+5 V Supply
V_{SS}	Ground
W̄	Write Enable

description

This series of static random-access memories is organized as 1024 words of 4 bits each. Static design results in reducing overhead costs by elimination of refresh-clocking circuitry and by simplification of timing requirements. Because this series is fully static, chip select may be tied low to further simplify system timing. Output data is always available during a read cycle.

All inputs and outputs are fully compatible with Series 74, 74S or 74LS TTL. No pull-up resistors are required. This 4K Static RAM series is manufactured using TI's reliable N-channel silicon-gate technology to optimize the cost/performance relationship.

The TMS 2114/2114L series is offered in the 18-pin dual-in-line plastic (NL suffix) package designed for insertion in mounting-hole rows on 300-mil (7.62 mm) centers. The series is guaranteed for operation from 0°C to 70°C.

operation

addresses (A0-A9)

The ten address inputs select one of the 1024 4-bit words in the RAM. The address inputs must be stable for the duration of a write cycle. The address inputs can be driven directly from standard Series 54/74 TTL with no external pull-up resistors.

chip select ($\overline{S}$)

The chip-select terminal, which can be driven directly from standard TTL circuits, affects the data-in and data-out terminals. When chip select is at a logic low level, both terminals are enabled. When chip select is high, data-in is inhibited and data-out is in the floating or high-impedance state.

write enable ($\overline{W}$)

The read or write mode is selected through the write enable terminal. A logic high selects the read mode; a logic low selects the write mode. $\overline{W}$ or $\overline{S}$ must be high when changing addresses to prevent erroneously writing data into a memory location. The $\overline{W}$ input can be driven directly from standard TTL circuits.

data-in/data-out (DQ1-DQ4)

Data can be written into a selected device when the write enable input is low. The DQ terminal can be driven directly from standard TTL circuits. The three-state output buffer provides direct TTL compatibility with a fan-out of two Series 74 TTL gates, one Series 74S TTL gate, or eight Series 74LS TTL gates. The DQ terminals are in the high-impedance state when chip select ($\overline{S}$) is high or whenever a write operation is being performed. Data-out is the same polarity as data-in.

TEXAS INSTRUMENTS
INCORPORATED
POST OFFICE BOX 225012 • DALLAS, TEXAS 75265

logic symbol†

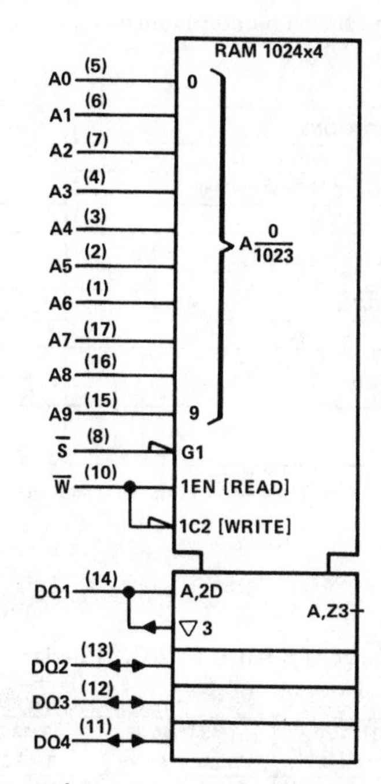

FUNCTION TABLE

W̄	S̄	DQ1-DQ4	MODE
L	L	VALID DATA	WRITE
H	L	DATA OUTPUT	READ
X	H	HI-Z	DEVICE DISABLED

† This symbol is in accordance with IEEE Std 91/ANSI Y32.14 and recent decisions by IEEE and IEC. See explanation on page 289.

absolute maximum ratings over operating free-air temperature (unless otherwise noted)*

Supply voltage, V_{CC} (see Note 1) . . . . −0.5 to 7 V
Input voltage (any input) (see Note 1) . . . . −1 to 7 V
Continuous power dissipation . . . . 1 W
Operating free-air temperature range . . . . 0°C to 70°C
Storage temperature range . . . . −55°C to 150°C

NOTE 1: Voltage values are with respect to the ground material.

* Stresses beyond those listed under "Absolute Maximum Ratings" may cause permanent damage to the device. This is a stress rating only and functional operation of the device at these or any other conditions beyond those indicated in the "Recommended Operating Conditions" section of this specification is not implied. Exposure to absolute-maximum-rated conditions for extended periods may affect device reliability.

recommended operating conditions

PARAMETER	TMS 2114-15, TMS 2114L-15 TMS 2114-20, TMS 2114L-20 TMS 2114-25, TMS 2114L-25			TMS 2114-45, TMS 2114L-45			UNIT
	MIN	NOM	MAX	MIN	NOM	MAX	
Supply voltage, V_{CC}	4.5	5	5.5	4.75	5	5.25	V
Supply voltage, V_{SS}		0			0		V
High-level input voltage, V_{IH}	2		5.5	2		5.25	V
Low-level input voltage, V_{IL} (see Note 2)	−1		0.8	−0.3		0.8	V
Operating free-air temperature, T_A	0		70	0		70	°C

NOTE 2: The algebraic convention, where the more negative (less positive) limit is designated as minimum, is used in this data sheet for logic voltage levels only.

TEXAS INSTRUMENTS
INCORPORATED
POST OFFICE BOX 225012 • DALLAS, TEXAS 75265

electrical characteristics over recommended operating free-air temperature range (unless otherwise noted)

PARAMETER		TEST CONDITIONS†			MIN	TYP‡	MAX	UNIT
V_{OH}	High-level voltage	$I_{OH} = -1$ mA**	V_{CC} = MIN (operating)		2.4			V
V_{OL}	Low-level voltage	$I_{OL} = 3.2$ mA**	V_{CC} = MIN (operating)				0.4	V
I_I	Input current	$V_I = 0$ V to MAX					10	μA
I_{OZ}	Off-state output current	$\overline{S}$ at 2 V or $\overline{W}$ at 0.8 V	$V_O = 0$ V to MAX				±10	μA
I_{CC}	Supply current from V_{CC}	$I_O = 0$ mA, $T_A = 0°C$ (worst case)	TMS 2114	V_{CC} = MAX		90	100	mA
			TMS 2114L	V_{CC} = MAX		50	60	
C_i	Input capacitance	$V_I = 0$ V, f = 1 MHz					8	pF
C_o	Output capacitance	$V_O = 0$ V, f = 1 MHz					8	pF

† For conditions shown as MIN or MAX, use the appropriate value specified under recommended operating conditions.
‡ All typical values are at $V_{CC} = 5$ V, $T_A = 25°C$.
* TMS 2114/TMS 2114L-15, -20, -25 only.
** TMS 2114/TMS 2114L-45: $I_{OH} = -200$ μA, $I_{OL} = 2$ mA.

timing requirements over recommended supply voltage range, $T_A = 0°C$ to $70°C$ 1 Series 74 TTL load $C_L = 100$ pF

PARAMETER		TMS 2114-15 TMS 2114L-15		TMS 2114-20 TMS 2114L-20		TMS 2114-25 TMS 2114L-25		TMS 2114-45 TMS 2114L-45		UNIT
		MIN	MAX	MIN	MAX	MIN	MAX	MIN	MAX	
$t_{c(rd)}$	Read cycle time	150		200		250		450		ns
$t_{c(wr)}$	Write cycle time	150		200		250		450		ns
$t_{w(W)}$	Write pulse width	80		100		100		200		ns
$t_{su(A)}$	Address set up time	0		0		0		0		ns
$t_{su(S)}$	Chip select set up time	80		100		100		200		ns
$t_{su(D)}$	Data set up time	80		100		100		200		ns
$t_{h(D)}$	Data hold time	0		0		0		0		ns
$t_{h(A)}$	Address hold time	0		0		0		20		ns

TEXAS INSTRUMENTS
INCORPORATED
POST OFFICE BOX 225012 • DALLAS, TEXAS 75265

switching characteristics over recommended voltage range, T_A = 0°C to 70°C, 1 Series 74 TTL load, C_L = 100 pF

PARAMETER		TMS 2114-15 TMS 2114L-15		TMS 2114-20 TMS 2114L-20		TMS 2114-25 TMS 2114L-25		TMS 2114-45 TMS 2114L-45		UNIT
		MIN	MAX	MIN	MAX	MIN	MAX	MIN	MAX	
$t_{a(A)}$	Access time from address		150		200		250		450	ns
$t_{a(S)}$	Access time from chip select (or output enable) low		70		85		100		120	ns
$t_{a(W)}$	Access time from write enable high		70		85		100		120	ns
$t_{v(A)}$	Output data valid after address change	20		20		20		20		ns
$t_{dis(S)}$	Output disable time after chip select (or output enable) high		50		60		60		100	ns
$t_{dis(W)}$	Output disable time after write enable low		50		60		60		100	ns

read cycle timing**

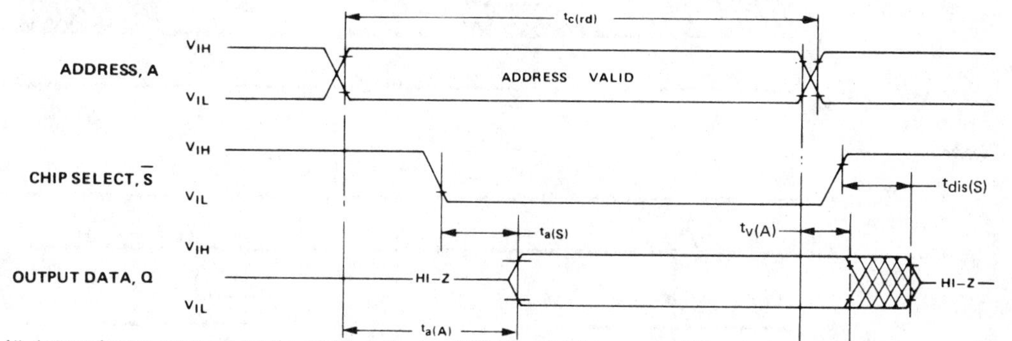

All timing reference points are 0.8 V and 2.0V on inputs and 0.6 V and 2.2 V on outputs (90% points). Input rise and fall times equal 10 nanoseconds.

**Write enable Is high for a read cycle.

early write cycle timing

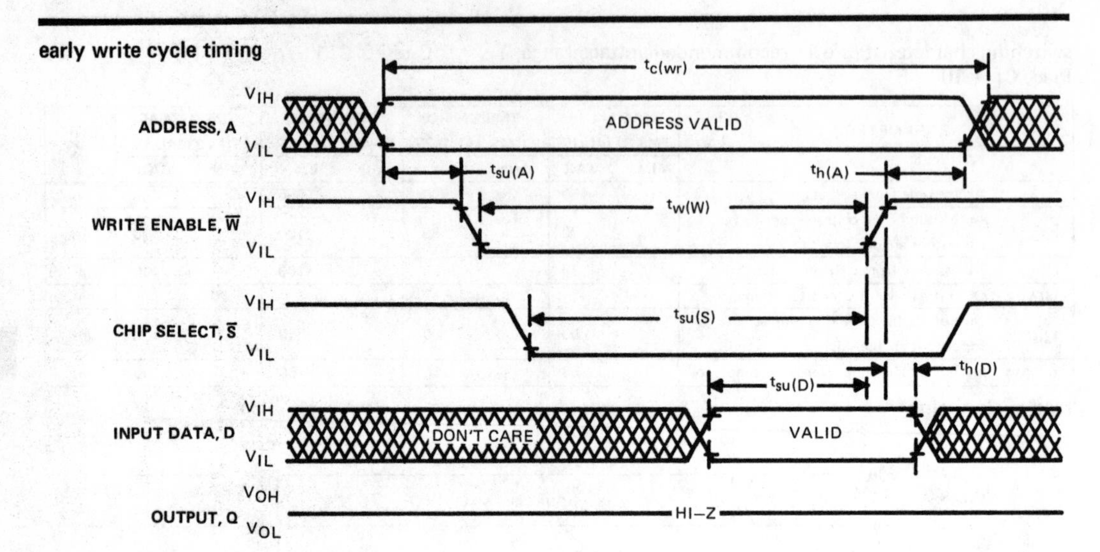

read-write cycle timing

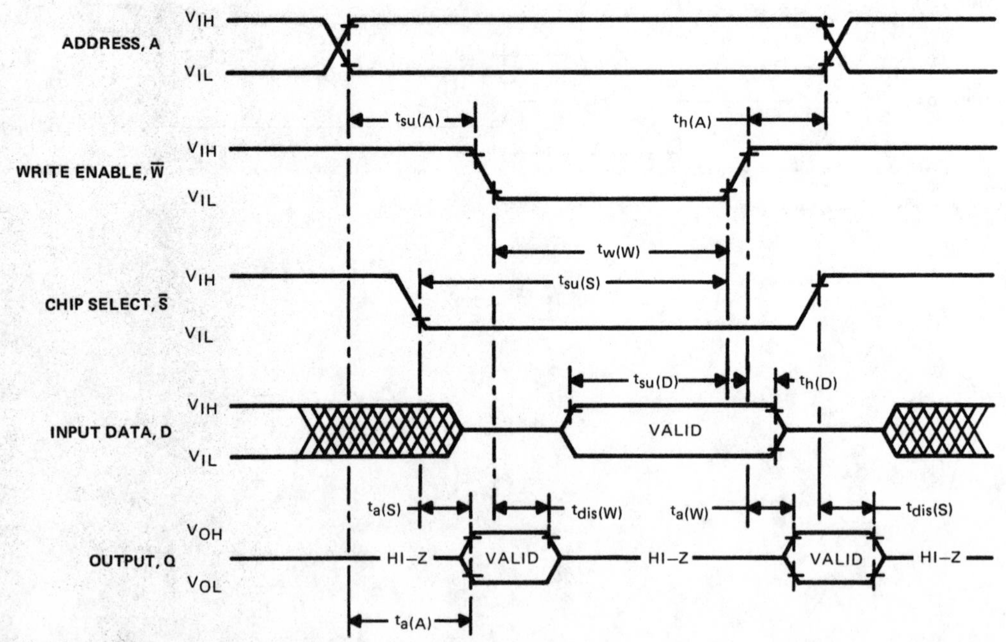

applications data

Early write cycle avoids DQ conflicts by controlling the write time with $\overline{S}$. On the diagram above, the write operation will be controlled by the leading edge of $\overline{S}$, not $\overline{W}$. Data can only be written when both $\overline{S}$ and $\overline{W}$ are low. Either $\overline{S}$ or $\overline{W}$ being high inhibits the write operation. To prevent erroneous data being written into the array, the addresses must be stable during the write cycle as defined by $t_{su(A)}$, $t_{w(W)}$, and $t_{h(A)}$.

TEXAS INSTRUMENTS
INCORPORATED
POST OFFICE BOX 225012 • DALLAS, TEXAS 75265

TMS 2147H JL, NL, FPL
FAST 4096-WORD BY 1-BIT STATIC RAM

FEBRUARY 1981 – REVISED MAY 1982

- 4096 X 1 Organization
- Single +5 V Supply (±10% Tolerance)
- High-Density 300-mil (7.62 mm) Packages
- Fully Static Operation (No Clocks, No Refresh, No Timing Strobe)
- Fast . . . 4 Performance Ranges:

	ACCESS TIME (MAX)	READ OR WRITE CYCLE (MIN)
TMS 2147H-3	35 ns	35 ns
TMS 2147H-4	45 ns	45 ns
TMS 2147H-5	55 ns	55 ns
TMS 2147H-7	70 ns	70 ns

- Inputs and Outputs TTL Compatible
- Common I/O Capability
- 3-State Outputs and Chip Enable Control for OR-Tie Capability
- Automatic Chip Enable/Power Down Operation
- Reliable SMOS (Scaled-MOS) N-Channel Technology
- Direct Performance Upgrade for Industry Standard 2147

**TMS 2147H
18-PIN PLASTIC AND CERAMIC
DUAL-IN-LINE PACKAGES
(TOP VIEW)**

Pin	Name	Pin	Name
A0	1	18	V_{CC}
A1	2	17	A6
A2	3	16	A7
A3	4	15	A8
A4	5	14	A9
A5	6	13	A10
Q	7	12	A11
$\overline{W}$	8	11	D
V_{SS}	9	10	$\overline{E}$

**18-PIN PLASTIC
CHIP CARRIER PACKAGE
(TOP VIEW)**

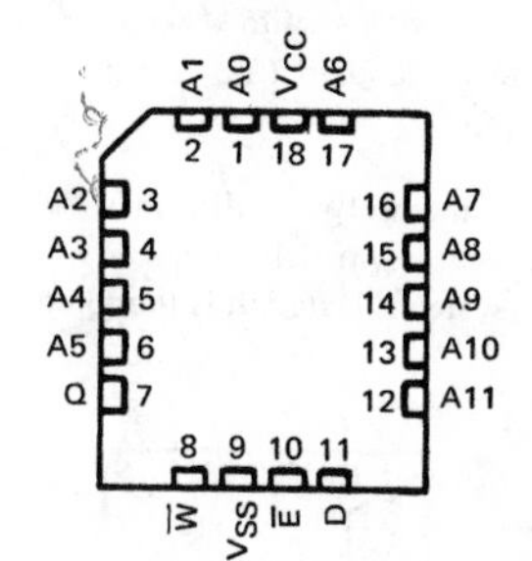

PIN NAMES

A0-A11	Addresses
D	Data In
Q	Data Out
$\overline{E}$	Chip Enable/Power Down
V_{CC}	+5 V Supply
V_{SS}	Ground
$\overline{W}$	Write Enable

description

These high-speed static random-access memories are organized as 4096 words of 1 bit. Static design results in reduced overhead costs by elimination of refresh-clocking circuitry and by simplification of timing requirements. Automatic chip enable/power down allows devices to be placed in the reduced-power mode whenever deselected.

All inputs and outputs are fully compatible with Series 74, 74S or 74LS TTL. No pull-up resistors are required. These 4K static RAM series are manufactured using TI's reliable state-of-the-art SMOS (scaled MOS) N-channel silicon-gate technology to optimize the cost/performance relationship.

The TMS 2147H is offered in 18-pin dual-in-line plastic (NL suffix) and ceramic (JL suffix) packages designed for insertion in mounting-hole rows on 300-mil (7.62 mm) centers. An 18-pin plastic chip carrier (FP suffix) is also available. The series is guaranteed for operation from 0°C to 70°C.

TEXAS INSTRUMENTS
INCORPORATED
POST OFFICE BOX 225012 • DALLAS, TEXAS 75265

operation

addresses (A0-A11)

The 12 address inputs select one of the 4096 storage locations in the RAM. The address inputs must be stable for the duration of a write cycle. The address inputs can be driven directly from standard Series 54/74 TTL with no external pull-up resistors.

chip enable/power down ($\overline{E}$)

The chip enable/power down terminal, which can be driven directly by standard TTL circuits, affects the data-in and data-out terminals and the internal functioning of the chip itself. Whenever the chip enable/power down is low (enabled), the device is operational, input and output terminals are enabled, and data can be read or written. When the chip enable/power down terminal is high (disabled), the device is deselected and put into a reduced-power standby mode. Data is retained during standby.

write-enable ($\overline{W}$)

The read or write mode is selected through the write-enable terminal. A logic high selects the read mode; a logic low selects the write mode. $\overline{W}$ must be high when changing addresses to prevent erroneously writing data into a memory location. The $\overline{W}$ input can be driven directly from standard TTL circuits.

data-in (D)

Data can be written into a selected device when the write-enable input is low. The data-in terminal can be driven directly from standard TTL circuits.

data-out (Q)

The three-state output buffer provides direct TTL compatibility. The output is in the high-impedance state when chip enable/power down ($\overline{E}$) is high or whenever a write operation is being performed, facilitating device operation in common I/O systems. Data-out is the same polarity as data-in.

logic symbol†

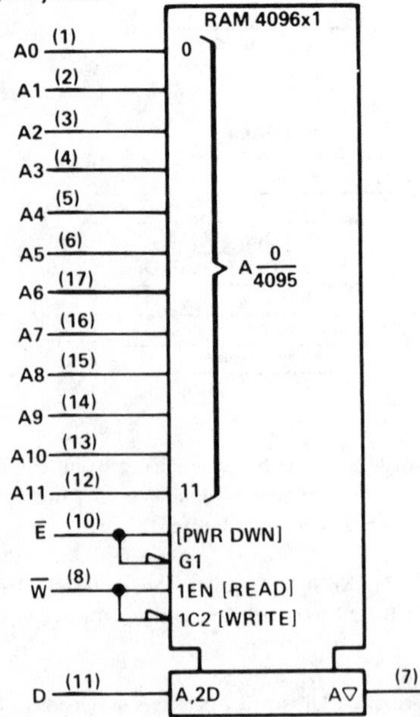

FUNCTION TABLE

INPUTS		OUTPUT	MODE
$\overline{E}$	$\overline{W}$	Q	
H	X	Hi-Z	POWER DOWN
L	L	Hi-Z	WRITE
L	H	DATA OUT	READ

† This symbol is in accordance with IEEE Std 91/ANSI Y32.14 and recent decisions by IEEE and IEC. See explanation on page 289.

TEXAS INSTRUMENTS
INCORPORATED
POST OFFICE BOX 225012 • DALLAS, TEXAS 75265

functional block diagram

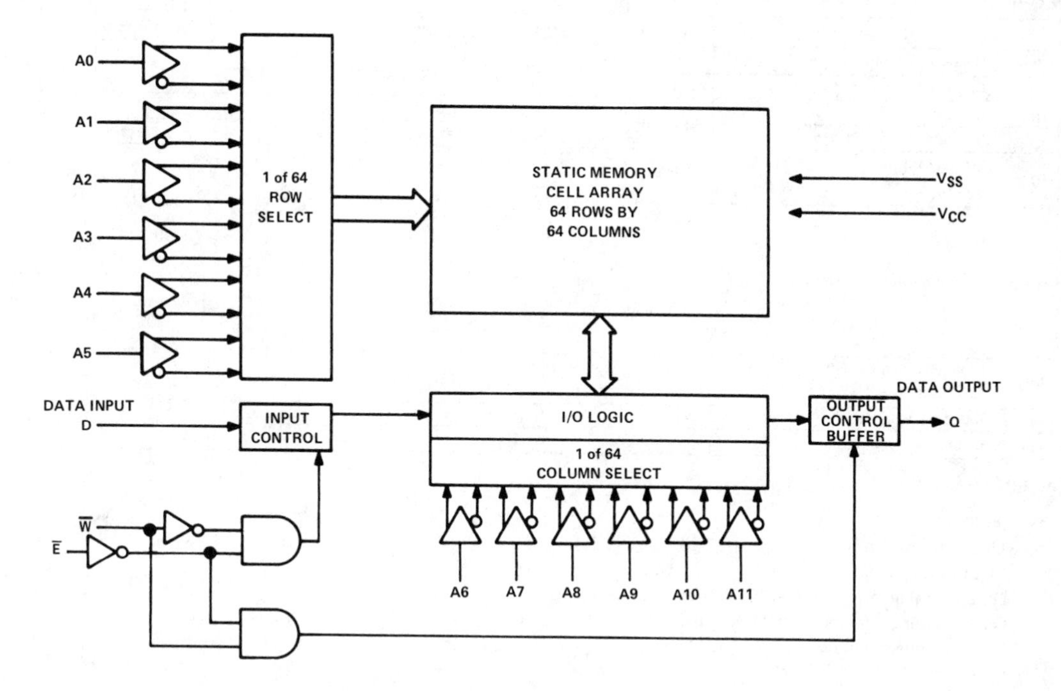

absolute maximum ratings over operating ambient temperature† range (unless otherwise noted)‡

Supply voltage, V_{CC} (see Note 1)	−1.5 V to 7 V
Input voltage (any input) (see Note 1)	−1.5 V to 7 V
Continuous power dissipation	1 W
Operating ambient temperature range	0°C to 70°C
Storage temperature range	−65°C to 150°C

recommended operating conditions

PARAMETER	MIN	NOM	MAX	UNIT
Supply voltage, V_{CC}	4.5	5	5.5	V
Supply voltage, V_{SS}		0		V
High-level input voltage, V_{IH}	2		6	V
Low-level input voltage, V_{IL}	−1§		0.8	V
Operating ambient temperature†, T_A	0		70	°C

†The ambient temperature conditions assume air moving perpendicular to the longitudinal axis and parallel to the seating plane of the device at a velocity of 400 ft/min (122 m/min) with the device under test soldered to a 4 X 6 X 0.062-inch (102 X 152 X 1.6-mm) double-sided 2-ounce copper-clad circuit board (plating thickness 0.07 mm).

‡Stresses beyond those listed under "Absolute Maximum Ratings" may cause permanent damage to the device. This is stress rating only and functional operation of the device at these or any other conditions beyond those indicated in the "Recommended Operating Conditions" section of this specification is not implied. Exposure to absolute-maximum-rated conditions for extended periods may affect device reliability.

§The algebraic convention, where the more negative limit is designated as minimum, is used in this data sheet for logic voltage levels only.

NOTE 1: Voltage values are with respect to the ground terminal.

TEXAS INSTRUMENTS
INCORPORATED
POST OFFICE BOX 225012 • DALLAS, TEXAS 75265

electrical characteristics over recommended operating ambient temperature† range (unless otherwise noted)

PARAMETER		TEST CONDITIONS	MIN	TYP‡	MAX	UNIT
V_{OH}	High-level output voltage	$I_{OH} = -4$ mA, $V_{CC} = 4.5$ V	2.4			V
V_{OL}	Low-level output voltage	$I_{OL} = 8$ mA, $V_{CC} = 4.5$ V			0.4	V
I_I	Input current	$V_I = 0$ V to 5.5 V			10	μA
I_{OZ}	Off-state output current	$\overline{E}$ at 2 V, $V_O = 0$ V to 4.5 V, $V_{CC} = 5.5$ V			±50	μA
I_{CC1}	Standby supply current from V_{CC}	$\overline{E}$ at V_{IH}		18	30	mA
I_{CC2}	Operating supply current from V_{CC}	$\overline{E}$ at V_{IL}, $I_O = 0$ mA, $T_A = 0°C$ (worst case)		90	120	mA
		$\overline{E}$ at V_{IL}, $I_O = 0$ mA, $T_A = 70°C$			100	mA
I_{PO}	Peak power-on current (see Note 2)	V_{CC} = GND to V_{CC} min, $\overline{E}$ at lower of V_{CC} or V_{IH} min			70	mA
C_i	Input capacitance	$V_I = 0$ V, f = 1 MHz			5	pF
C_O	Output capacitance	$V_O = 0$ V, f = 1 MHz			6	pF

ac test conditions

Input pulse levels . . . . GND to 3 V
Input rise and fall times . . . . 5 ns
Input timing reference levels . . . . 1.5 V
Output timing reference level (2147H-3) . . . . 1.5 V
Output timing reference high level (2147H-4, -5, -7) . . . . 2 V
Output timing reference, low level (2147H-4, -5, -7) . . . . 0.8 V
Output loading . . . . See Figure 1

timing requirements over recommended supply voltage range and operating ambient temperature† range

PARAMETER		TMS 2147H-3		TMS 2147H-4		TMS 2147H-5		TMS 2147H-7		UNIT
		MIN	MAX	MIN	MAX	MIN	MAX	MIN	MAX	
$t_{c(rd)}$	Read cycle time	35		45		55		70		ns
$t_{c(wr)}$	Write cycle time	35		45		55		70		ns
$t_{w(W)}$	Write pulse width	20		25		25		40		ns
$t_{su(A)}$	Address setup time	0		0		0		0		ns
$t_{su(E)}$	Chip enable setup time	35		45		45		55		ns
$t_{su(D)}$	Data setup time	20		25		25		30		ns
$t_{h(D)}$	Data hold time	10		10		10		10		ns
$t_{h(A)}$	Address hold time	0		0		10		15		ns
t_{AVWH}	Address valid to write enable high	35		45		45		55		ns

† The ambient temperature conditions assume air moving at a velocity of 400 ft/min (122 m/min).
‡ All typical values are at $V_{CC} = 5$, $T_A = 25°C$.

NOTE 2: I_{PO} exceeds I_{CC1} maximum during power on. A pull-up resistor to V_{CC} on the $\overline{E}$ input is required to keep the device deselected; otherwise, power-on current approaches I_{CC2}.

TEXAS INSTRUMENTS
INCORPORATED
POST OFFICE BOX 225012 • DALLAS, TEXAS 75265

switching characteristics over recommended supply voltage range and operating ambient temperature† range

PARAMETER		TEST CONDITIONS	TMS 2147H-3		TMS 2147H-4		TMS 2147H-5		TMS 2147H-7		UNIT
			MIN	MAX	MIN	MAX	MIN	MAX	MIN	MAX	
$t_{a(A)}$	Access time from address	R_L = 510 Ω, C_L = 30 pF, See Figure 1		35		45		55		70	ns
$t_{a(E)}$	Access time from chip enable			35		45		55		70	ns
$t_{v(A)}$	Output data valid after address change		5		5		5		5		ns
$t_{dis(W)}$	Output disable time from write enable ‡			20		25		25		35	ns
$t_{en(W)}$	Output enable time from write enable‡		0		0		0		0		ns
$t_{dis(E)}$	Output disable time from chip enable‡			30		30		30		40	ns
$t_{en(E)}$	Output enable time from chip enable‡		5		5		10		10		ns
t_{pwrdn}	Power down time from chip select			20		20		20		30	ns

† The ambient temperature conditions assume air moving at a velocity of 400 ft/min (122 m/min).
‡Transition is measured ±500 mV from steady state voltage with specified loading in Figure 1.

PARAMETER MEASUREMENT INFORMATION

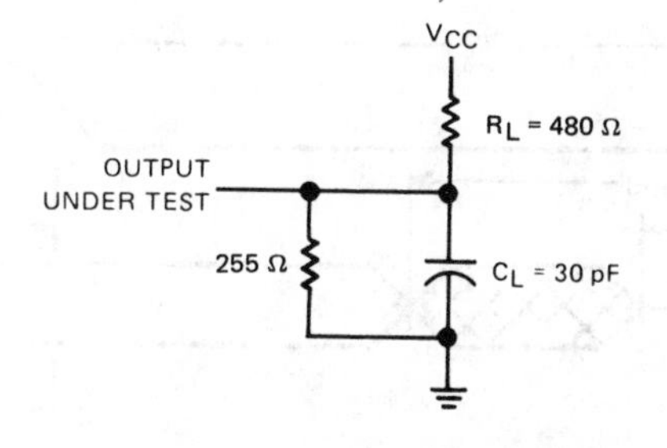

FIGURE 1 – LOAD CIRCUIT

read cycle timing

from address

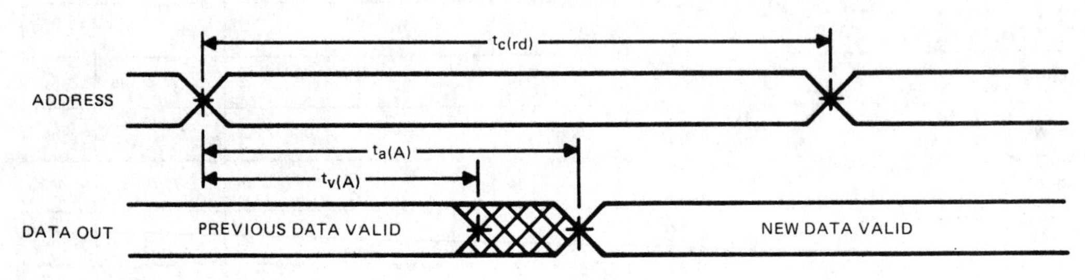

$\overline{W}$ is high, $\overline{E}$ is low.

from chip select

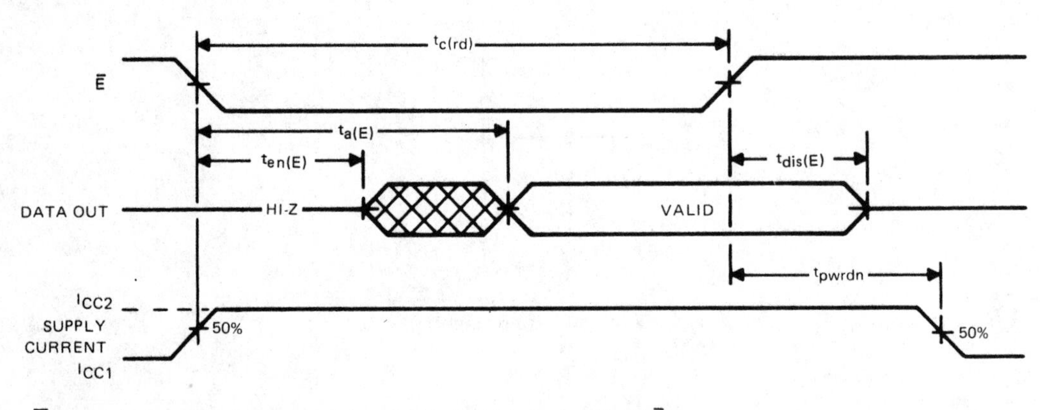

$\overline{W}$ is high, address is valid prior to or simultaneously with the high-to-low transition of $\overline{E}$.

TEXAS INSTRUMENTS
INCORPORATED
POST OFFICE BOX 225012 • DALLAS, TEXAS 75265

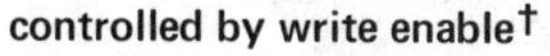
write cycle timing
controlled by write enable†

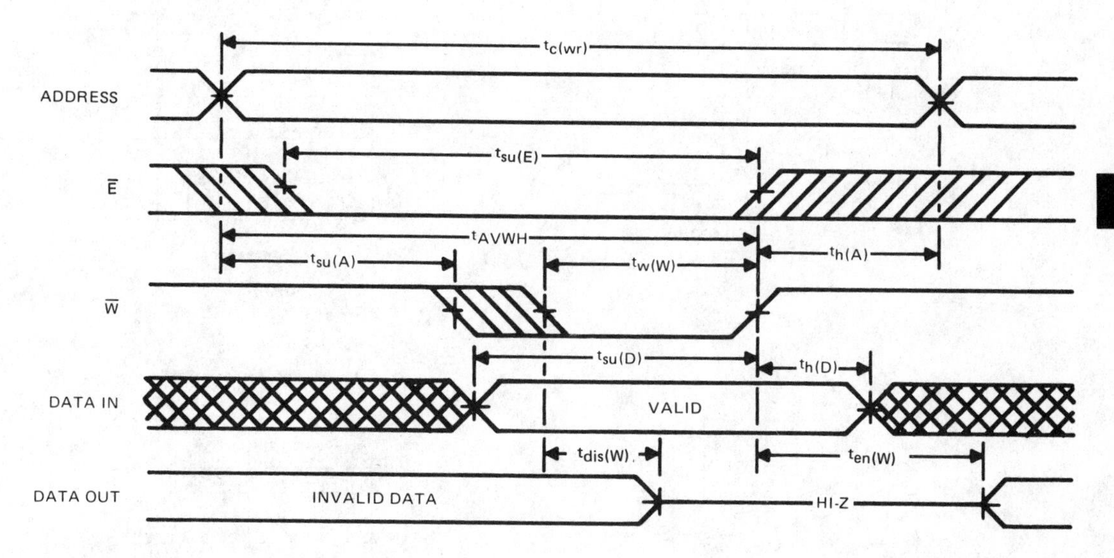

controlled by chip enable†

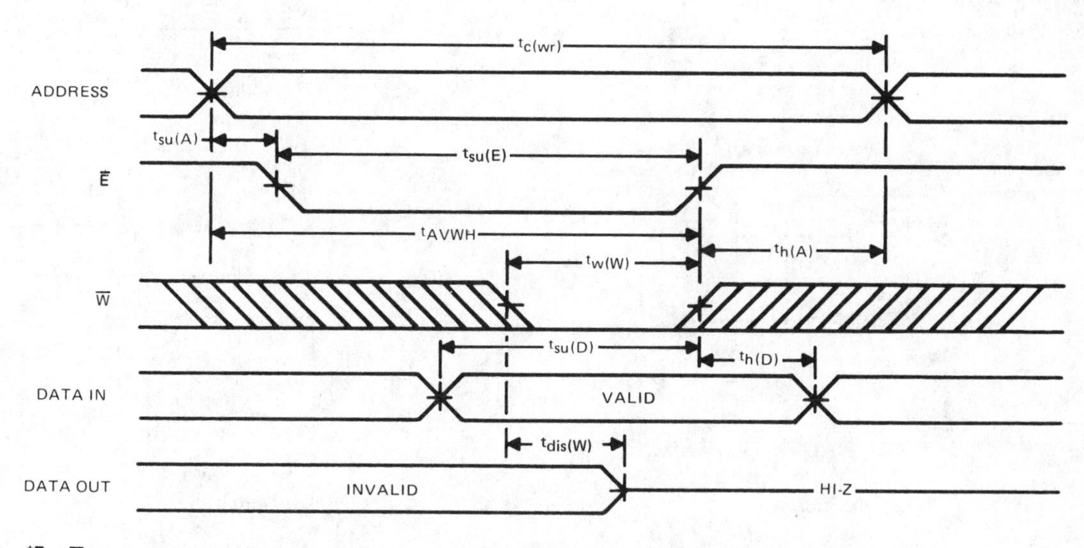

†$\overline{E}$ or $\overline{W}$ must be high during address transitions.
NOTE: If $\overline{E}$ goes high simultaneously with $\overline{W}$ going high, the output remains in the high-impedance state.

TYPICAL CHARACTERISTICS

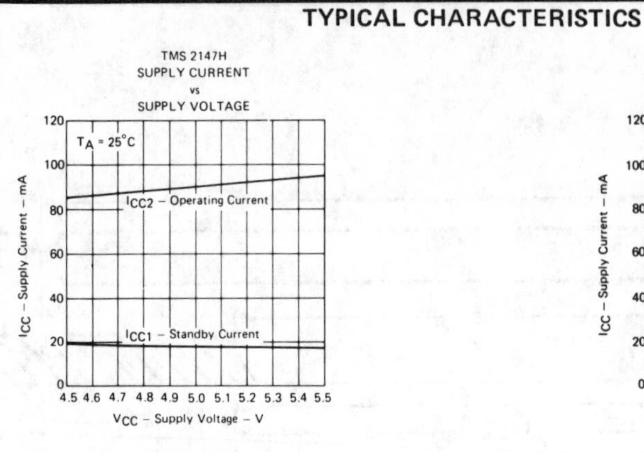

FIGURE 2

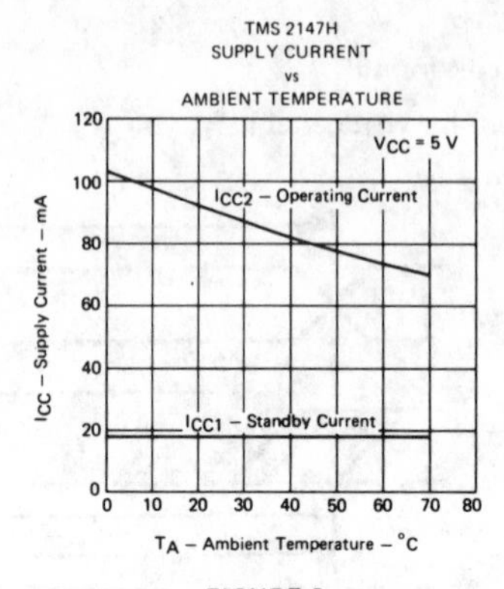

FIGURE 3

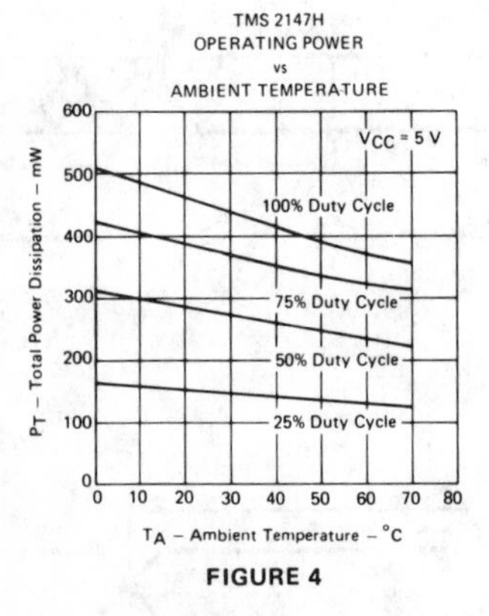

FIGURE 4

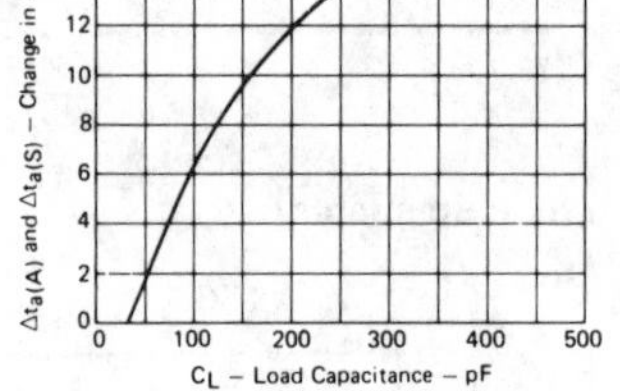

FIGURE 5

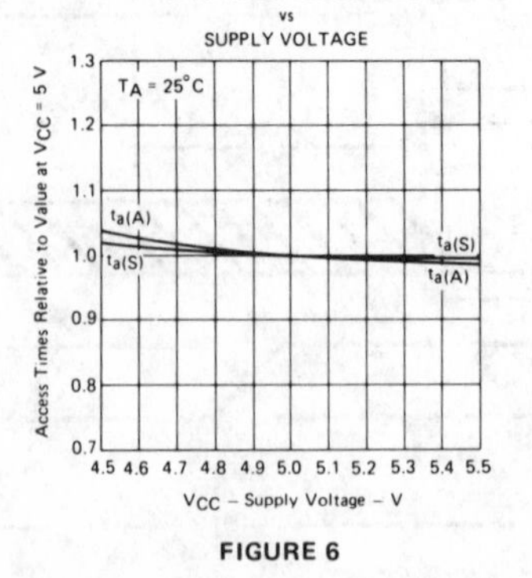

FIGURE 6

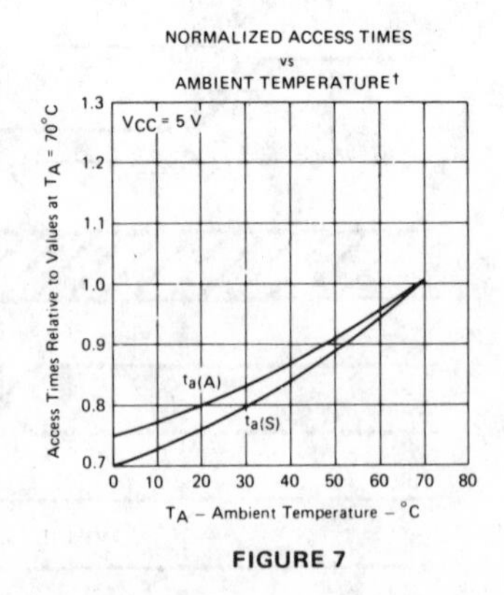

FIGURE 7

† The ambient temperature conditions assume air moving at a velocity of 400 feet per minute.

MOS
LSI

TMS 2149 JL, NL, FPL
FAST 1024-WORD BY 4-BIT STATIC RAM

JANUARY 1982 – REVISED MAY 1982

- 1024 X 4 Organization
- Single +5 V Supply (±10% Tolerance)
- High-Density 300 mil (7.62 mm) Packages
- Fully Static Operation (No Clocks, No Refresh, No Timing Strobe)
- Fast . . . 4 Performance Ranges:

	ADDRESS ACCESS TIME (MAX)	CS ACCESS TIME (MAX)	READ OR WRITE CYCLE (MIN)
TMS 2149-3	35 ns	15 ns	35 ns
TMS 2149-4	45 ns	20 ns	45 ns
TMS 2149-5	55 ns	25 ns	55 ns
TMS 2149-7	70 ns	30 ns	70 ns

- Inputs and Outputs TTL Compatible
- Common I/O
- 3-State Outputs
- Reliable SMOS (Scaled-MOS) N-Channel Technology
- Industry Standard 1K X 4 Pinout

TMS 2149
18-PIN PLASTIC AND CERAMIC DUAL-IN-LINE PACKAGES
(TOP VIEW)

Pin name	Pin	Pin	Pin name
A3	1	18	V_{CC}
A2	2	17	A8
A1	3	16	A4
A0	4	15	A5
A9	5	14	DQ1
A7	6	13	DQ2
A6	7	12	DQ3
$\overline{S}$	8	11	DQ4
V_{SS}	9	10	$\overline{W}$

18-PIN PLASTIC CHIP CARRIER PACKAGE
(TOP VIEW)

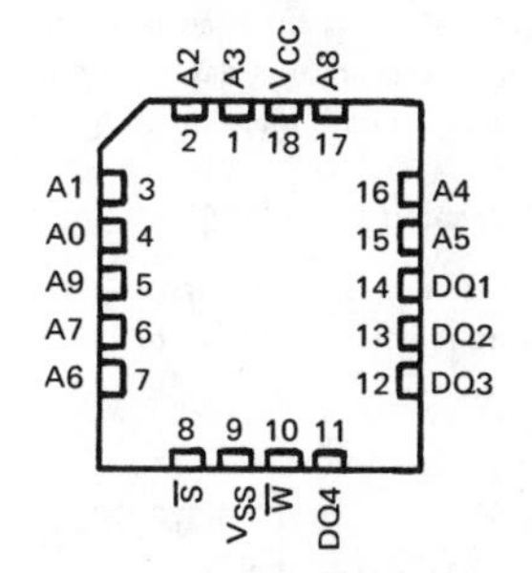

PIN NAMES

A0 - A9	Addresses
DQ	Data In/Data Out
$\overline{S}$	Chip Select
V_{CC}	+5 V Supply
V_{SS}	Ground
$\overline{W}$	Write Enable

description

These high-speed static random-access memories are organized as 1024 words of four bits each. Static design results in reduced overhead costs by elimination of refresh-clocking circuitry and by simplification of timing requirements.

All inputs and outputs are fully compatible with Series 74, 74S, or 74LS TTL. No pull-up resistors are required. These 4K static RAM series are manufactured using TI's reliable state-of-the-art SMOS (scaled MOS) N-channel silicon-gate technology to optimize the cost/performance relationship.

The TMS 2149 is offered in 18-pin dual-in-line plastic (NL suffix) and ceramic (JL suffix) packages designed for insertion in mounting-hole rows on 300-mil (7.62-mm) centers. An 18-pin plastic chip carrier (FP suffix) package is also available. The series is guaranteed for operation from 0°C to 70°C.

TEXAS INSTRUMENTS
INCORPORATED
POST OFFICE BOX 225012 • DALLAS, TEXAS 75265

operation

addresses (A0-A9)

The 10 address inputs select one of the 1024 4-bit words in the RAM. The address inputs must be stable for the duration of a write cycle. The address inputs can be driven directly from standard Series 54/74 TTL with no external pull-up resistors.

chip-select ($\overline{S}$)

The chip-select terminal, which can be driven directly by standard TTL circuits, affects the data-in/data-out (DQ) terminals and the internal functioning of the chip itself. Whenever the chip-select terminal is low (enabled), the device is operational. DQ terminals function as data-in or data-out depending on the level of the write enable terminal. When the chip-select terminal is high (disabled), the device is deselected, data-in is inhibited and data-out is in the floating or high impedance state.

write enable ($\overline{W}$)

The read or write mode is selected through the write enable terminal. If chip-select is low (enabled), a logic high on write enable selects the read mode and activates data-out on the DQ terminals. A logic low on write enable selects the write mode and accepts data-in from the DQ terminals. $\overline{W}$ or $\overline{S}$ must be high when changing addresses to prevent erroneously writing data into a memory location.

data-in/data-out (DQ1-DQ4)

The DQ terminals can be driven directly from standard TTL circuits. The DQ terminals are in the high impedance state when chip-select ($\overline{S}$) is high. Data-out is the same polarity as data-in.

logic symbol†

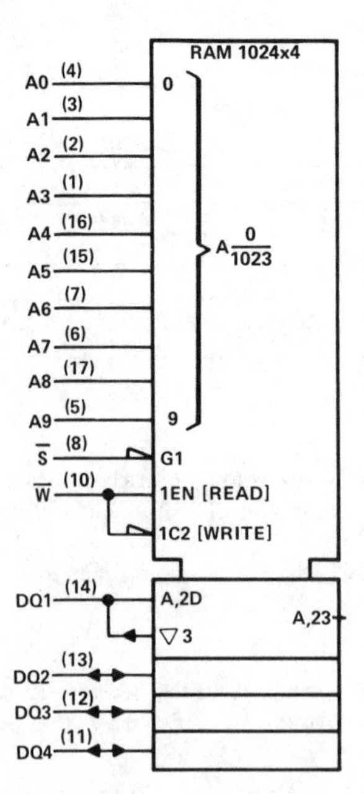

FUNCTION TABLE

$\overline{W}$	$\overline{S}$	DQ1-DQ4	MODE
L	L	VALID DATA	WRITE
H	L	DATA OUTPUT	READ
X	H	HI-Z	DEVICE DISABLED

† This symbol is in accordance with IEEE Std 91/ANSI Y32.14 and recent decisions by IEEE and IEC. See explanation on page 289.

functional block diagram

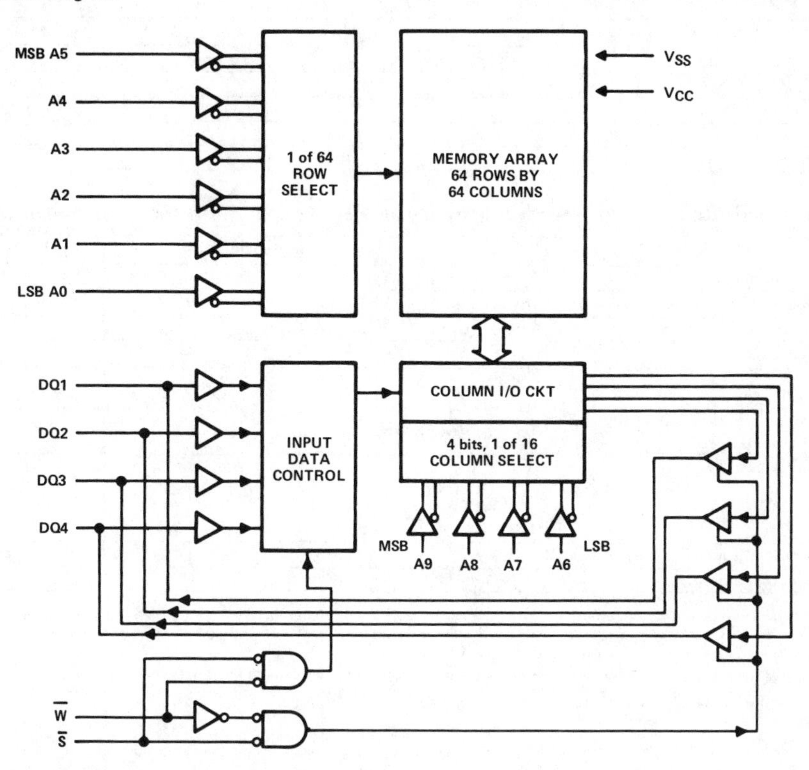

4

absolute maximum ratings over operating ambient temperature† range (unless otherwise noted) ‡

Supply voltage, V_{CC} (see Note 1)	−1.5 V to 7 V
Input voltage (any input) (see Note 1)	−1.5 V to 7 V
Continuous power dissipation	1 W
Operating ambient temperature range	0°C to 70°C
Storage temperature range	−65°C to 150°C

†The ambient temperature conditions assume air moving perpendicular to the longitudinal axis and parallel to the seating plane of the device at a velocity of 400 ft/min (122 m/min) with the device under test soldered to a 4 X 6 X 0.062-inch (102 X 152 X 1.6-mm) double-sided 2-ounce copper-clad circuit board (plating thickness 0.07 mm).

‡Stresses beyond those listed under "Absolute Maximum Ratings" may cause permanent damage to the device. This is stress rating only and functional operation of the device at these or any other conditions beyond those indicated in the "Recommended Operating Conditions" section of this specification is not implied. Exposure to absolute-maximum-rated conditions for extended periods may affect device reliability.

NOTE 1: Voltage values are with respect to the ground terminal.

recommended operating conditions

PARAMETER	MIN	NOM	MAX	UNIT
Supply voltage, V_{CC}	4.5	5	5.5	V
Supply voltage, V_{SS}		0		V
High-level input voltage, V_{IH}	2		6	V
Low-level input voltage, V_{IL}	−1‡		0.8	V
Operating ambient temperature†, T_A	0		70	°C

electrical characteristics over recommended operating ambient temperature† range (unless otherwise noted)

PARAMETER		TEST CONDITIONS	MIN	TYP§	MAX	UNIT
V_{OH}	High-level output voltage	I_{OH} = −4 mA, V_{CC} = 4.5 V	2.4			V
V_{OL}	Low-level output voltage	I_{OL} = 8 mA, V_{CC} = 4.5 V			0.4	V
I_I	Input current	V_I = 0 V to 5.5 V			10	μA
I_{OZ}	Off-state output current	$\overline{S}$ at 2 V, V_O = 0.2 V to 4.5 V V_{CC} = 5.5 V			±50	μA
I_{CC2}	Operating supply current from V_{CC}	$\overline{S}$ at V_{IL}, I_O = 0 mA, T_A = 0°C (worst case)		90	120	mA
		$\overline{S}$ at V_{IL}, I_O = 0 mA, T_A = 70°C			100	mA
C_i	Input capacitance	V_I = 0 V, f = 1 MHz			5	pF
C_O	Output capacitance	V_O = 0 V, f = 1 MHz			7	pF

† The ambient temperature conditions assume air moving perpendicular to the longitudinal axis and parallel to the seating plane of the device at a velocity of 400 ft/min (122 m/min) with the device under test soldered to a 4 X 6 X 0.062-inch (102 X 152 X 1.6-mm) double-sided 2-ounce copper-clad circuit board (plating thickness 0.07 mm).

‡ The algebraic convention, where the more negative limit is designated as minimum, is used in this data sheet for logic voltage levels only.

§ All typical values are at V_{CC} = 5, T_A = 25°C.

ac test conditions

Input pulse levels . . . . 0 V to 3 V
Input rise and fall times . . . . 5 ns
Input timing reference levels . . . . 1.5 V
Output timing reference level . . . . 1.5 V
Output loading . . . . See Figure 1

TEXAS INSTRUMENTS
INCORPORATED
POST OFFICE BOX 225012 • DALLAS, TEXAS 75265

timing requirements over recommended supply voltage range and operating ambient temperature† range

PARAMETER		TMS 2149-3		TMS 2149-4		TMS 2149-5		TMS 2149-7		UNIT
		MIN	MAX	MIN	MAX	MIN	MAX	MIN	MAX	
$t_{c(rd)}$	Read cycle time	35		45		55		70		ns
$t_{c(wr)}$	Write cycle time	35		45		55		70		ns
$t_{w(W)}$	Write pulse width	30		30		40		50		ns
$t_{su(A)}$	Address setup time	0		0		0		0		ns
$t_{su(S)}$	Chip select setup time	30		30		40		50		ns
$t_{su(D)}$	Data setup time	20		20		20		25		ns
$t_{h(D)}$	Data hold time	5		5		5		5		ns
$t_{h(A)}$	Address hold time	0		5		5		5		ns
t_{AVWH}	Address valid to write enable high	35		40		50		65		ns

switching characteristics over recommended supply voltage range and operating ambient temperature† range

PARAMETER		TEST CONDITIONS	TMS 2149-3		TMS 2149-4		TMS 2149-5		TMS 2149-7		UNIT
			MIN	MAX	MIN	MAX	MIN	MAX	MIN	MAX	
$t_{a(A)}$	Access time from address	$R_L = 480\ \Omega$, $C_L = 30$ pF, See Figure 1		35		45		55		70	ns
$t_{a(S)}$	Access time from chip select			15		20		25		30	ns
$t_{v(A)}$	Output data valid after address change		5		5		5		5		ns
$t_{dis(W)}$	Output disable time from write enable‡	$R_L = 480\ \Omega$, $C_L = 5$ pF, See Figure 2		10		10		20		25	ns
$t_{en(W)}$	Output enable time from write enable‡		5		5		5		5		ns
$t_{dis(S)}$	Output disable time from chip select‡			10		10		15		15	ns
$t_{en(S)}$	Output enable time from chip select‡		5		5		5		5		ns

† The ambient temperature conditions assume air moving at a velocity of 400 ft/min (122 m/min).
‡Transition is measured ±500 mV from steady state voltage with specified loading in Figure 2. This parameter is sampled and not 100% tested.

PARAMETER MEASUREMENT INFORMATION

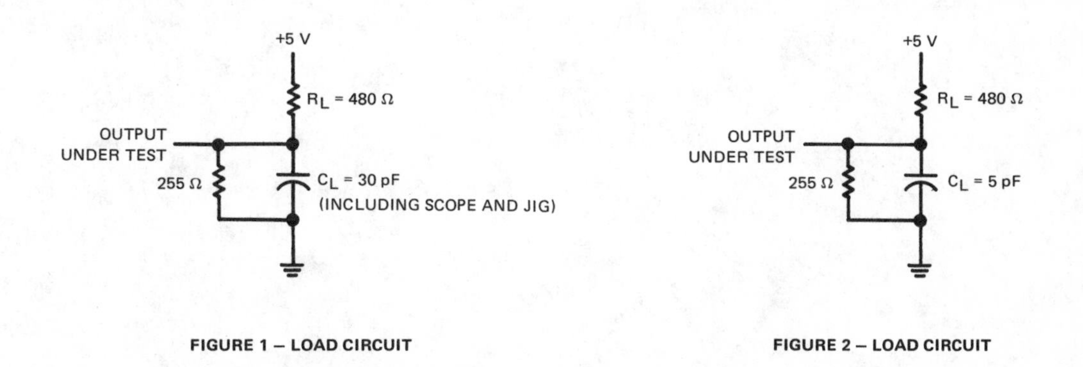

FIGURE 1 – LOAD CIRCUIT

FIGURE 2 – LOAD CIRCUIT

TEXAS INSTRUMENTS
INCORPORATED
POST OFFICE BOX 225012 • DALLAS, TEXAS 75265

AC CHARACTERISTICS

read cycle timing

from address

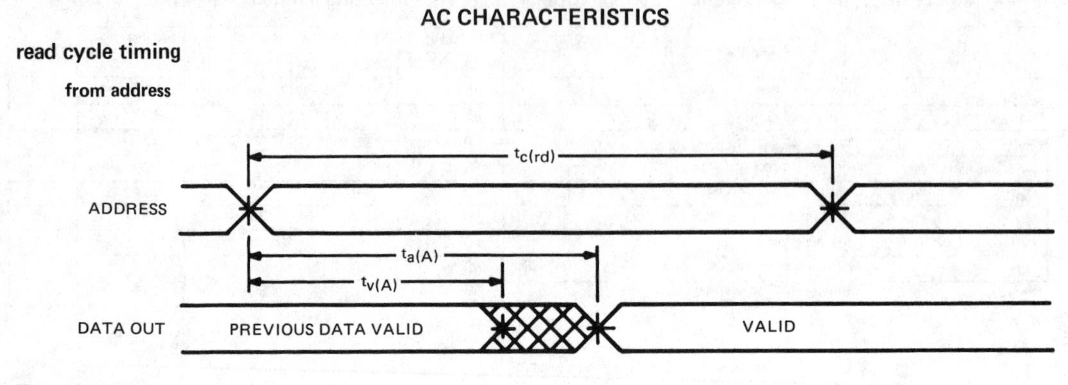

$\overline{W}$ is high, $\overline{S}$ is low.

from chip select

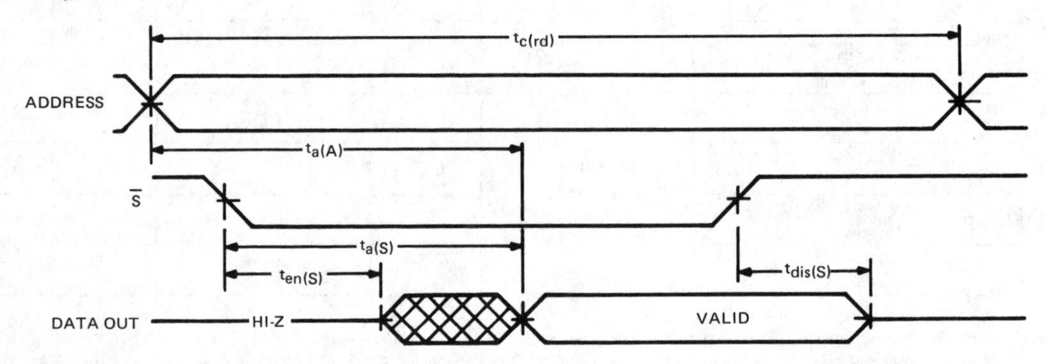

$\overline{W}$ is high.

TEXAS INSTRUMENTS
INCORPORATED
POST OFFICE BOX 225012 • DALLAS, TEXAS 75265

AC CHARACTERISTICS

write cycle timing

controlled by write enable†

controlled by chip select†

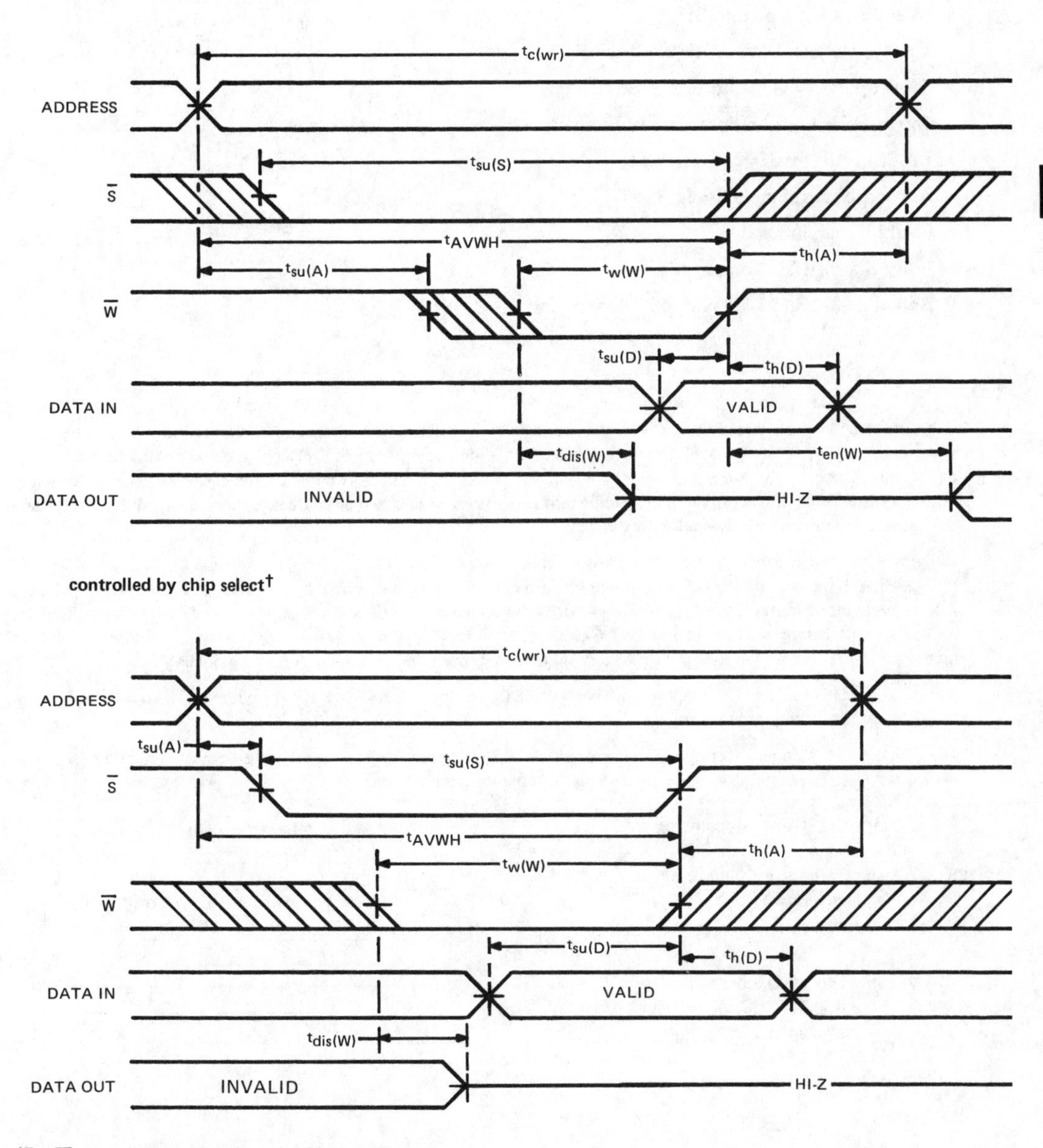

†S̄ or W̄ must be high during address transitions

NOTE: If S̄ goes high simultaneously with W̄ going high, the output remains in the high-impedance state.

TEXAS INSTRUMENTS
INCORPORATED
POST OFFICE BOX 225012 • DALLAS, TEXAS 75265

TMS 2150 JL
CACHE ADDRESS COMPARATOR

MARCH 1982

- Fast Address to Match Valid Delay — Four Speed Ranges: 45 ns, 55 ns, 70 ns, 90 ns
- 512 X 9 Internal RAM
- 300-Mil 24-Pin Ceramic DIP
- Max Power Dissipation: 660 mW
- On-Chip Parity Generation and Checking
- Parity Error Output/Force Parity Error Input
- On-Chip Address/Data Comparator
- Asynchronous, Single-Cycle Reset
- Easily Expandable
- Fully Static, TTL Compatible
- Reliable SMOS (Scaled NMOS) Technology

J PACKAGE
(TOP VIEW)

1	$\overline{RESET}$	24	V_{CC}
2	A5	23	A1
3	A4	22	A0
4	A3	21	A8
5	A2	20	A7
6	D3	19	A6
7	D0	18	D5
8	D1	17	D4
9	D2	16	D7
10	$\overline{W}$	15	D6
11	$\overline{PE}$	14	MATCH
12	V_{SS}	13	$\overline{S}$

description

The 8-bit-slice cache address comparator consists of a high-speed 512 X 9 static RAM array, parity generator, and parity checker, and 9-bit high-speed comparator. It is fabricated using N-channel silicon gate technology for high speed and simple interface with MOS and bipolar TTL circuits. The cache address comparator is easily cascadable for wider tag addresses or deeper tag memories. Significant reductions in cache memory component count, board area, and power dissipation can be achieved with this device.

When $\overline{S}$ is low and $\overline{W}$ is high, the cache address comparator compares the contents of the memory location addressed by A0-A8 with the data on D0-D7 plus generated parity. An equality is indicated by a high level on the MATCH output. A low-level output from $\overline{PE}$ signifies a parity error in the internal RAM data. $\overline{PE}$ is an N-channel open-drain output for easy OR-tieing. During a write cycle ($\overline{S}$ and $\overline{W}$ low), data on D0-D7 plus generated even parity are written in the 9-bit memory location addressed by A0-A8. Also during write, a parity error may be forced by holding $\overline{PE}$ low.

A $\overline{RESET}$ input is provided for initialization. When $\overline{RESET}$ goes low, all 512 X 9 RAM locations will be cleared and the MATCH output will be forced high.

The cache address comparator operates from a single +5 V supply and is offered in a 24-pin 300-mil CERPAK. The device is fully TTL compatible and is guaranteed to operate from 0 °C to 70 °C.

MATCH OUTPUT DESCRIPTION

MATCH = V_{OH} if: [A0-A8] = D0-D7 + parity,
or: $\overline{RESET} = V_{IL}$,
or: $\overline{S} = V_{IH}$,
or: $\overline{W} = V_{IL}$

MATCH = V_{OL} if: [A0-A8] ≠ D0-D7 + parity,
with $\overline{RESET} = V_{IH}$,
$\overline{S} = V_{IL}$, and $\overline{W} = V_{IH}$

FUNCTION TABLE

OUTPUT		FUNCTION
MATCH	$\overline{PE}$	DESCRIPTION
L	L	Parity Error
L	H	Not Equal
H	L	Undefined Error
H	H	Equal

Where $\overline{S} = V_{IL}$, $\overline{W} = V_{IH}$, $\overline{RESET} = V_{IH}$

PRODUCT PREVIEW
This document contains information on a product under development. Texas Instruments reserves the right to change or discontinue this product without notice.

TEXAS INSTRUMENTS
INCORPORATED
POST OFFICE BOX 225012 • DALLAS, TEXAS 75265

functional block diagram (positive logic)

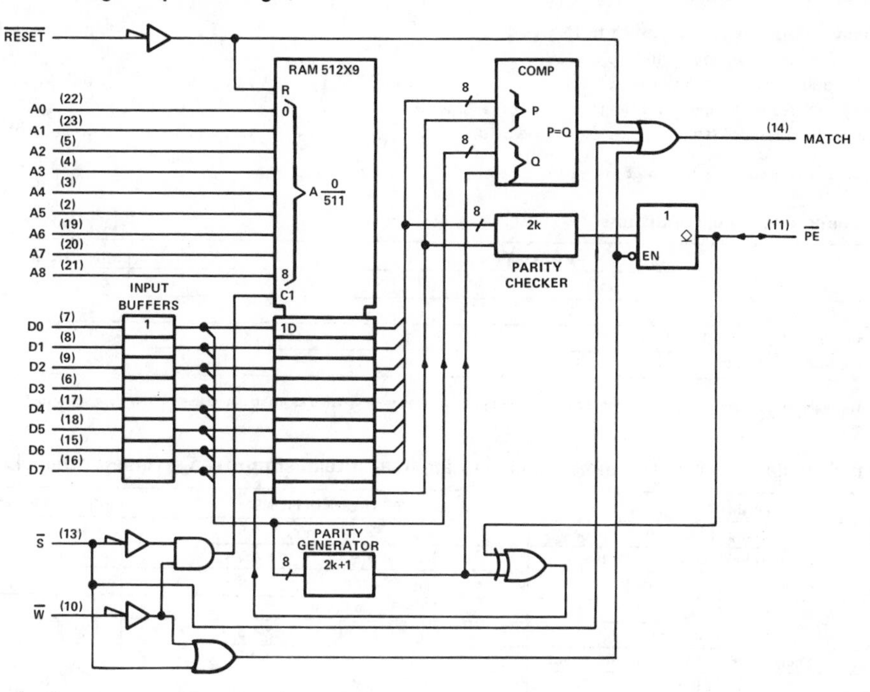

PIN FUNCTION	DESCRIPTION
A0-A8, Address Inputs	Address 1 of 512-by-9-bit random-access memory locations.
D0-D7, Data Inputs	Compared with memory location addressed by A0-A8 when $\overline{W}$ = V_{IH} and $\overline{S}$ = V_{IL}. Provides input data to RAM when $\overline{W}$ = V_{IL} and $\overline{S}$ = V_{IL}.
$\overline{RESET}$, Input	Asynchronously clears entire RAM array and forces MATCH high when $\overline{RESET}$ = V_{IL} and $\overline{W}$ = V_{IH}.
$\overline{S}$, Chip Select Input	Enables device when $\overline{S}$ = V_{IL}. Deselects device and forces MATCH high when $\overline{S}$ = V_{IH}.
$\overline{W}$, Write Control Input	Writes D0-D7 + generated parity into RAM and forces MATCH high when $\overline{W}$ = V_{IL} with $\overline{S}$ = V_{IL}. Places selected device in compare mode if $\overline{W}$ = V_{IH}.
$\overline{PE}$, Parity Error Input/Output	During write cycles $\overline{PE}$ can force a parity error into the 9-bit location specified by A0-A8 when $\overline{PE}$ = V_{IL}. For compare cycles, $\overline{PE}$ = V_{OL} indicates a parity error in the stored data. $\overline{PE}$ is an open-drain output so an external pull-up resistor is required.
MATCH, Output	When MATCH = V_{OH} during a compare cycle, D0-D7 + parity equal the contents of the 9-bit memory location addressed by A0-A8.
V_{SS}	Circuit GND potential.
V_{CC}	+5 V circuit power supply.

absolute maximum ratings over operating free-air temperature range (unless otherwise specified)

Supply voltage range, V_{CC} (see Note 1) .... −1.5 V to 7 V
Input voltage range, any input .... −1.5 V to 7 V
Continuous power dissipation .... 1 W
Operating free-air temperature range .... 0°C to 70°C
Storage temperature range .... −65°C to 150°C

NOTE 1: All voltage values are with respect to V_{SS}.

recommended operating conditions

PARAMETER	MIN	NOM	MAX	UNIT
Supply voltage, V_{CC}	4.5	5	5.5	V
High-level input voltage, V_{IH}	2		6	V
Low-level input voltage, V_{IL} (See Note 2)	−1		0.8	V
Operating free-air temperature, T_A	0		70	°C

NOTE 2: The algebraic convention, where the more negative (less positive) limit is designated as minimum, is used in this data sheet for logic voltage levels only.

electrical characteristics over recommended operating free-air temperature range (unless otherwise noted)

PARAMETER		TEST CONDITIONS	MIN	TYP	MAX	UNIT
$V_{OH(M)}$	MATCH high-level output voltage	$I_{OH} = -2$ mA, $V_{CC} = 4.5$ V	2.4			V
$V_{OL(M)}$	MATCH low-level output voltage	$I_{OL} = 4$ mA, $V_{CC} = 4.5$ V			0.4	V
$V_{OL(PE)}$	$\overline{PE}$ low-level output voltage	$I_{OL} = 12$ mA, $V_{CC} = 4.5$ V			0.4	V
I_I	Input current	$V_I = 0$ V to 5.5 V			10	μA
$I_{OL(PE)}$	$\overline{PE}$ output sink current	$V_{OL} = 0.4$ V, $V_{CC} = 4.5$ V	12			mA
I_{OS}	Short-circuit MATCH output current	$V_{CC} = 5.5$ V, V_O = GND			−150	mA
I_{CC1}	Supply current (operative)	$\overline{RESET} = V_{IH}$		85	120	mA
I_{CC2}	Supply current (reset)	$\overline{RESET} = V_{IL}$		110	140	mA
C_i	Input capacitance	$V_I = 0$ V, f = 1 MHz			5	pF
C_o	Output capacitance	$V_O = 0$ V, f = 1 MHz			6	pF

ac test conditions

Input pulse levels .... GND to 3 V
Input rise and fall times .... 5 ns
Input timing reference levels .... 1.5 V
Output timing reference level .... 1.5 V
Output loading .... See Figures 1A and 1B

TEXAS INSTRUMENTS
INCORPORATED
POST OFFICE BOX 225012 • DALLAS, TEXAS 75265

switching characteristics over recommended ranges of supply voltage and operating free-air temperature

PARAMETER		TMS 2150-4		TMS 2150-5		TMS 2150-7		TMS 2150-9		UNIT
		MIN	MAX	MIN	MAX	MIN	MAX	MIN	MAX	
$t_{a(A)}$	Access time from address to MATCH		45		55		70		90	ns
$t_{a(A-P)}$	Access time from address to $\overline{PE}$		55		65		80		90	ns
$t_{a(S)}$	Access time from $\overline{S}$ to MATCH		25		35		45		60	ns
$t_{p(D)}$	Propagation time, data inputs to MATCH		35		45		55		70	ns
$t_{p(R-MH)}$	Propagation time, $\overline{RESET}$ low to MATCH high		30		40		50		60	ns
$t_{p(S-MH)}$	Propagation time, $\overline{S}$ high to MATCH high		30		40		50		60	ns
$t_{p(W-MH)}$	Propagation time, $\overline{W}$ low to MATCH high		25		35		45		55	ns
$t_{p(W-PH)}$	Propagation time, $\overline{W}$ low to $\overline{PE}$ high		25		35		45		55	ns
$t_{v(A)}$	MATCH valid time after change of address	5		5		5		5		ns
$t_{v(A-P)}$	$\overline{PE}$ valid time after change of address	15		15		15		15		ns

timing requirements over recommended ranges of supply voltage and operating free-air temperature

PARAMETER		TMS 2150-4		TMS 2150-5		TMS 2150-7		TMS 2150-9		UNIT
		MIN	MAX	MIN	MAX	MIN	MAX	MIN	MAX	
$t_{c(W)}$	Write cycle time	45		55		70		90		ns
$t_{c(rd)}$	Read cycle time	45		55		70		90		ns
$t_{w(RL)}$	Pulse duration, $\overline{RESET}$ low	35		45		55		55		ns
$t_{w(WL)}$	Pulse duration, $\overline{W}$ low	25		35		45		55		ns
$t_{su(A)}$	Address setup time before $\overline{W}$ low	0		0		0		0		ns
$t_{su(D)}$	Data setup time before $\overline{W}$ high	25		30		35		40		ns
$t_{su(P)}$	$\overline{PE}$ setup time before $\overline{W}$ high	25		30		35		40		ns
$t_{su(S)}$	Chip select setup time before $\overline{W}$ high	25		35		45		50		ns
$t_{su(RH)}$	$\overline{RESET}$ inactive setup time before first tag cycle	0		0		0		0		ns
$t_{h(A)}$	Address hold time after $\overline{W}$ high	0		5		10		15		ns
$t_{h(D)}$	Data hold time after $\overline{W}$ high	5		10		20		25		ns
$t_{h(P)}$	$\overline{PE}$ hold time after $\overline{W}$ high	0		5		10		15		ns
$t_{h(S)}$	Chip select hold time after $\overline{W}$ high	0		0		0		0		ns
t_{AVWH}	Address valid to write enable	45		50		60		75		ns

PARAMETER MEASUREMENT INFORMATION

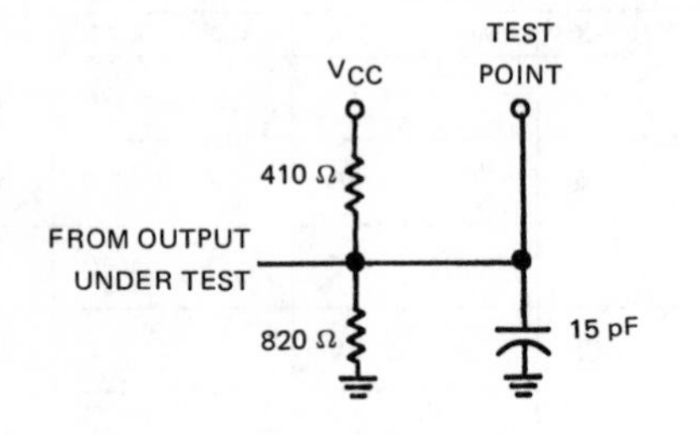

FIGURE 1A – $\overline{PE}$ OUTPUT LOAD CIRCUIT

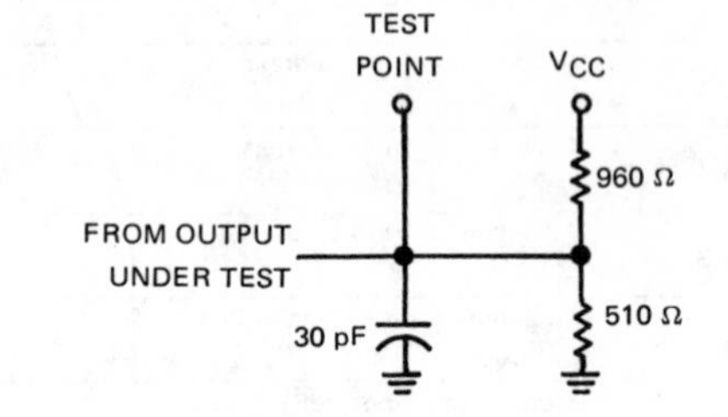

FIGURE 1B – MATCH OUTPUT LOAD CIRCUIT

compare cycle timing

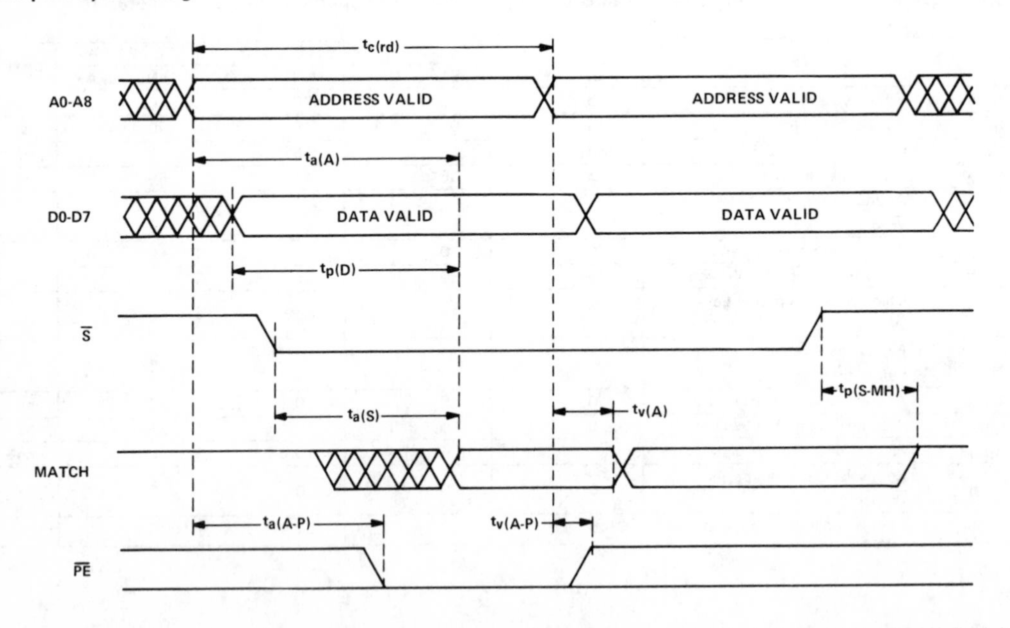

NOTE: Input pulse levels are 0 V and 3 V, with rise and fall times of 5 ns. The timing reference levels on the input pulses are 0.8 V and 2.0 V. The timing reference level for output pulses is 1.5 V. See Figures 1A and 1B for output loading.

TEXAS INSTRUMENTS
INCORPORATED
POST OFFICE BOX 225012 • DALLAS, TEXAS 75265

write cycle timing

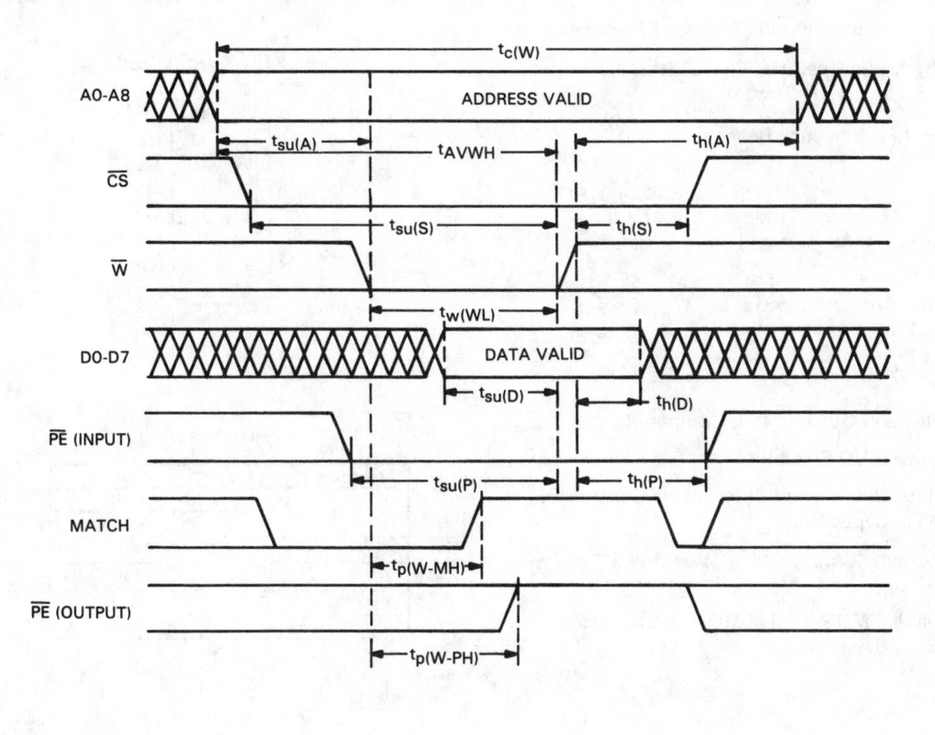

reset cycle timing

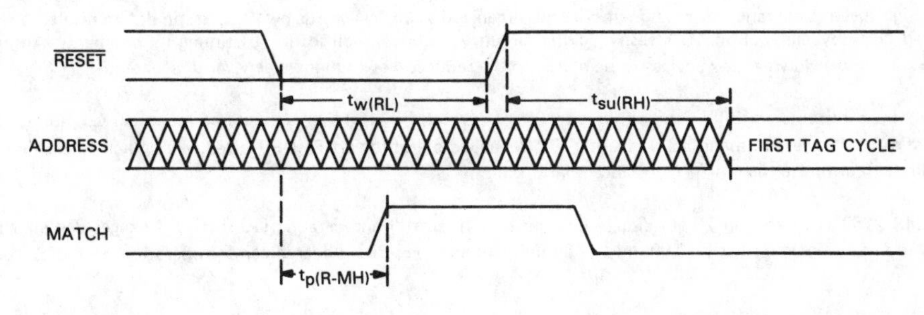

NOTE: Input pulse levels are 0 V and 3 V, with rise and fall times of 5 ns. The timing reference levels on the input pulses are 0.8 V and 2.0 V. The timing reference level for output pulses is 1.5 V. See Figures 1A and 1B for output loading.

TMS 2167 JL, TMS 2167 NL
FAST 16,384-WORD BY 1-BIT STATIC RAM

MAY 1982

- 16K X 1 Organization
- Single +5 V Supply (± 10% Tolerance)
- High-Density 300-mil (7.62 mm) Packages
- Fully Static Operation (No Clocks, No Refresh, No Timing Strobe)
- **3 Fast Performance Ranges:**

	ACCESS TIME (MAX)	READ OR WRITE CYCLE (MIN)
TMS 2167-4	45 ns	45 ns
TMS 2167-5	55 ns	55 ns
TMS 2167-7	70 ns	70 ns

- Inputs and Outputs TTL Compatible
- Common I/O Capability
- 3-State Outputs and Chip-Enable Control for OR-Tie Capability
- Automatic Chip-Enable/Power-Down Operation
- Reliable SMOS (Scaled-MOS) N-Channel Technology

TMS 2167
20-PIN PLASTIC AND CERAMIC
DUAL-IN-LINE PACKAGES
(TOP VIEW)

1	A3	20	V_{CC}
2	A4	19	A2
3	A5	18	A1
4	A6	17	A0
5	A7	16	A13
6	A8	15	A12
7	A9	14	A11
8	Q	13	A10
9	$\overline{W}$	12	D
10	V_{SS}	11	$\overline{E}$

PIN NAMES

A0 - A13	Addresses
D	Data In
Q	Data Out
$\overline{E}$	Chip-Enable/Power-Down
V_{CC}	+5 V Supply
V_{SS}	Ground
$\overline{W}$	Write Enable

description

These high-speed static random-access memories are organized as 16,384 words by 1 bit. Static design results in reduced overhead costs by elimination of refresh-clocking circuitry and by simplification of timing requirements. Automatic chip-enable/power-down allows devices to be placed in the reduced-power mode whenever deselected.

All inputs and outputs are fully compatible with Series 74, 74S or 74LS TTL. No pull-up resistors are required. These 16K static RAM series are manufactured using TI's reliable state-of-the-art SMOS (scaled MOS) N-channel silicon-gate technology to optimize the cost/performance relationship.

The TMS 2167 is offered in 20-pin dual-in-line plastic (NL suffix) and ceramic (JL suffix) packages designed for insertion in mounting-hole rows on 300-mil (7.62 mm) centers. The series is guaranteed for operation from 0°C to 70°C.

PRODUCT PREVIEW
This document contains information on a product under development. Texas Instruments reserves the right to change or discontinue this product without notice.

TEXAS INSTRUMENTS
INCORPORATED
POST OFFICE BOX 225012 • DALLAS, TEXAS 75265

operation

addresses (A0-A13)

The 14 address inputs select one of the 16,384 storage locations in the RAM. The address inputs must be stable for the duration of a write cycle. The address inputs can be driven directly from standard Series 54/74 TTL with no external pull-up resistors.

chip-enable/power-down ($\overline{E}$)

The chip-enable/power-down terminal, which can be driven directly by standard TTL circuits, affects the data-in and data-out terminals and the internal functioning of the chip itself. When the chip-enable/power-down terminal is low (enabled), the device is operational, input and output terminals are enabled, and data can be read or written. When the chip-enable/power-down terminal is high (disabled), the device is deselected and put into a reduced power standby mode. Data is retained during standby.

write-enable ($\overline{W}$)

The read or write mode is selected through the write-enable terminal. A logic high selects the read mode; a logic low selects the write mode. $\overline{W}$ must be high when changing addresses to prevent erroneously writing data into a memory location. The $\overline{W}$ input can be driven directly from standard TTL circuits.

data-in (D)

Data can be written into a selected device when the write-enable input is low. The data-in terminal can be driven directly from standard TTL circuits.

data-out (Q)

The three-state output buffer provides direct TTL compatibility. The output is in the high-impedance state when chip-enable/power-down ($\overline{E}$) is high or whenever a write operation is being performed, facilitating device operation in common I/O systems. Data-out is the same polarity as data-in.

logic symbol†

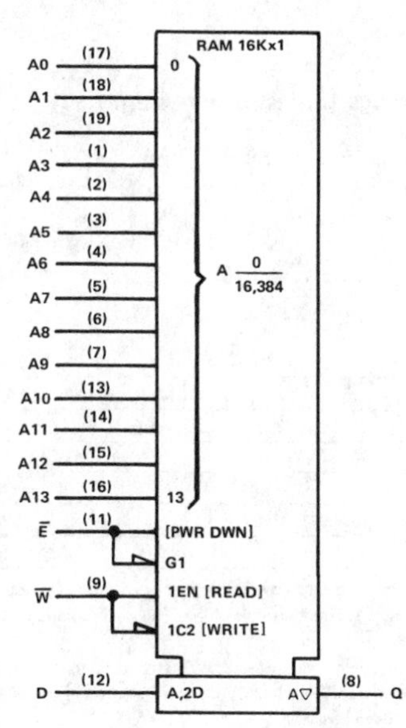

FUNCTION TABLE

INPUTS		OUTPUT	MODE
$\overline{E}$	$\overline{W}$	Q	
H	X	Hi-Z	POWER DOWN
L	L	Hi-Z	WRITE
L	H	DATA OUT	READ

† This symbol is in accordance with IEEE Std 91/ANSI Y32.14 and recent decisions by IEEE and IEC. See explanation on page 289.

functional block diagram

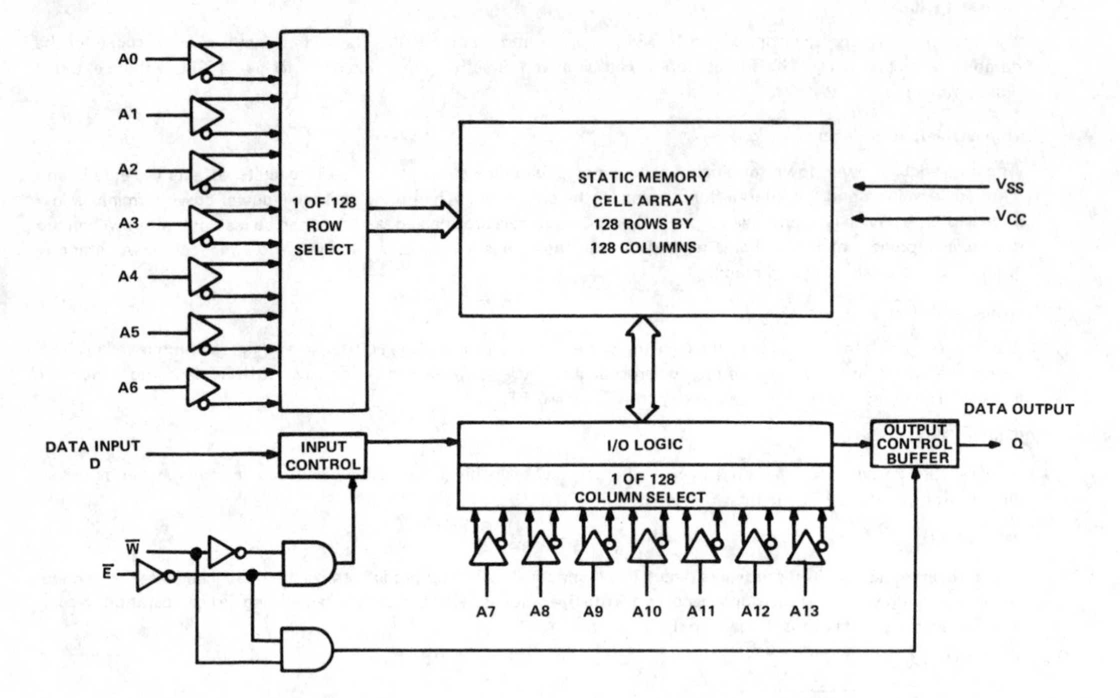

absolute maximum ratings over operating ambient temperature† range (unless otherwise noted)‡

Supply voltage, V_{CC} (see Note 1)	−1.5 V to 7 V
Input voltage (any input) (see Note 1)	−1.5 V to 7 V
Continuous power dissipation	1 W
Operating ambient temperature range	0°C to 70°C
Storage temperature range	−65°C to 150°C

recommended operating conditions

PARAMETER	MIN	NOM	MAX	UNIT
Supply voltage, V_{CC}	4.5	5	5.5	V
Supply voltage, V_{SS}		0		V
High-level input voltage, V_{IH}	2		6	V
Low-level input voltage, V_{IL}	−1§		0.8	V
Operating ambient temperature†, T_A	0		70	°C

† The ambient temperature conditions assume air moving perpendicular to the longitudinal axis and parallel to the seating plane of the device at a velocity of 400 ft/min (122 m/min) with the device under test soldered to a 4 × 6 × 0.062-inch (102 × 152 × 1.6-mm) double-sided 2-ounce copper-clad circuit board (plating thickness 0.07 mm).

‡ Stresses beyond those listed under "Absolute Maximum Ratings" may cause permanent damage to the device. This is stress rating only and functional operation of the device at these or any other conditions beyond those indicated in the "Recommended Operating Conditions" section of this specification is not implied. Exposure to absolute-maximum-rated conditions for extended periods may affect device reliability.

§ The algebraic convention, where the morenegative limit is designated as minimum, is used in this data sheet for logic voltage levels only.

NOTE 1: Voltage values are with respect to the ground terminal.

TEXAS INSTRUMENTS
INCORPORATED
POST OFFICE BOX 225012 • DALLAS, TEXAS 75265

electrical characteristics over recommended operating ambient temperature† range (unless otherwise noted)

PARAMETER		TEST CONDITIONS	MIN	TYP‡	MAX	UNIT
V_{OH}	High-level output voltage	$I_{OH} = -4$ mA, $V_{CC} = 4.5$ V	2.4			V
V_{OL}	Low-level output voltage	$I_{OL} = 8$ mA, $V_{CC} = 4.5$ V			0.4	V
I_I	Input current	$V_I = 0$ V to 5.5 V			10	μA
I_{OZ}	Off-state output current	$\overline{E}$ at 2 V, $V_O = 0$ V to 4.5 V, $V_{CC} = 5.5$ V			±50	μA
I_{CC1}	Standby supply current from V_{CC}	$\overline{E}$ at V_{IH}		9	20	mA
I_{CC2}	Operating supply current from V_{CC}	$\overline{E}$ at V_{IL}, $I_O = 0$ mA, $T_A = 0°C$ (worst case)		70	120	mA
		$\overline{E}$ at V_{IL}, $I_O = 0$ mA, $T_A = 70°C$			90	mA
I_{PO}	Peak power-on current (see Note 2)	V_{CC} = GND to V_{CC} min, $\overline{E}$ at lower of V_{CC} or V_{IH} min			70	mA
C_i	Input capacitance	$V_I = 0$ V, f = 1 MHz			5	pF
C_O	Output capacitance	$V_O = 0$ V, f = 1 MHz			6	pF

ac test conditions

Input pulse levels . . . . GND to 3 V
Input rise and fall times . . . . 5 ns
Input timing reference levels . . . . 1.5 V
Output timing reference, high level (2167-4, -5, -7) . . . . 2 V
Output timing reference, low level (2167-4, -5, -7) . . . . 0.8 V
Output loading . . . . See Figure 1

timing requirements over recommended supply voltage range and operating ambient temperature† range

PARAMETER		TMS 2167-4		TMS 2167-5		TMS 2167-7		UNIT
		MIN	MAX	MIN	MAX	MIN	MAX	
$t_{c(rd)}$	Read cycle time	45		55		70		ns
$t_{c(wr)}$	Write cycle time	45		55		70		ns
$t_{w(W)}$	Write pulse width	30		40		50		ns
$t_{su(A)}$	Address setup time	0		0		0		ns
$t_{su(E)}$	Chip-enable setup time	30		40		50		ns
$t_{su(D)}$	Data setup time	15		15		25		ns
$t_{h(D)}$	Data hold time	5		5		5		ns
$t_{h(A)}$	Address hold time	0		5		5		ns
t_{AVWH}	Address valid to write enable high	35		45		55		ns

† The ambient temperature conditions assume air moving at a velocity of 400 ft/min (122 m/min.).
‡ All typical values are at $V_{CC} = 5$, $T_A = 25°C$.
NOTE 2: I_{PO} exceeds I_{CC1} maximum during power on. A pull-up resistor to V_{CC} on the $\overline{E}$ input is required to keep the device deselected; otherwise, power-on current approaches I_{CC2}.

4

switching characteristics over recommended supply voltage range and operating ambient temperature† range

PARAMETER		TEST CONDITIONS	TMS 2167-4		TMS 2167-5		TMS 2167-7		UNIT
			MIN	MAX	MIN	MAX	MIN	MAX	
$t_{a(A)}$	Access time from address	R_L = 510 Ω, C_L = 30 pF, See Figure 1		45		55		70	ns
$t_{a(E)}$	Access time from chip enable			45		55		70	ns
$t_{v(A)}$	Output data valid after address change		5		5		5		ns
$t_{dis(W)}$	Output disable time from write enable‡			20		25		25	ns
$t_{en(W)}$	Output enable time from write enable‡		5		5		5		ns
$t_{dis(E)}$	Output disable time from chip enable‡			20		25		25	ns
$t_{en(E)}$	Output enable time from chip enable‡		5		5		5		ns
t_{pwrdn}	Power down time from chip enable			20		20		30	ns

† The ambient temperature conditions assume air moving at a velocity of 400 ft/min (122 m/min.).
‡ Transition is measured ±500 mV from steady state voltage with specified loading in Figure 1.

PARAMETER MEASUREMENT INFORMATION

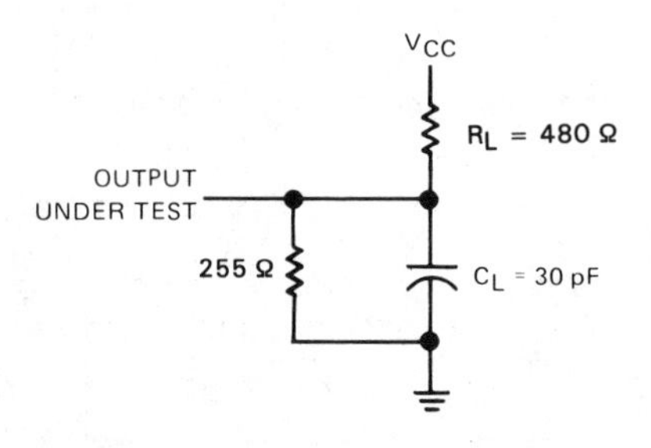

FIGURE 1 – LOAD CIRCUIT

TEXAS INSTRUMENTS
INCORPORATED
POST OFFICE BOX 225012 • DALLAS, TEXAS 75265

read cycle timing

from address

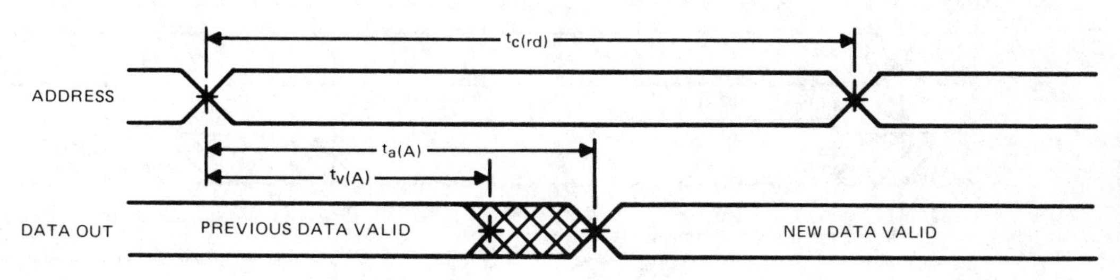

$\overline{W}$ is high, $\overline{E}$ is low.

from chip enable

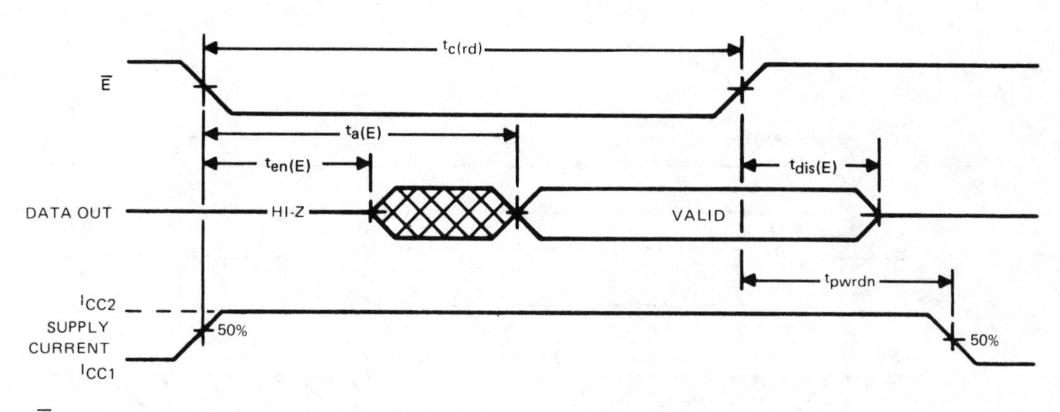

$\overline{W}$ is high, address is valid prior to or simultaneously with the high-to-low transition of $\overline{E}$.

write cycle timing

controlled by write enable†

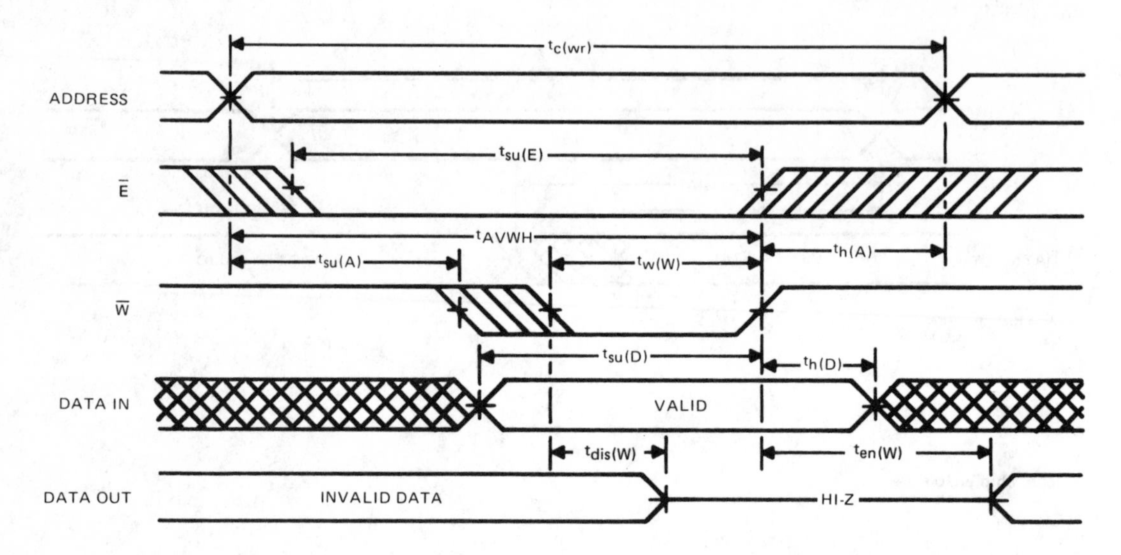

controlled by chip enable†

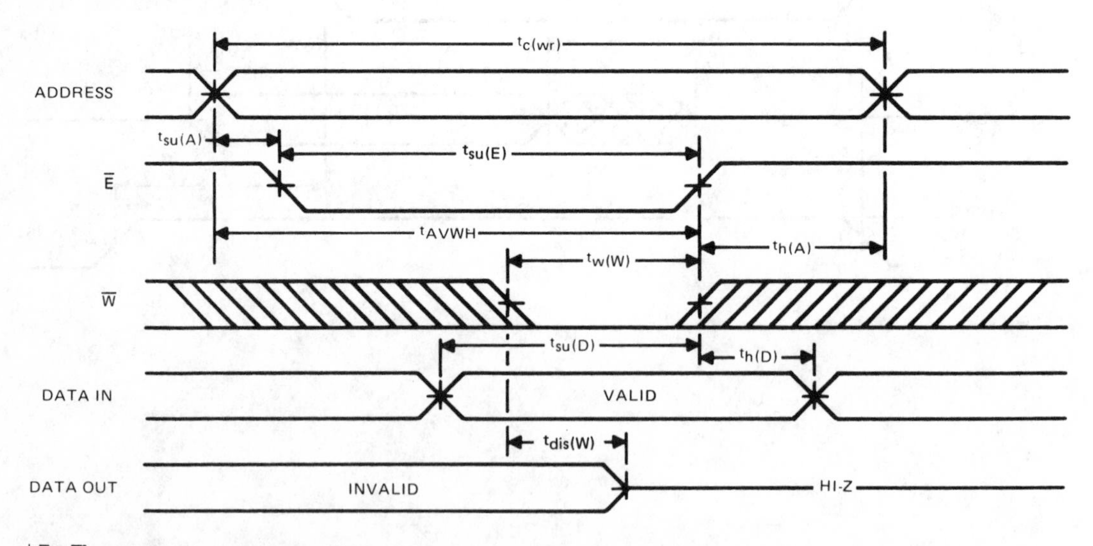

† $\overline{E}$ or $\overline{W}$ must be high during address transitions.

NOTE: If $\overline{E}$ goes high simultaneously with $\overline{W}$ going high, the output remains in the high-impedance state.

TEXAS INSTRUMENTS
INCORPORATED
POST OFFICE BOX 225012 • DALLAS, TEXAS 75265

TMS 2168 JL, TMS 2168 NL
FAST 4096-WORD BY 4-BIT STATIC RAM

MARCH 1982

- **4096 X 4 Organization**
- **Single +5 V Supply (±10% Tolerance)**
- **High-Density 300 mil (7.62 mm) Packages**
- **Fully Static Operation (No Clocks, No Refresh, No Timing Strobe)**
- **Fast . . . 3 Performance Ranges:**

	ADDRESS ACCESS TIME (MAX)	READ OR WRITE CYCLE (MIN)
TMS 2168-4	45 ns	45 ns
TMS 2168-5	55 ns	55 ns
TMS 2168-7	70 ns	70 ns

- **Inputs and Outputs TTL Compatible**
- **Automatic Chip-Enable/Power-Down Operation**
- **3-State Outputs**
- **Reliable SMOS (Scaled-MOS) N-Channel Technology**
- **Industry Standard 4K X 4 Pinout**

TMS 2168
20-PIN PLASTIC AND CERAMIC DUAL-IN-LINE PACKAGES
(TOP VIEW)

1	A7	20	V_{CC}
2	A6	19	A8
3	A5	18	A9
4	A4	17	A10
5	A0	16	A11
6	A1	15	DQ1
7	A2	14	DQ2
8	A3	13	DQ3
9	$\overline{E}$	12	DQ4
10	V_{SS}	11	$\overline{W}$

PIN NAMES

A0-A11	Addresses
DQ	Data In/Data Out
$\overline{E}$	Chip Enable/Power Down
V_{CC}	+5 V Supply
V_{SS}	Ground
W	Write Enable

description

These static random-access memories are organized as 4096 words of four bits each. Static design results in reduced overhead costs by elimination of refresh-clocking circuitry and by simplification of timing requirements. Automatic chip enable/power down allows devices to be placed in the reduced-power mode whenever deselected.

All inputs and outputs are fully compatible with Series 74, 74S, or 74LS TTL. No pull-up resistors are required. These 16K static RAM series are manufactured using TI's reliable state-of-art SMOS (scaled MOS) N-channel silicon gate technology to optimize the cost/performance relationship.

The TMS 2168 is offered in 20-pin dual-in-line plastic (NL suffix) and ceramic (JL suffix) packages designed for insertion in mounting-hole rows on 300-mil (7.62 mm) centers. The series is guaranteed for operation from 0 °C to 70 °C.

operation

addresses (A0-A11)

The 12 address inputs select one of the 4096 4-bit words in the RAM. The address inputs must be stable for the duration of a write cycle. The address inputs can be driven directly from standard Series 54/74 TTL with no external pull-up resistors.

TEXAS INSTRUMENTS
INCORPORATED
POST OFFICE BOX 225012 • DALLAS, TEXAS 75265

logic symbol†

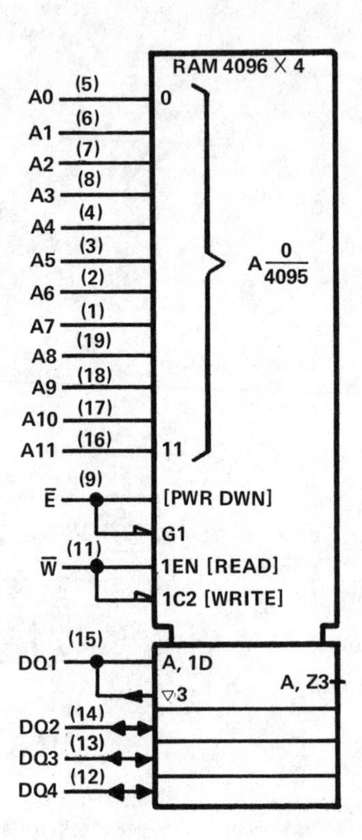

† This symbol is in accordance with IEEE Std 91/ANSI Y32.14 and recent decisions by IEEE and IEC. See explanation on page 289.

TEXAS INSTRUMENTS
INCORPORATED
POST OFFICE BOX 225012 • DALLAS, TEXAS 75265

TMS 2169 JL, TMS 2169 NL
FAST 4096-WORD BY 4-BIT STATIC RAM

MARCH 1982

- 4096 X 4 Organization
- Single +5 V Supply (± 10% Tolerance)
- High-Density 300 mil (7.62 mm) Packages
- Fully Static Operation (No Clocks, No Refresh, No Timing Strobe)
- Fast . . . 3 Performance Ranges:

	ADDRESS ACCESS TIME (MAX)	CS ACCESS TIME (MAX)	READ OR WRITE CYCLE (MIN)
TMS 2169-4	45 ns	25 ns	45 ns
TMS 2169-5	55 ns	30 ns	55 ns
TMS 2169-7	70 ns	35 ns	70 ns

- Inputs and Outputs TTL Compatible
- Common I/O
- 3-State Outputs
- Reliable SMOS (Scaled-MOS) N-Channel Technology
- Industry Standard 4K X 4 Pinout

TMS 2169
20-PIN PLASTIC AND CERAMIC
DUAL-IN-LINE PACKAGES
(TOP VIEW)

A7	1	20	V_{CC}
A6	2	19	A8
A5	3	18	A9
A4	4	17	A10
A0	5	16	A11
A1	6	15	DQ1
A2	7	14	DQ2
A3	8	13	DQ3
$\overline{S}$	9	12	DQ4
V_{SS}	10	11	$\overline{W}$

PIN NAMES

A0-A11	Addresses
DQ	Data In/Data Out
$\overline{S}$	Chip Select
V_{CC}	+5 V Supply
V_{SS}	Ground
$\overline{W}$	Write Enable

description

These high-speed static random-access memories are organized as 4096 words of four bits each. Static design results in reduced overhead costs by elimination of refresh-clocking circuitry and by simplification of timing requirements.

All inputs and outputs are fully compatible with Series 74, 74S, or 74LS TTL. No pull-up resistors are required. These 16K static RAM series are manufactured using TI's reliable state-of-art SMOS (scaled MOS) N-channel silicon gate technology to optimize the cost/performance relationship.

The TMS 2169 is offered in 20-pin dual-in-line plastic (NL suffix) and ceramic (JL suffix) packages designed for insertion in mounting-hole rows on 300-mil (7.62 mm) centers. The series is guaranteed for operation from 0 °C to 70 °C.

operation

addresses (A0-A11)

The 12 address inputs select one of the 4096 4-bit words in the RAM. The address inputs must be stable for the duration of a write cycle. The address inputs can be driven directly from standard Series 54/74 TTL with no external pull-up resistors.

chip-select ($\overline{S}$)

The chip-select terminal, which can be driven directly by standard TTL circuits, affects the data-in/data-out (DQ) terminals and the internal functioning of the chip itself. Whenever the chip-select terminal is low (enabled), the device is operational. DQ terminals function as data-in or data-out depending on the level of the write enable terminal. When the chip-select terminal is high (disabled), the device is deselected, data-in is inhibited and data-out is in the floating or high impedance state.

logic symbol†

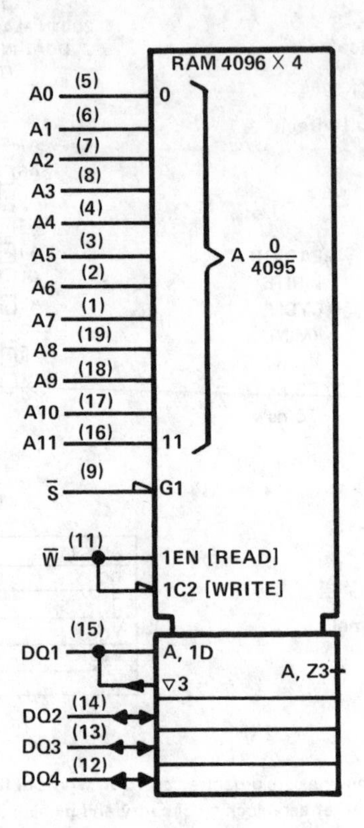

† This symbol is in accordance with IEEE Std 91/ANSI Y32.14 and recent decisions by IEEE and IEC. See explanation on page 289.

TMS 4016 NL
2048-WORD BY 8-BIT STATIC RAM

FEBRUARY 1981 – REVISED DECEMBER 1981

- 2K X 8 Organization, Common I/O
- Single +5 V Supply
- Fully Static Operation (No Clocks, No Refresh)
- JEDEC Standard Pinout
- 24-Pin 600 Mil (15.2 mm) Package Configuration
- Plug-in Compatible with 16K 5 V EPROMs
- 8-Bit Output for Use in Microprocessor-Based Systems
- 3-State Outputs with $\overline{S}$ for OR-ties
- $\overline{G}$ Eliminates Need for External Bus Buffers
- All Inputs and Outputs Fully TTL Compatible
- Fanout to Series 74, Series 74S or Series 74LS TTL Loads
- N-Channel Silicon-Gate Technology
- Power Dissipation Under 385 mW Max
- Guaranteed dc Noise Immunity of 400 mV with Standard TTL Loads
- 4 Performance Ranges:

	ACCESS TIME (MAX)
TMS 4016-12	120 ns
TMS 4016-15	150 ns
TMS 4016-20	200 ns
TMS 4016-25	250 ns

TMS 4016
24-PIN PLASTIC
DUAL-IN-LINE PACKAGE
(TOP VIEW)

1	A7	24	V_{CC}
2	A6	23	A8
3	A5	22	A9
4	A4	21	$\overline{W}$
5	A3	20	$\overline{G}$
6	A2	19	A10
7	A1	18	$\overline{S}$
8	A0	17	DQ8
9	DQ1	16	DQ7
10	DQ2	15	DQ6
11	DQ3	14	DQ5
12	V_{SS}	13	DQ4

PIN NOMENCLATURE	
A0-A10	Addresses
DQ1-DQ8	Data In/Data Out
$\overline{S}$	Chip Select
$\overline{G}$	Output Enable
$\overline{W}$	Write Enable
V_{SS}	Ground
V_{CC}	+5 V Supply

4

description

The TMS 4016 static random-access memory is organized as 2048 words of 8 bits each. Fabricated using proven N-channel, silicon-gate MOS technology, the TMS 4016 operates at high speeds and draws less power per bit than 4K static RAMs. It is fully compatible with Series 74, 74S, or 74LS TTL. Its static design means that no refresh clocking circuitry is needed and timing requirements are simplified. Access time is equal to cycle time. A chip select control is provided for controlling the flow of data-in and data-out and an output enable function is included in order to eliminate the need for external bus buffers.

Of special importance is that the TMS 4016 static RAM has the same standardized pinout as TI's compatible EPROM family. This, along with other compatible features, makes the TMS 4016 plug-in compatible with the TMS 2516 (or other 16K 5 V EPROMs). Minimal, if any modifications are needed. This allows the microprocessor system designer complete flexibility in partitioning his memory board between read/write and non-volatile storage.

The TMS 4016 is offered in the plastic (NL suffix) 24-pin dual-in-line package designed for insertion in mounting hole rows on 600-mil (15.2 mm) centers. It is guaranteed for operation from 0°C to 70°C.

See page 253 for explanation of TMS 4016-16K 5 V EPROM compatibility.

operation

addresses (A0-A10)

The eleven address inputs select one of the 2048 8-bit words in the RAM. The address-inputs must be stable for the duration of a write cycle. The address inputs can be driven directly from standard Series 54/74 TTL with no external pull-up resistors.

output enable ($\overline{G}$)

The output enable terminal, which can be driven directly from standard TTL circuits, affects only the data-out terminals. When output enable is at a logic high level, the output terminals are disabled to the high-impedance state. Output enable provides greater output control flexibility, simplifying data bus design.

chip select ($\overline{S}$)

The chip-select terminal, which can be driven directly from standard TTL circuits, affects the data-in/data-out terminals. When chip select and output enable are at a logic low level, the D/Q terminals are enabled. When chip select is high, the D/Q terminals are in the floating or high-impedance state and the input is inhibited.

write enable ($\overline{W}$)

The read or write mode is selected through the write enable terminal. A logic high selects the read mode; a logic low selects the write mode. $\overline{W}$ must be high when changing addresses to prevent erroneously writing data into a memory location. The $\overline{W}$ input can be driven directly from standard TTL circuits.

data-in/data-out (DQ1-DQ8)

Data can be written into a selected device when the write enable input is low. The D/Q terminal can be driven directly from standard TTL circuits. The three-state output buffer provides direct TTL compatibility with a fan-out of one Series 74 TTL gate, one Series 74S TTL gate, or five Series 74LS TTL gates. The D/Q terminals are in the high impedance state when chip select ($\overline{S}$) is high, output enable ($\overline{G}$) is high, or whenever a write operation is being performed. Data-out is the same polarity as data-in.

logic symbol†

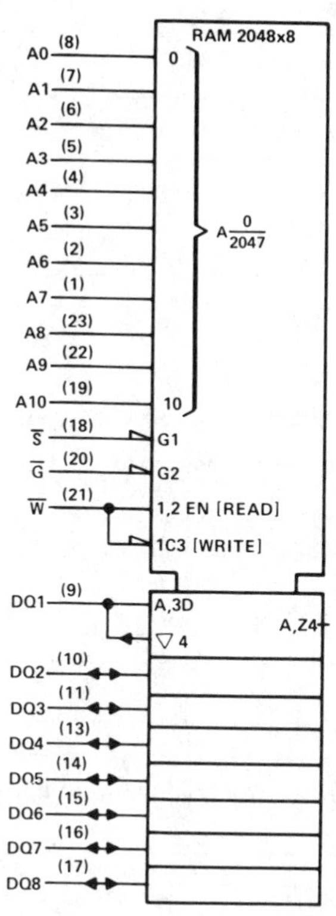

4

$\overline{W}$	$\overline{S}$	$\overline{G}$	DQ1-DQ8	MODE
L	L	X	VALID DATA	WRITE
H	L	L	DATA OUTPUT	READ
X	H	X	HI-Z	DEVICE DISABLED
H	L	H	HI-Z	OUTPUT DISABLED

† This symbol is in accordance with IEEE Std 91/ANSI Y32.14 and recent decisions by IEEE and IEC. See explanation on page 289.

absolute maximum ratings over operating free-air temperature range (unless otherwise noted)*

Supply voltage, V_{CC} (see Note 1)	−0.5 to 7 V
Input voltage (any input) (see Note 1)	−1 to 7 V
Continuous power dissipation	1 W
Operating free-air temperature range	0°C to 70°C
Storage temperature range	−55°C to 150°C

NOTE 1: Voltage values are with respect to the V_{SS} terminal.

* Stresses beyond those listed under "Absolute Maximum Ratings" may cause permanent damage to the device. This is a stress rating only and functional operation of the device at these or any other conditions beyond those indicated in the "Recommended Operating Conditions" section of this specification is not implied. Exposure to absolute-maximum-rated conditions for extended periods may affect device reliability.

recommended operating conditions

PARAMETER	MIN	NOM	MAX	UNIT
Supply voltage, V_{CC}	4.5	5	5.5	V
Supply voltage, V_{SS}		0		V
High-level input voltage, V_{IH}	2		5.5	V
Low-level input voltage, V_{IL} (algebraic limits)	−1		0.8	V
Operating free-air temperature, T_A	0		70	°C

electrical characteristics over recommended operating free-air temperature range (unless otherwise noted)

PARAMETER		TEST CONDITIONS		MIN	TYP†	MAX	UNIT
V_{OH}	High level voltage	$I_{OH} = -1$ mA,	$V_{CC} = 4.5$ V	2.4			V
V_{OL}	Low level voltage	$I_{OL} = 2.1$ mA,	$V_{CC} = 4.5$ V			0.4	V
I_I	Input current	$V_I = 0$ V to 5.5 V				10	μA
I_{OZ}	Off-state output current	$\overline{S}$ or $\overline{G}$ at 2 V or $\overline{W}$ at 0.8 V, $V_O = 0$ V to 5.5 V				10	μA
I_{CC}	Supply current from V_{CC}	$I_O = 0$ mA, $T_A = 0°C$ (worst case)	$V_{CC} = 5.5$ V,		40	70	mA
C_i	Input capacitance	$V_I = 0$ V,	f = 1 MHz			8	pF
C_o	Output capacitance	$V_O = 0$ V,	f = 1 MHz			12	pF

† All typical values are at $V_{CC} = 5$ V, $T_A = 25°C$.

timing requirements over recommended supply voltage range and operating free-air temperature range

PARAMETER		TMS 4016-12		TMS 4016-15		TMS 4016-20		TMS 4016-25		UNIT
		MIN	MAX	MIN	MAX	MIN	MAX	MIN	MAX	
$t_{c(rd)}$	Read cycle time	120		150		200		250		ns
$t_{c(wr)}$	Write cycle time	120		150		200		250		ns
$t_{w(W)}$	Write pulse width	60		80		100		120		ns
$t_{su(A)}$	Address setup time	20		20		20		20		ns
$t_{su(S)}$	Chip select setup time	60		80		100		120		ns
$t_{su(D)}$	Data setup time	50		60		80		100		ns
$t_{h(A)}$	Address hold time	0		0		0		0		ns
$t_{h(D)}$	Data hold time	5		10		10		10		ns

switching characteristics over recommended voltage range, $T_A = 0°C$ to $70°C$ with output loading of Figure 1 (see notes 2, 3)

PARAMETER		TMS 4016-12		TMS 4016-15		TMS 4016-20		TMS 4016-25		UNIT
		MIN	MAX	MIN	MAX	MIN	MAX	MIN	MAX	
$t_{a(A)}$	Access time from address		120		150		200		250	ns
$t_{a(S)}$	Access time from chip select low		60		75		100		120	ns
$t_{a(G)}$	Access time from output enable low		50		60		80		100	ns
$t_{v(A)}$	Output data valid after address change	10		15		15		15		ns
$t_{dis(S)}$	Output disable time after chip select high		40		50		60		80	ns
$t_{dis(G)}$	Output disable time after output enable high		40		50		60		80	ns
$t_{dis(W)}$	Output disable time after write enable low		50		60		60		80	ns
$t_{en(S)}$	Output enable time after chip select low	5		5		10		10		ns
$t_{en(G)}$	Output enable time after output enable low	5		5		10		10		ns
$t_{en(W)}$	Output enable time after write enable high	5		5		10		10		ns

NOTES: 2. $C_L = 100$ pF for all measurements except t_{dis}.
$C_L = 5$ pF for t_{dis}.
3. t_{dis} and t_{en} parameters are sampled and not 100% tested.

TEXAS INSTRUMENTS
INCORPORATED
POST OFFICE BOX 225012 • DALLAS, TEXAS 75265

timing waveform of read cycle (see note 4)

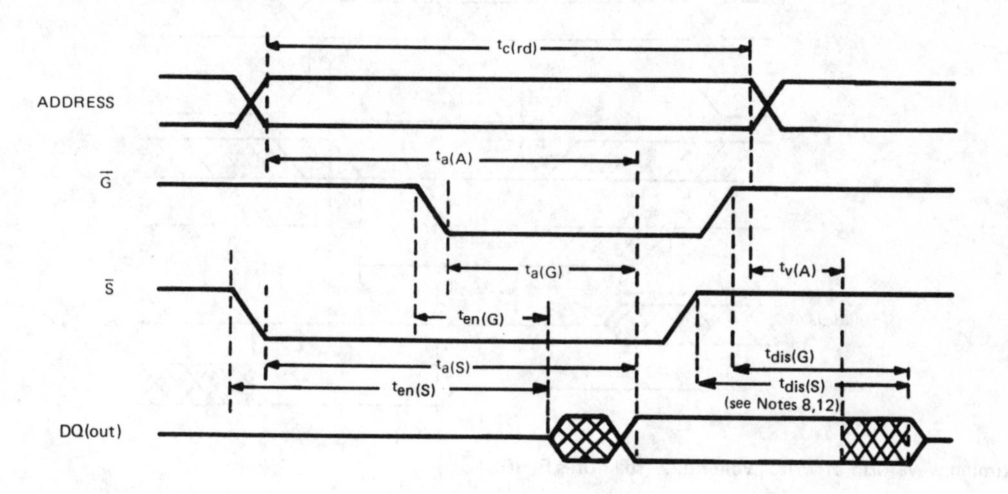

All timing reference points are 0.8 V and 2.0 V on inputs and 0.6 V and 2.2 V on outputs (90% points). Input rise and fall times equal 10 ns.

NOTE 4: $\overline{W}$ is high for Read Cycle.

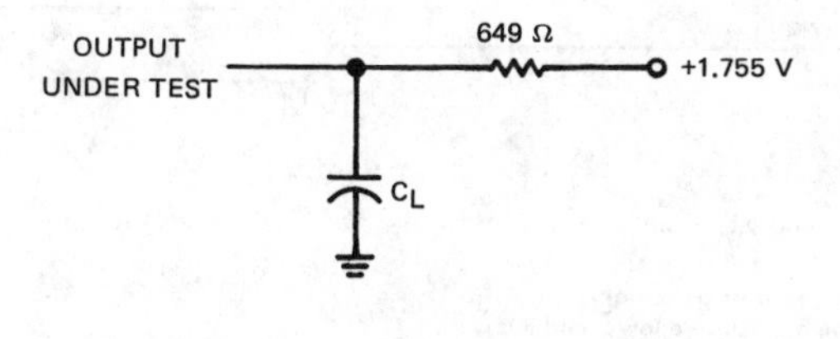

FIGURE 1 – OUTPUT LOAD

timing waveform of write cycle no. 1 (see note 5)

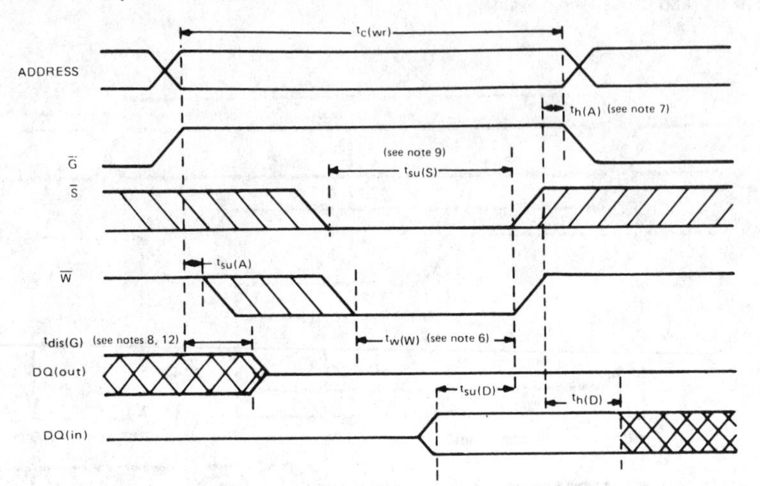

timing waveform of write cycle no. 2 (see notes 5, 10)

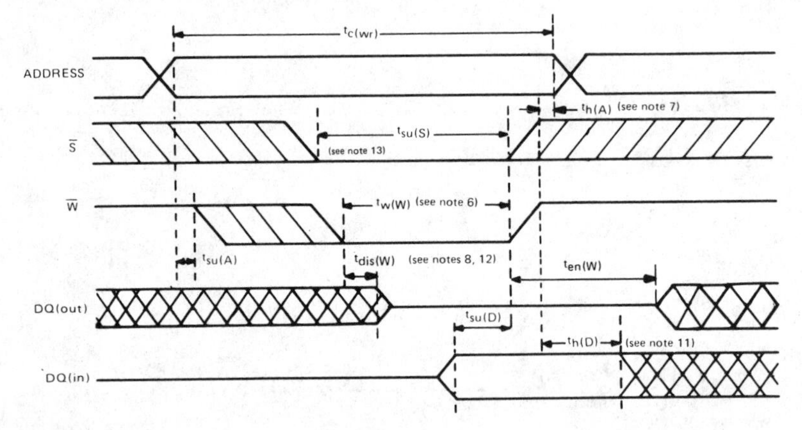

All timing reference points are 0.8 V and 2.0 V on inputs and 0.6 V and 2.2 V on outputs (90% points). Input rise and fall times equal 10 nanoseconds.

NOTES:
5. $\overline{W}$ must be high during all address transitions.
6. A write occurs during the overlap of a low $\overline{S}$ and a low $\overline{W}$.
7. $t_{h(A)}$ is measured from the earlier of $\overline{S}$ or $\overline{W}$ going high to the end of the write cycle.
8. During this period, I/O pins are in the output state so that the input signals of opposite phase to the outputs must not be applied.
9. If the $\overline{S}$ low transition occurs simultaneously with the $\overline{W}$ low transitions or after the $\overline{W}$ transition, output remains in a high impedance state.
10. $\overline{G}$ is continuously low ($\overline{G} = V_{IL}$).
11. If $\overline{S}$ is low during this period, I/O pins are in the output state. Data input signals of opposite phase to the outputs must not be applied.
12. Transition is measured ±200 mV from steady-state voltage.
13. If the $\overline{S}$ low transition occurs before the $\overline{W}$ low transition, then the data input signals of opposite phase to the outputs must not be applied for the duration of $t_{dis(W)}$ after the $\overline{W}$ low transition.

TMS 4044 NL; TMS 40L44 NL
4096-WORD BY 1-BIT STATIC RAMS

DECEMBER 1977 – REVISED MAY 1982

- Single +5 V Supply (±10% Tolerance)
- High Density 300-mil (7.62 mm) 18-Pin Package
- Fully Static Operation (No Clocks, No Refresh, No Timing Strobe)
- 4 Performance Ranges:

	ACCESS TIME (MAX)	READ OR WRITE CYCLE (MIN)
TMS 4044-12, TMS 40L44-12	120 ns	120 ns
TMS 4044-20, TMS 40L44-20	200 ns	200 ns
TMS 4044-25, TMS 40L44-25	250 ns	250 ns
TMS 4044-45, TMS 40L44-45	450 ns	450 ns

- 400-mV Guranteed DC Noise Immunity with Standard TTL Loads – No Pull-Up Resistors Required
- Common I/O Capability
- 3-State Outputs and Chip Select Control for OR-Tie Capability
- Fan-Out to 2 Series 74, 1 Series 74S, or 8 Series 74LS TTL Loads
- Low Power Dissipation

	MAX (OPERATING)	MAX (STANDBY)
TMS 4044	303 mW	84 mW
TMS 40L44	220 mW	60 mW

TMS 4044/TMS 40L44
18-PIN PLASTIC
DUAL-IN-LINE PACKAGE
(TOP VIEW)

1	A0	18	V_{CC}
2	A1	17	A6
3	A2	16	A7
4	A3	15	A8
5	A4	14	A9
6	A5	13	A10
7	Q	12	A11
8	$\overline{W}$	11	D
9	V_{SS}	10	$\overline{S}$

PIN NAMES

A0-A11	Addresses
D	Data In
Q	Data Out
$\overline{S}$	Chip Select
V_{CC}	+5 V Supply
V_{SS}	Ground
$\overline{W}$	Write Enable

description

This series of static random-access memories is organized as 4096 words of 1 bit. Static design results in reduced overhead costs by elimination of refresh-clocking circuitry and by simplification of timing requirements. Because this series is fully static, chip select may be tied low to further simplify system timing. Output data is always available during a read cycle.

All inputs and outputs are fully compatible with Series 74, 74S or 74LS TTL. No pull-up resistors are required. This 4K Static RAM series is manufactured using TI's reliable N-channel silicon-gate technology to optimize the cost/performance relationship. All versions are characterized to retain data at V_{CC} = 2.4 V to reduce power dissipation.

The TMS 4044/40L44 series is offered in the 18-pin dual-in-line plastic (NL suffix) packages designed for insertion in mounting-hole rows on 300-mil (7.62 mm) centers. The series is guaranteed for operation from 0°C to 70°C.
for operation from 0°C to 70°C.

operation

addresses (A0-A11)

The twelve address inputs select one of the 4096 storage locations in the RAM. The address inputs must be stable for the duration of a write cycle. The address inputs can be driven directly from standard Series 54/74 TTL with no external pull-up resistors.

chip select ($\overline{S}$)

The chip-select terminal, which can be driven directly from standard TTL circuits, affects the data-in and data-out terminals. When chip select is at a logic low level, both terminals are enabled. When chip select is high, data-in is inhibited and data-out is in the floating or high-impedance state.

write enable ($\overline{W}$)

The read or write mode is selected through the write enable terminal. A logic high selects the read mode; a logic low selects the write mode. $\overline{W}$ must be high when changing addresses to prevent erroneously writing data into a memory location. The $\overline{W}$ input can be driven directly from standard TTL circuits.

data-in (D)

Data can be written into a selected device when the write enable input is low. The data-in terminal can be driven directly from standard TTL circuits.

data-out (Q)

The three-state output buffer provides direct TTL compatibility with a fan-out of two Series 74 TTL gates, one Series 74S TTL gate, or eight Series 74LS TTL gates. The output is in the high-impedance state when chip select ($\overline{S}$) is high or whenever a write operation is being performed, facilitating device operation in common I/O systems. Data-out is the same polarity as data-in.

standby operation

The standby mode, which will retain data while reducing power consumption, is attained by recuding the V_{CC} supply from 5 volts to 2.4 volts. When reducing supply voltage during the standby mode, $\overline{S}$ and $\overline{W}$ must be high to retain data. The V_{CC} transition rate should not exceed 26 mV/ms. During standby operation, data can not be read or written into the memory. When resuming normal operation, five cycle times must be allowed after normal supplies are returned for the memory to resume steady state operating conditions.

TEXAS INSTRUMENTS
INCORPORATED
POST OFFICE BOX 225012 • DALLAS, TEXAS 75265

logic symbol†

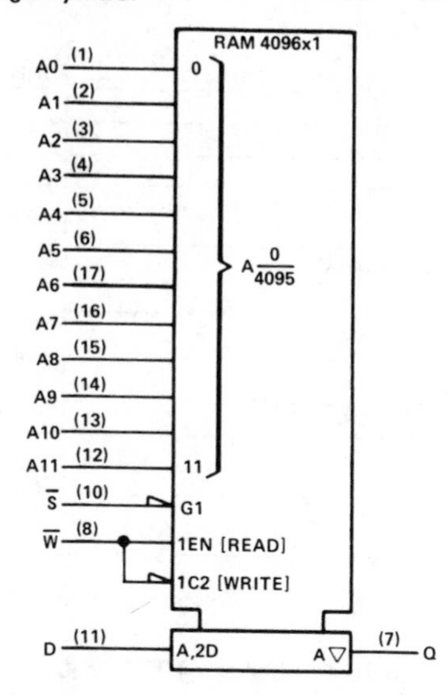

FUNCTION TABLE

INPUTS		OUTPUT	MODE
$\overline{S}$	$\overline{W}$	Q	
H	X	HI-Z	DEVICE DISABLED
L	L	HI-Z	WRITE
L	H	DATA OUT	READ

† This symbol is in accordance with IEEE Std 91/ANSI Y32.14 and recent decisions by IEEE and IEC. See explanation on page 289.

absolute maximum ratings over operating free-air temperature (unless otherwise noted)*

Supply voltage, V_{CC} (see Note 1)	−0.5 to 7 V
Input voltage (any input) (see Note 1)	−1 to 7 V
Continuous power dissipation	1 W
Operating free-air temperature range	0°C to 70°C
Storage temperature range	−55°C to 150°C

NOTE 1: Voltage values are with respect to the ground terminal.

* Stresses beyond those listed under "Absolute Maximum Ratings" may cause permanent damage to the device. This is a stress rating only and functional operation of the device at these or any other conditions beyond those indicated in the "Recommended Operating Conditions" section of this specification is not implied. Exposure to absolute-maximum-rated conditions for extended periods may affect device reliability.

recommended operating conditions

PARAMETER			MIN	NOM	MAX	UNIT
Supply voltage, V_{CC}	TMS 4044-12 TMS 40L44-12	Operating	4.5	5	5.5	V
		Standby	2.4		5.5	V
	TMS 4044-20 TMS 40L44-20	Operating	4.5		5.5	V
		Standby	2.4		5.5	V
	TMS 4044-25 TMS 40L44-25 TMS 40L44-45	Operating	4.5		5.5	V
		Standby	2.4		5.5	V
	TMS 4044-45	Operating	4.5		5.5	V
Supply voltage, V_{SS}				0		V
High-level input voltage, V_{IH}			2		5.5	V
Low-level input voltage, V_{IL} (see Note 2)			−1		0.8	V
Operating free-air temperature, T_A			0		70	°C

NOTE 2: The algebraic convention, where the more negative (less positive) limit is designated as minimum, is used in this data sheet for logic voltage levels only.

TEXAS INSTRUMENTS
INCORPORATED
POST OFFICE BOX 225012 • DALLAS, TEXAS 75265

electrical characteristics over recommended operating free-air temperature ranges (unless otherwise noted)

PARAMETER		TEST CONDITIONS			MIN	TYP‡	MAX	UNIT
V_{OH}	High level voltage	I_{OH} = −1.0 mA	V_{CC} = 4.5 V		2.4			V
V_{OL}	Low level voltage	I_{OL} = 3.2 mA	V_{CC} = 4.5 V				0.4	V
I_I	Input current	V_I = 0 V to 5.5 V					10	μA
I_{OZ}	Off-state output current	$\overline{S}$ at 2 V or $\overline{W}$ at 0.8 V	V_O = 0 V to 5.5 V				±10	μA
I_{CC}	Supply current from V_{CC}	I_O = 0 mA, T_A = 0°C (worst case)	TMS 40L44	V_{CC} = MAX		25	40	mA
				V_{CC} = 2.4 V		15	25	
			TMS 4044-12, TMS 4044-20, TMS 4044-25	V_{CC} = MAX		50	55	
				V_{CC} = 2.4 V		25	35	
			TMS 4044-45	V_{CC} = MAX		50	55	
C_i	Input capacitance	V_I = 0 V, f = 1 MHz					8	pF
C_o	Output capacitance	V_O = 0 V, f = 1 MHz					8	pF

‡ All typical values are at V_{CC} = 5 V, T_A = 25°C.

timing requirements over recommended supply voltage range, T_A = 0°C to 70°C 1 Series 74 TTL load, C_L = 100 pF

PARAMETER		TMS 4044-12 TMS 40L44-12		TMS 4044-20 TMS 40L44-20		TMS 4044-25 TMS 40L44-25		TMS 4044-45 TMS 40L44-45		UNIT
		MIN	MAX	MIN	MAX	MIN	MAX	MIN	MAX	
$t_{c(rd)}$	Read cycle time	120		200		250		450		ns
$t_{c(wr)}$	Write cycle time	120		200		250		450		ns
$t_{v(w)}$	Address valid to end of write	110		180		230		230		ns
$t_{w(W)}$	Write pulse width	60		60		75		200		ns
$t_{su(A)}$	Address set up time	0		0		0		0		ns
$t_{su(S)}$	Chip select set up time	60		60		75		200		ns
$t_{su(D)}$	Data set up time	50		60		75		200		ns
$t_{h(D)}$	Data hold time	0		0		0		0		ns
$t_{h(A)}$	Address hold time	0		0		0		0		ns

switching characteristics over recommended voltage range, $T_A = 0°C$ to $70°C$ 1 Series 74 TTL load, $C_L = 100$ pF

PARAMETER		TMS 4044-12 TMS 40L44-12		TMS 4044-20 TMS 40L44-20		TMS 4044-25 TMS 40L44-25		TMS 4044-45 TMS 40L44-45		UNIT
		MIN	MAX	MIN	MAX	MIN	MAX	MIN	MAX	
$t_{a(A)}$	Access time from address		120		200		250		450	ns
$t_{a(S)}$	Access time from chip select low		70		70		100		100	ns
$t_{a(W)}$	Access time from write enable high		70		70		100		100	ns
$t_{v(A)}$	Output data valid after address change	20		20		20		20		ns
$t_{dis(S)}$	Output disable time after chip select high		50		60		60		80	ns
$t_{dis(W)}$	Output disable time after write enable low		50		60		60		80	ns

read cycle timing**

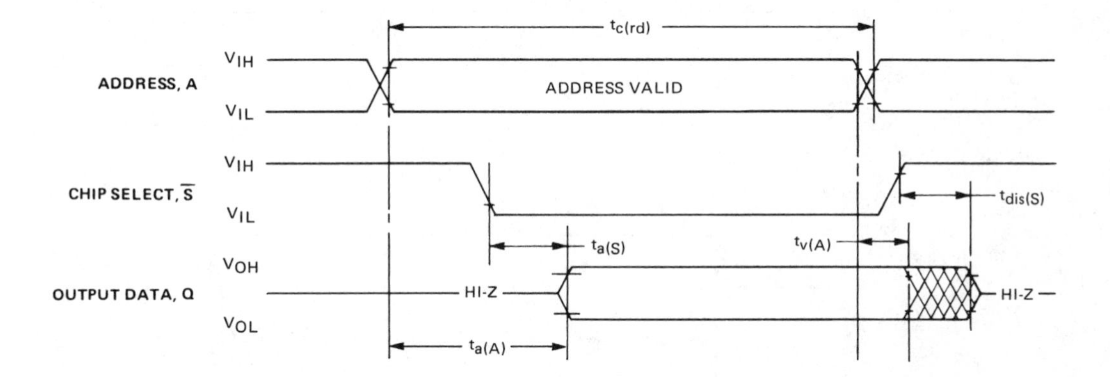

All timing reference points are 0.8 V and 2.0 V on inputs and 0.6 V and 2.2 V on outputs (90% points). Input rise and fall times = 10 ns.
**Write enable is high for a read cycle.

early write cycle timing

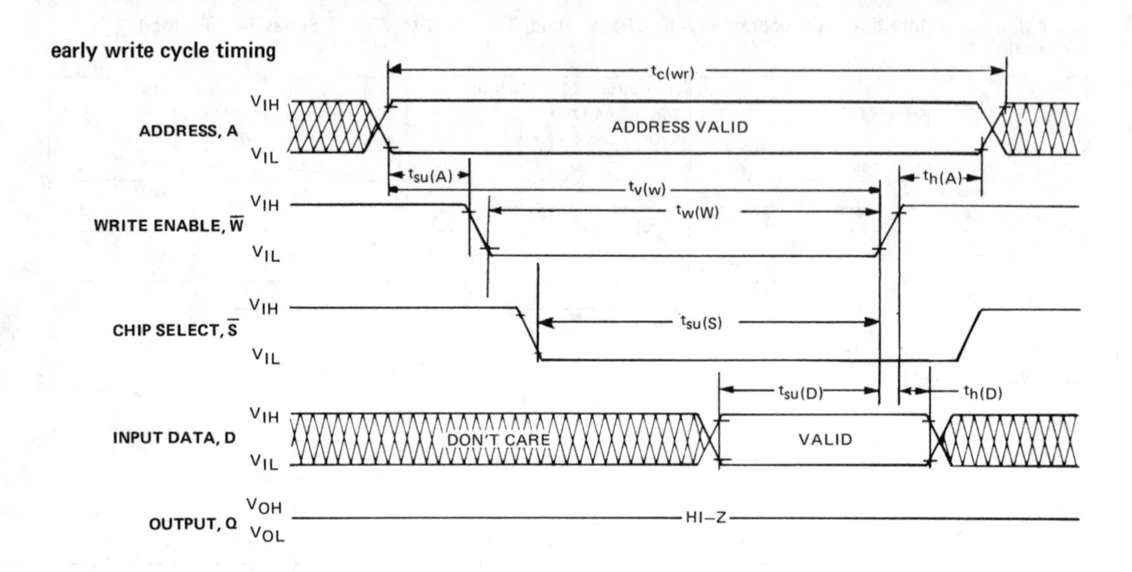

read-write cycle timing

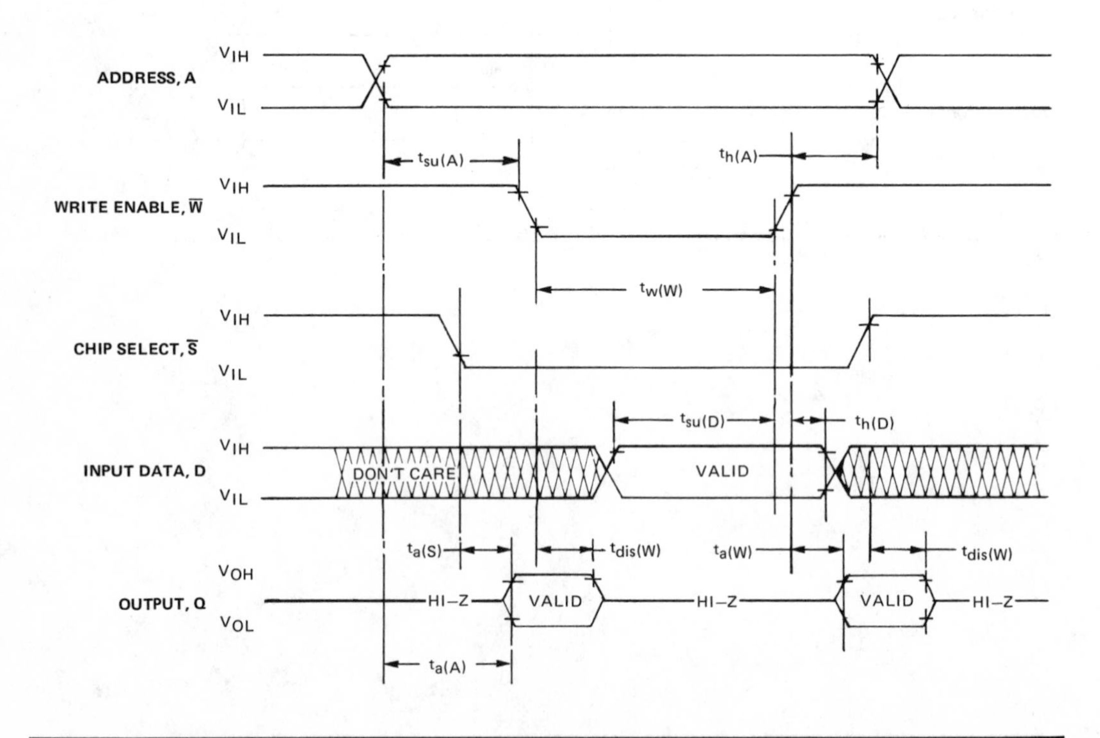

582

TEXAS INSTRUMENTS
INCORPORATED
POST OFFICE BOX 225012 • DALLAS, TEXAS 75265

EPROM
Data Sheets

TMS 2516-25 JL, TMS 2516-35 JL AND TMS 2516-45 JL 16,384-BIT ERASABLE PROGRAMMABLE READ-ONLY MEMORIES

DECEMBER 1979—REVISED MAY 1982

- Organization . . . 2048 X 8
- Single +5 V Power Supply
- Pin Compatible with Existing ROMs and EPROMs (16K, 32K, and 64K)
- JEDEC Standard Pinout
- All Inputs/Outputs Fully TTL Compatible
- Static Operation (No Clocks, No Refresh)
- Max Access/Min Cycle Time
  - TMS 2516-25 . . . 250 ns
  - TMS 2516-35 . . . 350 ns
  - TMS 2516-45 . . . 450 ns
- 8-Bit Output for Use in Microprocessor-Based Systems
- N-Channel Silicon-Gate Technology
- 3-State Output Buffers
- Low Power
  - Active . . . 285 mW Typical
  - Standby . . . 50 mW Typical
- Guaranteed DC Noise Immunity with Standard TTL Loads
- No Pull-Up Resistors Required

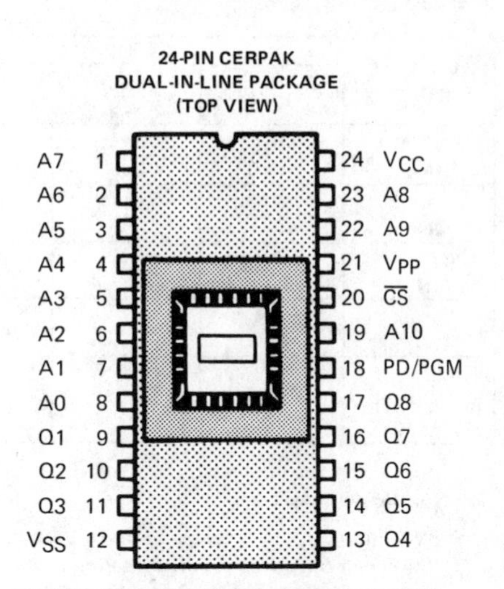

PIN NOMENCLATURE	
A(N)	Address inputs
$\overline{CS}$	Chip Select
PD/PGM	Power Down/Program
Q(N)	Input/Output
V_{CC}	+5 V Power Supply
V_{PP}	+25 V Power Supply
V_{SS}	0 V Ground

description

The TMS 2516 series are 16,384-bit, ultraviolet-light erasable, electrically programmable read-only memories. These devices are fabricated using N-channel silicon-gate technology for high speed and simple interface with MOS and bipolar circuits. All inputs (including program data inputs) can be driven by Series 74 TTL circuits without the use of external pull-up resistors, and each output can drive one Series 74 TTL circuit without external resistors. The data outputs are three-state for connecting multiple devices to a common bus. The TMS 2516 is plug-in compatible with the TMS 4016 16K static RAM. It is offered in a dual-in-line cerpak package (JL suffix) rated for operation from 0 °C to 70 °C.

Since these EPROMs operate from a single +5 V supply (in the read mode), they are ideal for use in microprocessor systems. One other (+25 V) supply is needed for programming but all programming signals are TTL level, requiring a single 50-ms pulse. For programming outside of the system, existing EPROM programmers can be used. Locations may be programmed singly, in blocks, or at random. Total programming time for all bits is 100 seconds.

TEXAS INSTRUMENTS
INCORPORATED
POST OFFICE BOX 225012 • DALLAS, TEXAS 75265

operation

FUNCTION (PINS)	MODE					
	Read	Output Disable	Power Down	Start Programming	Inhibit Programming	Program Verification
PD/PGM (18)	V_{IL}	Don't Care	V_{IH}	Pulsed V_{IL} to V_{IH}	V_{IL}	V_{IL}
$\overline{CS}$ (20)	V_{IL}	V_{IH}	Don't Care	V_{IH}	V_{IH}	V_{IL}
V_{PP} (21)	+5 V	+5 V	+5 V	+25 V	+25 V	+25 V (or +5 V)
V_{CC} (24)	+5 V	+5 V	+5 V	+5 V	+5 V	+5 V
Q (9 to 11, 13 to 17)	Q	HI-Z	HI-Z	D	HI-Z	Q

read/output disable

When the outputs of two or more TMS 2516's are connected to the same bus, the output of any particular device in the circuit can be read with no interference from the competing outputs of the other devices. The device whose output is to be read should have a low-level TTL signal applied to the $\overline{CS}$ and PD/PGM pins. All other devices in the circuit should have their outputs disabled by applying high-level signals to these same pins. PD/PGM can be left low, but it may be advantageous to power down the device during output disable. Output data is accessed at pins Q1 through Q8. On the TMS 2516-45 data can be accessed in 450 ns and access time from $\overline{CS}$ is 150 ns. On the TMS 2516-35 and TMS 2516-25 data can be accessed in 350 and 250 (respectively) and access time from $\overline{CS}$ is 120 ns. These access times assume that the addresses are stable.

power down

Active power dissipation can be cut by 80% by applying a high TTL signal to the PD/PGM pin. In this mode all outputs are in a high-impedance state.

erasure

Before programming, the TMS 2516 is erased by exposing the chip through the transparent lid to high-intensity ultraviolet light having a wavelength of 253.7 nm (2537 angstroms). The recommended minimum exposure dose (UV intensity times exposure time) is fifteen watt-seconds per square centimeter. Thus, a typical 12-milliwatt per-square-centimeter, filterless UV lamp will erase the device in a minimum of 21 minutes. The lamp should be located about 2.5 centimeters above the chip during erasure. After erasure, all bits are in the "1" state (assuming a high-level output corresponds to logic "1").

start programming

After erasure (all bits in logic "1" state), logic "0's" are programmed into the desired locations. A "0" can be erased only by ultraviolet light. The programming mode is achieved when V_{PP} is 25 V and $\overline{CS}$ is at V_{IH}. Data is presented in parallel (8 bits) on pins Q1 through Q8. Once addresses and data are stable, a 50-millisecond TTL high-level pulse should be applied to the PGM pin at each address location to be programmed. Maximum pulse width is 55 milliseconds. Locations can be programmed in any order. Several TMS 2516's can be programmed simultaneously when the devices are connected in parallel.

TEXAS INSTRUMENTS
INCORPORATED
POST OFFICE BOX 225012 • DALLAS, TEXAS 75265

inhibit programming

When two or more devices are connected in parallel, data can be programmed into all devices or only chosen devices. TMS 2516's not intended to be programmed (i.e., inhibited) should have a low level applied to the PD/PGM pin and a high-level applied to the CS pin

program verification

A verification is done to see if the device was programmed correctly. A verification can be done at any time. It can be done on each location immediately after that location is programmed. To do a verification, V_{PP} may be kept at +25 V.

logic symbol†

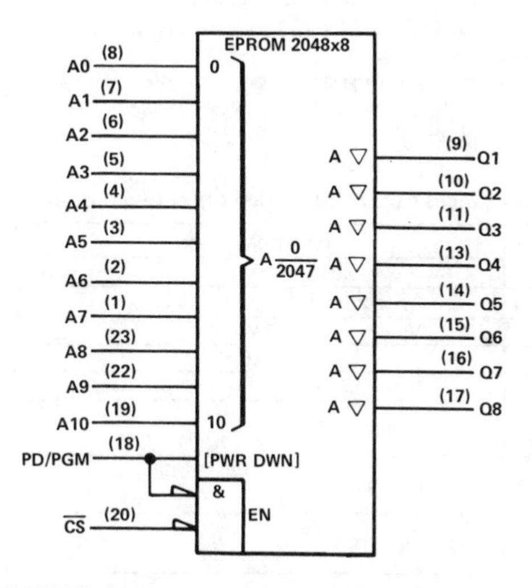

† This symbol is in accordance with IEEE Std 91/ANSI Y32.14 and recent decisions by IEEE and IEC. See explanation on page 289.

absolute maximum ratings over operating free-air temperature range (unless otherwise noted)*

Supply voltage, V_{CC} (see Note 1)	−0.3 to 6 V
Supply voltage, V_{PP} (see Note 1)	−0.3 to 28 V
All input voltages (see Note 1)	−0.3 to 6 V
Output voltage (operating with respect to V_{SS})	−0.3 to 6 V
Operating free-air temperature range	0 °C to 70 °C
Storage temperature range	−55 °C to 125 °C

NOTE 1: Under absolute maximum ratings, voltage values are with respect to the most negative supply voltage, V_{SS} (substrate).

* Stresses beyond those listed under "Absolute Maximum Ratings" may cause permanent damage to the device. This is a stress rating only and functional operation of the device at these or any other conditions beyond those indicated in the "Recommended Operating Conditions" section of this specification is not implied. Exposure to absolute-maximum-rated conditions for extended periods may affect device reliability.

recommended operating conditions

PARAMETER	TMS 2516-25 MIN	TMS 2516-25 NOM	TMS 2516-25 MAX	TMS 2516-35 MIN	TMS 2516-35 NOM	TMS 2516-35 MAX	TMS 2516-45 MIN	TMS 2516-45 NOM	TMS 2516-45 MAX	UNIT
Supply voltage, V_{CC} (see Note 2)	4.75	5	5.25	4.75	5	5.25	4.75	5	5.25	V
Supply voltage, V_{PP} (see Note 3)		VCC			V_{CC}			V_{CC}		V
Supply voltage, V_{SS}		0			0			0		V
High-level input voltage, V_{IH}	2		$V_{CC}+1$	2		$V_{CC}+1$	2		$V_{CC}+1$	V
Low-level input voltage, V_{IL}	−0.1		0.8	−0.1		0.8	−0.1		0.8	V
Read cycle time, $t_{c(rd)}$	250			350			450			ns
Operating free-air temperature, T_A	0		70	0		70	0		70	°C

NOTES: 2. V_{CC} must be applied before or at the same time as V_{PP} and removed after or at the same time as V_{PP}. The device must not be inserted into or removed from the board when V_{PP} or V_{CC} is applied.

3. V_{PP} can be connected to V_{CC} directly (except in the program mode). V_{CC} supply current in this case would be $I_{CC} + I_{PP}$. During programming, V_{PP} must be maintained at 25 V (± 1V).

electrical characteristics over full ranges of recommended operating conditions

PARAMETER		TEST CONDITIONS	MIN	TYP†	MAX	UNIT
V_{OH}	High-level output voltage*	$I_{OH} = -400\ \mu A$	2.4			V
V_{OL}	Low-level output voltage*	$I_{OL} = 2.1$ mA			0.45	V
I_I	Input current (leakage)	V_I = 0 V to 5.25 V			10	μA
I_O	Output current (leakage)	V_O = 0.4 V to 5.25 V			10	μA
I_{PP1}	V_{PP} supply current	V_{PP} = 5.25 V, PD/PGM = V_{IL}			6	mA
I_{PP2}	V_{PP} supply current (during program pulse)	PD/PGM = V_{IH}			30	mA
I_{CC1}	V_{CC} supply current (standby)	PD/PGM = V_{IH}		10	25	mA
I_{CC2}	V_{CC} supply current (active)	$\overline{CS}$ = PD/PGM = V_{IL}		57	100	mA

† Typical values are at T_A = 25 °C and nominal voltage.
* All AC and DC measurements are made at 10% and 90% points with a 50% pattern.

capacitance over recommended supply voltage and operating free-air temperature ranges, f = 1 MHz*

PARAMETER		TEST CONDITIONS	TYP†	MAX	UNIT
C_i	Input capacitance	V_I = 0 V, f = 1 MHz	4	6	pF
C_o	Output capacitance	V_O = 0 V, f = 1 MHz	8	12	pF

†All typical values are T_A = 25°C and nominal voltage
*Capacitive measurements are made on sample basis only

TEXAS INSTRUMENTS
INCORPORATED
POST OFFICE BOX 225012 • DALLAS, TEXAS 75265

switching characteristics over full ranges of recommended operating conditions (see Note 4)

PARAMETER		TEST CONDITIONS (SEE NOTES 4 AND 5)	TMS 2516-25			TMS 2516-35			TMS 2516-45			UNIT
			MIN	TYP†	MAX	MIN	TYP†	MAX	MIN	TYP†	MAX	
$t_{a(A)}$	Access time from address	C_L = 100 pF, 1 Series 74 TTL load, $t_r \leq 20$ ns, $t_f \leq 20$ ns		230	250		250	350		280	450	ns
$t_{a(CS)}$	Access time from chip select				120			120			150	ns
$t_{a(PR)}$	Access time from PD/PGM			230	250		250	350		280	450	ns
$t_{v(A)}$	Output data valid after address change		0			0			0			ns
$t_{dis(CS)}$	Output disable time from chip select during read only‡		0		100	0		100	0		100	ns
$t_{dis(CS)}$	Output disable time from chip select during program and program verify‡				120			120			120	ns
$t_{dis(PR)}$	Output disable time from PD/PGM‡		0		100	0		100			100	ns

† All typical values are at T_A = 25°C and nominal voltages.

‡ Value calculated from 0.5 volt delta to measured output level.

recommended timing requirements for programming T_A = 25°C (see Note 4)

PARAMETER		MIN	TYP†	MAX	UNIT
$t_{w(PR)}$	Pulse width, program pulse	45	50	55	ms
$t_{r(PR)}$	Rise time, program pulse	5			ns
$t_{f(PR)}$	Fall time, program pulse	5			ns
$t_{su(A)}$	Address setup time	2			μs
$t_{su(CS)}$	Chip-select setup time	2			μs
$t_{su(D)}$	Data setup time	2			μs
$t_{su(VPP)}$	Setup time from V_{PP}	0			ns
$t_{h(A)}$	Address hold time	2			μs
$t_{h(CS)}$	Chip-select hold time	2			μs
$t_{h(D)}$	Data hold time	2			μs

† Typical values are at nominal voltages.

NOTES: 4. For all switching characteristics and timing measurements, input pulse levels are 0.65 V to 2.2 V and V_{PP} = 25 V ± 1 V during programming. All AC and DC measurements are made at 10% and 90% points with a 50% pattern.

5. Common test conditions apply for t_{dis} except during programming. For $t_{a(A)}$, $t_{a(CS)}$, and t_{dis}, PD/PGM = $\overline{CS}$ = V_{IL}.

PARAMETER MEASUREMENT INFORMATION

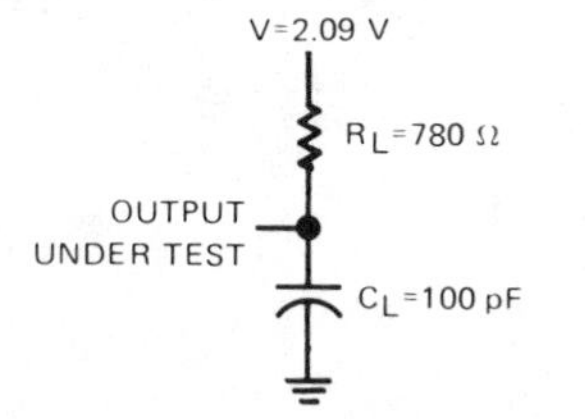

FIGURE 1 – TYPICAL OUTPUT LOAD CIRCUIT

read cycle timing

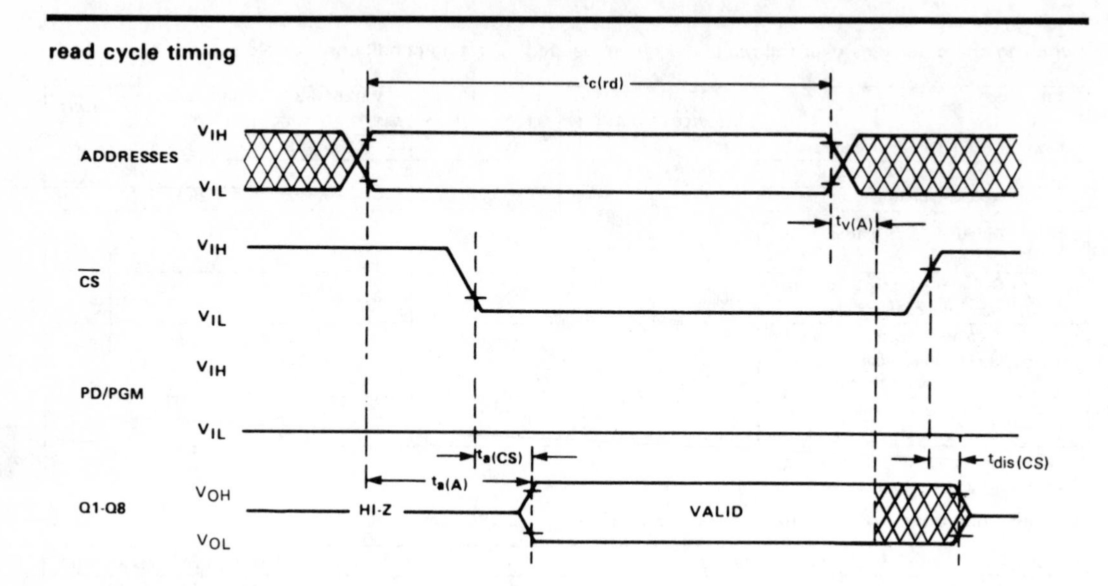

standby mode

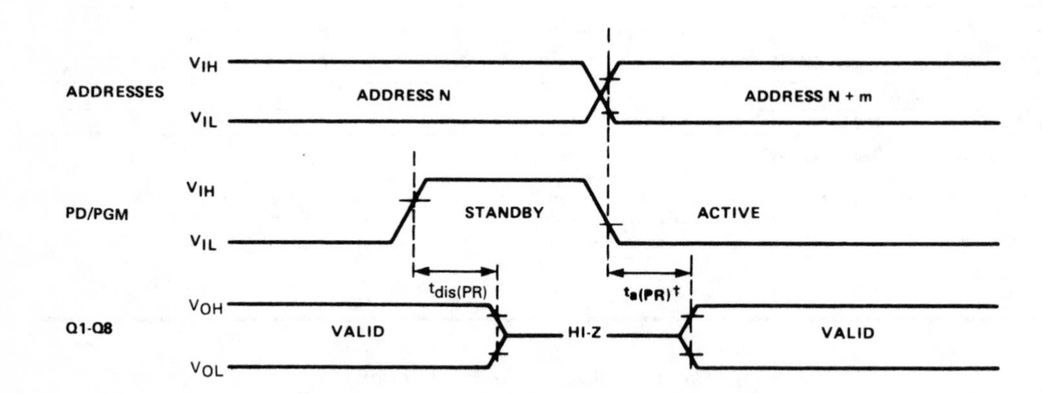

NOTE: $\overline{CS}$ must be in low state during Active Mode, "Don't Care" otherwise.
† $t_{a(PR)}$ referenced to PD/PGM or the address, whichever occurs last.

All timing reference points in this data sheet (inputs and outputs) are 90% points.

TEXAS INSTRUMENTS
INCORPORATED
POST OFFICE BOX 225012 • DALLAS, TEXAS 75265

program cycle timing

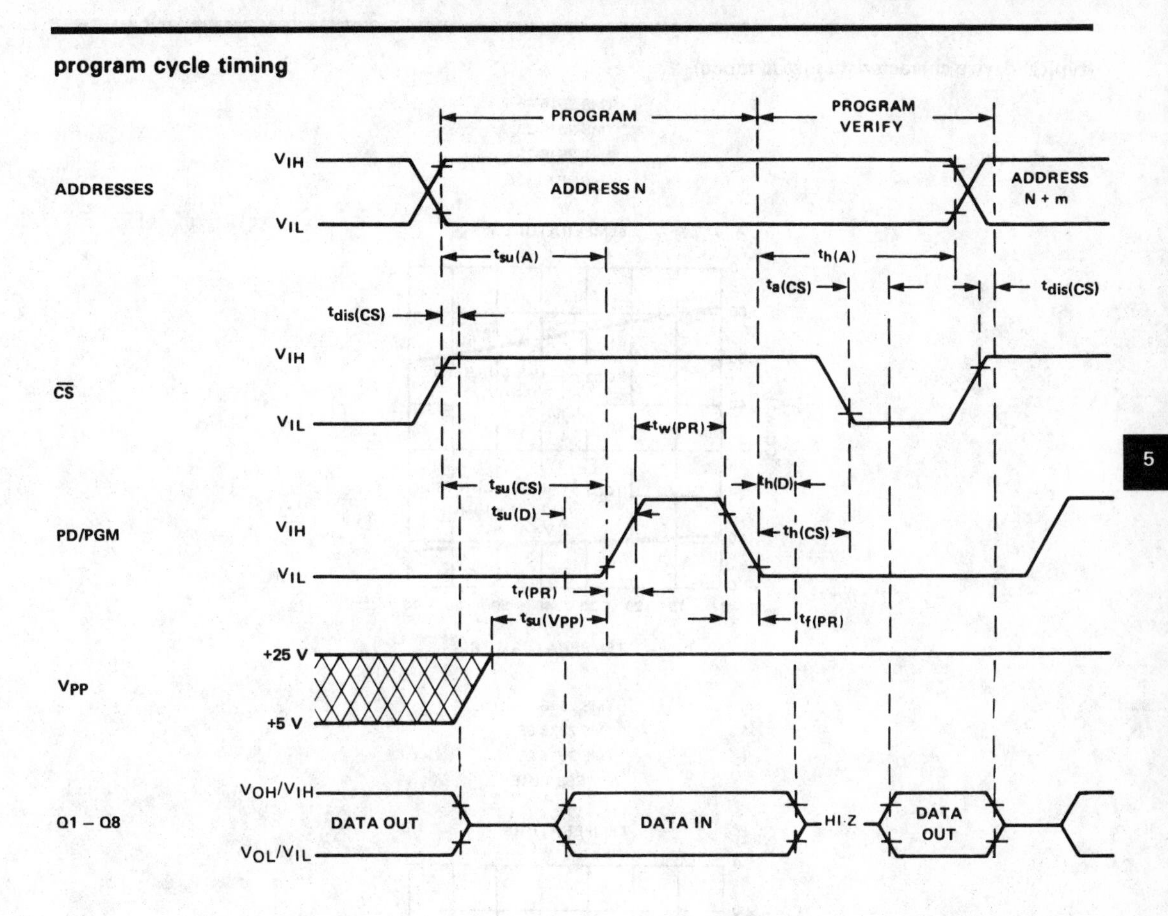

typical device characteristics (read mode)

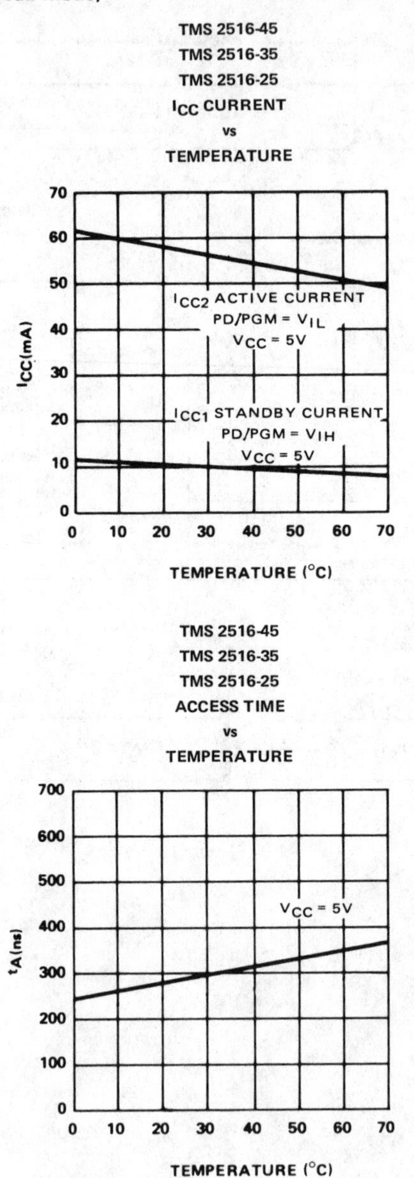

TMS 2532-30 JL, TMS 2532-35 JL, TMS 2532-45 JL, TMS 25L32-45 JL
32,768-BIT ERASABLE PROGRAMMABLE READ-ONLY MEMORIES

DECEMBER 1979–REVISED MAY 1982

- **Organization . . . 4096 X 8**
- **Single +5 V Power Supply**
- **Pin Compatible with Existing ROMs and EPROMs (8K, 16K, 32K, and 64K)**
- **JEDEC Standard Pinout**
- **All Inputs/Outputs Fully TTL Compatible**
- **Static Operation (No Clocks, No Refresh)**
- **Max Access/Min Cycle Time:**

TMS 2532-30	300 ns
TMS 2532-35	350 ns
TMS 2532-45	450 ns
TMS 25L32-45	450 ns

- **8-Bit Output for Use in Microprocessor-Based Systems**
- **N-Channel Silicon-Gate Technology**
- **3-State Output Buffers**
- **40% Lower Power**
  TMS 25L32 . . . 500 mW Max Active
  TMS 2532 . . . 840 mW Max Active
- **Guaranteed DC Noise Immunity with Standard TTL Loads**
- **No Pull-Up Resistors Required**

24-PIN CERAMIC
DUAL-IN-LINE PACKAGE
(TOP VIEW)

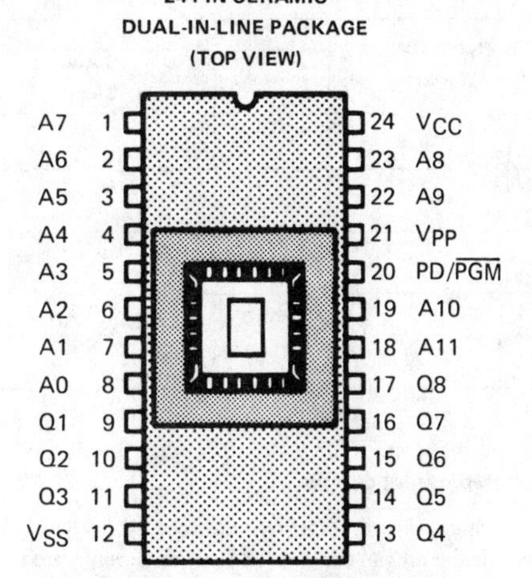

PIN NOMENCLATURE	
A(N)	Address inputs
PD/$\overline{PGM}$	Power Down/Program
Q(N)	Input/Output
V_{CC}	+5 V Power Supply
V_{PP}	+25 V Power Supply
V_{SS}	0 V Ground

description

The TMS 2532 series (TMS 2532-30 JL, TMS 2532-35 JL, TMS 2532-45 JL, and TMS 25L32-45 JL) are 32,768-bit, ultraviolet-light-erasable, electrically programmable read-only memories. These devices are fabricated using N-channel silicon-gate technology for high speed and simple interface with MOS and bipolar circuits. All inputs (including program data inputs) can be driven by Series 74 TTL circuits without the use of external pull-up resistors, and each output can drive one Series 74 TTL circuit without external resistors. The data outputs are three-state for connecting multiple devices to a common bus. The TMS 2532 series are plug-in compatible with the TMS 4732 32K ROM. The devices are offered in a dual-in-line ceramic package (JL suffix) rated for operation from 0 °C to 70 °C.

Since these EPROMs operate from a single +5 V supply (in the read mode), they are ideal for use in microprocessor systems. One other (+25 V) supply is needed for programming but all programming signals are TTL level, requiring a single 50 ms pulse. For programming outside of the system, existing EPROM programmers can be used. Locations may be programmed singly, in blocks, or at random. Total programming time for all bits is 200 seconds.

TEXAS INSTRUMENTS
INCORPORATED
POST OFFICE BOX 225012 • DALLAS, TEXAS 75265

operation

FUNCTION (PINS)	MODE				
	Read	Output Disable	Power Down	Start Programming	Inhibit Programming
PD/$\overline{PGM}$ (20)	V_{IL}	V_{IH}	V_{IH}	Pulsed V_{IH} to V_{IL}	V_{IH}
V_{PP} (21)	+5 V	+5 V	+5 V	+25 V	+25 V
V_{CC} (24)	+5 V	+5 V	+5 V	+5 V	+5 V
Q (9 to 11, 13 to 17)	Q	HI-Z	HI-Z	D	HI-Z

read/output disable

When the outputs of two or more TMS 2532's are connected on the same bus, the output of any particular device in the circuit can be read with no interference from the competing outputs of the other devices. The device whose output is to be read should have a low-level TTL signal applied to the PD/$\overline{PGM}$ Pin. All other devices in the circuit should have their outputs disabled by applying a high-level signal to this pin. Output data is accessed at pins Q1 through Q8. Data can be accessed in 450 ns = $t_{a(A)}$.

power down

Active power dissipation can be cut by over 70% by applying a high TTL signal to the PD/$\overline{PGM}$ pin. In this mode all outputs are in a high-impedance state.

erasure

Before programming, the TMS 2532 is erased by exposing the chip through the transparent lid to high-intensity ultraviolet light having a wavelength of 253.7 nm (2537 angstroms). The recommended minimum exposure dose (UV intensity times exposure time) is fifteen watt-seconds per square centimeter. Thus, a typical 12 milliwatt per square centimeter, filterless UV lamp will erase the device in a minimum of 21 minutes. The lamp should be located about 2.5 centimeters above the chip during erasure. After erasure, all bits are in the "1" state (assuming high-level output corresponds to logic "1").

start programming

After erasure (all bits in logic "1" state), logic "0's" are programmed into the desired locations. A "0" can be erased only by ultraviolet light. The programming mode is achieved when V_{PP} is 25 V. Data is presented in parallel (8 bits) on pins Q1 through Q8. Once addresses and data are stable, a 50-millisecond TTL low-level pulse should be applied to the $\overline{PGM}$ pin at each address location to be programmed. Maximum pulse width is 55-milliseconds. Locations can be programmed in any order. Several TMS 2532's can be programmed simultaneously when the devices are connected in parallel.

inhibit programming

When two or more devices are connected in parallel, data can be programmed into all devices or only chosen devices. Any TMS 2532's not intended to be programmed should have a high level applied to PD/$\overline{PGM}$.

program verification

The TMS 2532 program verification is simply the read operation, which can be performed as soon as V_{PP} returns to +5 V ending the program cycle.

TEXAS INSTRUMENTS
INCORPORATED
POST OFFICE BOX 225012 • DALLAS, TEXAS 75265

logic symbol†

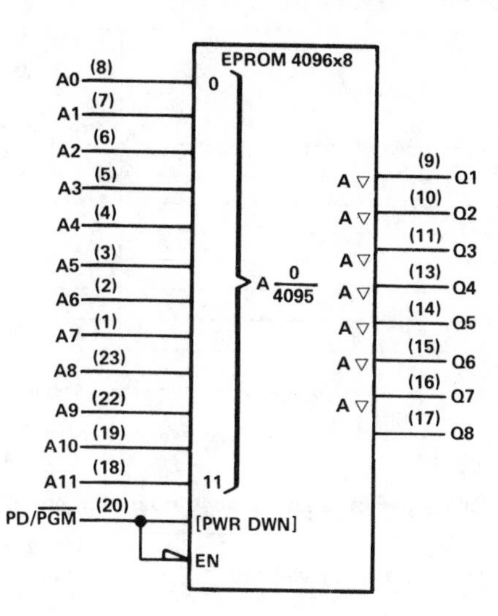

† This symbol is in accordance with IEEE Std 91/ANSI Y32.14 and recent decisions by IEEE and IEC. See explanation on page 289.

absolute maximum ratings over operating free-air temperature range (unless otherwise noted)*

Supply voltage, V_{CC} (see Note 1)	−0.3 to 6 V
Supply voltage, V_{PP} (see Note 1)	−0.3 to 28 V
All input voltages (see Note 1)	−0.3 to 6 V
Output voltage (operating with respect to V_{SS})	−0.3 to 6 V
Operating free-air temperature range	0 °C to 70 °C
Storage temperature range	−55 °C to 125 °C

NOTE 1: Under absolute maximum ratings, voltage values are with respect to the most negative supply voltage, V_{SS} (substrate).

*Stresses beyond those (listed under Absolute Maximum Ratings'' may cause permanent damage to the device. This is a stress rating only and functional operation of the device at these or any other conditions beyond those indicated in the ''Recommended Operating Conditions'' section of this specification is not implied. Exposure to absolute-maximum-rated conditions for extended periods may affect device reliability.

recommended operating conditions

	TMS 2532-45			TMS 2532-30 TMS 2532-35 TMS 25L32-45			UNIT
	MIN	NOM	MAX	MIN	NOM	MAX	
Supply voltage, V_{CC} (see Note 2)	4.75	5	5.25	4.75	5	5.25	V
Supply voltage, V_{PP} (see Note 3)		V_{CC}			V_{CC}		V
Supply voltage, V_{SS}		0			0		V
High-level input voltage, V_{IH}	2.2		$V_{CC}+1$	2		$V_{CC}+1$	V
Low-level input voltage, V_{IL}	−0.1		0.65	−0.1		0.8	V
Read cycle time, $t_{c(rd)}$	450			450			ns
Operating free-air temperature, T_A	0		70	0		70	°C

NOTES 2. V_{CC} must be applied before or at the same time as V_{PP} and removed after or at the same time as V_{PP}. The device must not be inserted into or removed from the board when V_{PP} is applied.
3. V_{PP} can be connected to V_{CC} directly (except in the program mode). V_{CC} supply current in this case would be $I_{CC} + I_{PP}$. During programming, V_{PP} must be maintained at 25 V (±1 V).

electrical characteristics over full ranges of recommended operating conditions

PARAMETER		TEST CONDITIONS	TMS 2532-30 TMS 2532-35 TMS 2532-45			TMS 25L32-45			UNIT
			MIN	TYP†	MAX	MIN	TYP†	MAX	
V_{OH}	High-level output voltage*	$I_{OH} = -400\ \mu A$	2.4			2.4			V
V_{OL}	Low-level output voltage*	$I_{OL} = 2.1$ mA			0.45			0.45	V
I_I	Input current (leakage)	$V_I = 0$ V or 5.25 V			±10			±10	μA
I_O	Output current (leakage)	$V_O = 0.4$ V or 5.25 V			±10			±10	μA
I_{PP1}	V_{PP} supply current	$V_{PP} = 5.25$ V, $PD/\overline{PGM} = V_{IL}$			12			12	mA
I_{PP2}	V_{PP} supply current (during program pulse)	$PD/\overline{PGM} = V_{IL}$			30			30	mA
I_{CC1}	V_{CC} supply current (standby)	$PD/\overline{PGM} = V_{IH}$		10	25		10	25	mA
I_{CC2}	V_{CC} supply current (active)	$PD/\overline{PGM} = V_{IL}$		80	160		65	95	mA

*AC and DC measurements are made at 10% and 90% points using a 50% pattern.

capacitance over recommended supply voltage and operating free-air temperature ranges, f = 1 MHz‡

PARAMETER		TEST CONDITIONS	TYP†	MAX	UNIT
C_i	Input capacitance	$V_I = 0$ V, f = 1 MHz	4	6	pF
C_o	Output capacitance	$V_O = 0$ V, f = 1 MHz	8	12	pF

†Typical values are $T_A = 25\ °C$ and nominal voltages.
‡Capacitance measurements are made on a sample basis only.

TEXAS INSTRUMENTS
INCORPORATED
POST OFFICE BOX 225012 • DALLAS, TEXAS 75265

switching characteristics over full ranges of recommended operating conditions (see note 4)

PARAMETER		TEST CONDITIONS (See Notes 4 & 5)	TMS 2532-30 MIN	TMS 2532-30 TYP†	TMS 2532-30 MAX	TMS 2532-35 MIN	TMS 2532-35 TYP†	TMS 2532-35 MAX	TMS 25L32-45 TMS 2532-45 MIN	TMS 25L32-45 TMS 2532-45 TYP†	TMS 25L32-45 TMS 2532-45 MAX	UNIT
$t_{a(A)}$	Access time from address	$C_L = 100$ pF, 1 Series 74 TTL load, $t_r \leq 20$ ns, $t_f \leq 20$ ns, See Figure 1			300			350		280	450	ns
$t_{a(PR)}$	Access time from PD/$\overline{PGM}$				300			350		280	450	ns
$t_{v(A)}$	Output data valid after address change								0			ns
t_{dis}	Output disable time from PD/$\overline{PGM}$ ‡				100			100	0		100	ns

†All typical values are at $T_A = 25\,°C$ and nominal voltages.
‡Value calculated from 0.5 volt delta to measured output level.

recommended timing requirements for programming $T_A = 25\,°C$ (see note 4)

PARAMETER		MIN	TYP†	MAX	UNIT
$t_{w(PR)}$	Pulse width, program pulse	45	50	55	ms
$t_{r(PR)}$	Rise time, program pulse	5			ns
$t_{f(PR)}$	Fall time, program pulse	5			ns
$t_{su(A)}$	Address setup time	2			μs
$t_{su(D)}$	Data setup time	2			μs
$t_{su(VPP)}$	Setup time from V_{PP}	0			ns
$t_{h(A)}$	Address hold time	2			μs
$t_{h(D)}$	Data hold time	2			μs
$t_{h(PR)}$	Program pulse hold time	0			ns
$t_{h(VPP)}$	V_{PP} hold time	0			ns

†Typical values are at nominal voltages.

NOTES: 4. For all switching characteristics and timing measurements, input pulse levels are 0.65 V to 2.2 V and $V_{PP} = 25\text{ V} \pm 1\text{ V}$ during programming. All AC and DC measurements are made at 10% and 90% points with a 50% pattern.
5. Common test conditions apply for t_{dis} except during programming. For $t_{a(A)}$ and t_{dis}, PD/$\overline{PGM} = V_{IL}$.

PARAMETER MEASUREMENT INFORMATION

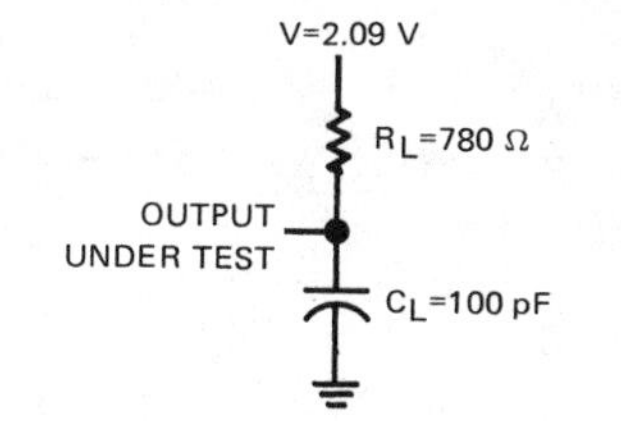

FIGURE 1 – TYPICAL OUTPUT LOAD CIRCUIT

read cycle timing

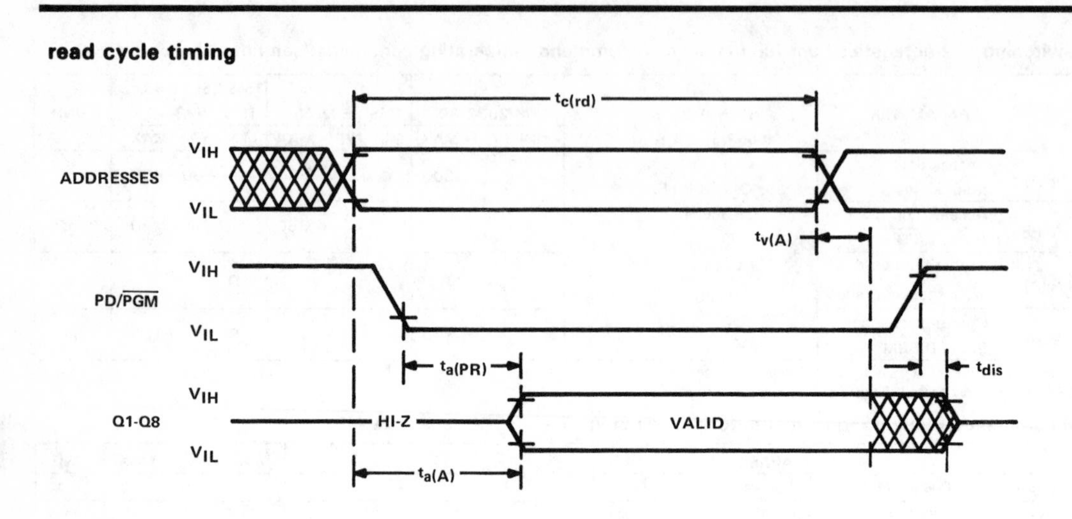

NOTE: There is no chip select pin on the TMS 2532.
The chip-select function is incorporated in the power-down mode.

standby mode

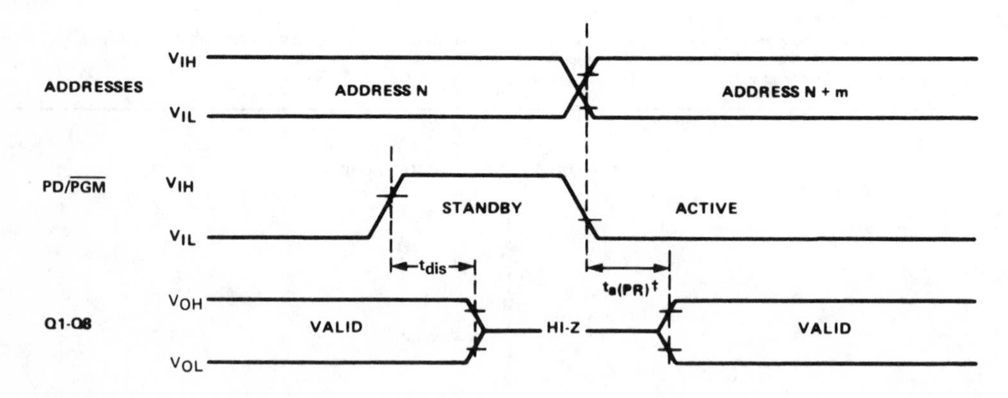

†$t_{a(PR)}$ referenced to PD/$\overline{PGM}$ or the address, whichever occurs last.

All timing reference points in this data sheet (inputs and outputs) are 90% points.

TEXAS INSTRUMENTS
INCORPORATED
POST OFFICE BOX 225012 • DALLAS, TEXAS 75265

program cycle timing

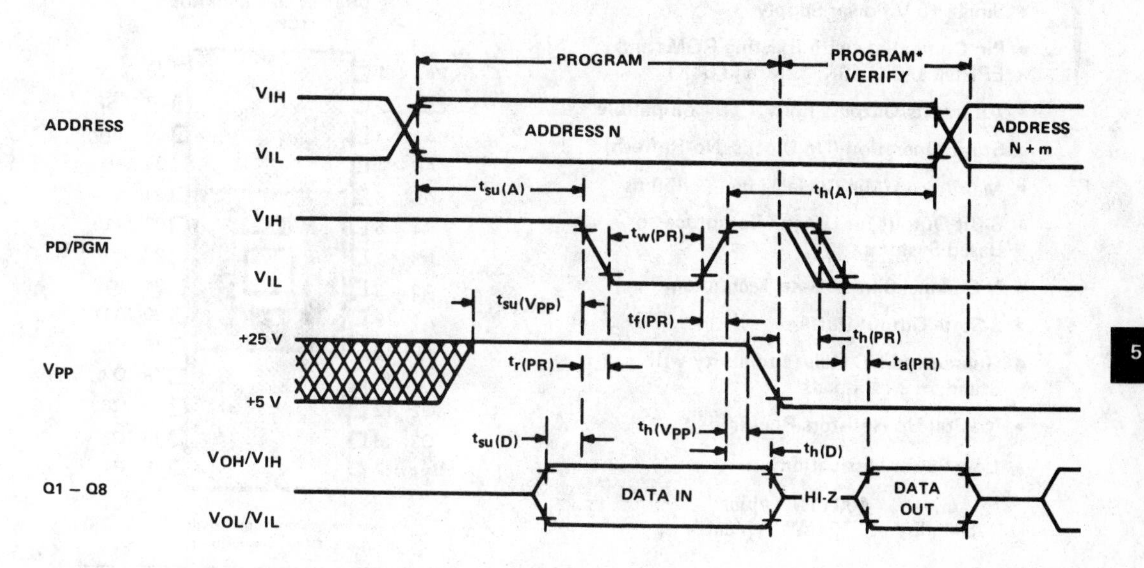

*Program verify equivalent to read mode.

TMS 2564-45 JL
65,536-BIT ERASABLE PROGRAMMABLE READ-ONLY MEMORY

MAY 1981—REVISED MAY 1982

- Organization . . . 8K X 8
- Single +5 V Power Supply
- Pin Compatible with Existing ROMs and EPROMs (8K, 16K, 32K, and 64K)
- All Inputs/Outputs Fully TTL Compatible
- Static Operation (No Clocks, No Refresh)
- Max Access/Min Cycle Time . . . 450 ns
- 8-Bit Output for Use in Microprocessor-Based Systems
- N-Channel Silicon-Gate Technology
- 3-State Output Buffers
- Guaranteed DC Noise Immunity with Standard TTL Loads
- No Pull-Up Resistors Required
- Low Power Dissipation:

  Active . . . 400 mW Typical
  Standby . . . 75 mW Typical

TMS 2564
28-PIN CERAMIC
DUAL-IN-LINE PACKAGE
(TOP VIEW)

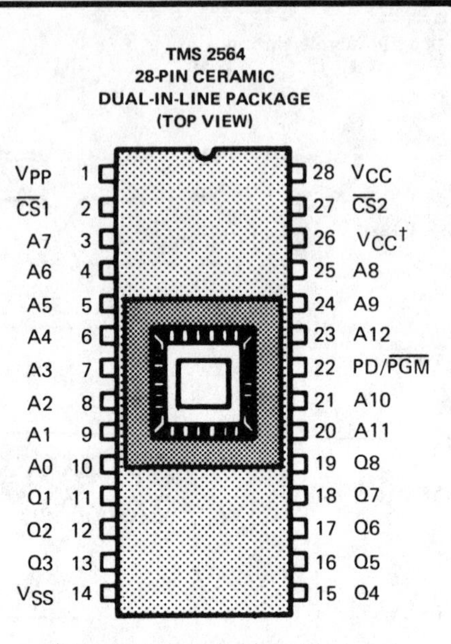

†V_{CC} may be connected to pin 26 for 24-pin ROM compatibility.

PIN NOMENCLATURE	
A(N)	Address inputs
$\overline{CS}$(N)	Chip Selects
PD/$\overline{PGM}$	Power Down/Program
Q(N)	Input/Output
V_{CC}	+5 V Power Supply
V_{PP}	+25 V Power Supply
V_{SS}	0 V Ground

description

The TMS 2564 is a 65,536-bit, ultraviolet-light-erasable, electrically programmable read-only memory. This device is fabricated using N-channel silicon-gate technology for high-speed and simple interface with MOS and bipolar circuits. All inputs (including program data inputs) can be driven by Series 74 TTL circuits without the use of external pull-up resistors, and each output can drive one Series 74 TTL circuit without external resistors. The data outputs are three-state for connecting multiple devices to a common bus. The TMS 2564 is offered in a dual-in-line ceramic package (JL or JDL suffix)* rated for operation from 0°C to 70°C.

Since this EPROM operates from a single +5 V supply (in the read mode), it is ideal for use in microprocessor systems. One other supply (+25 V) is needed for programming. Programming requires a single TTL level pulse per location. For programming outside of the system, existing EPROM programmers can be used. Locations may be programmed singly, in blocks, or at random.

The TMS 2564 is compatible with other 5-volt ROMs and EPROMs, including those in a 24-pin package.

POST OFFICE BOX 225012 • DALLAS, TEXAS 75265

operation

FUNCTION (PINS)	MODE								
	Read	Output Disable			Power Down	Start Programming	Inhibit Programming		
PD/$\overline{PGM}$ (22)	V_{IL}	V_{IH}	X	X	V_{IH}	Pulsed V_{IH} to V_{IL}	V_{IH}	X	X
$\overline{CS1}$ (21)	V_{IL}	X	V_{IH}	X	X	V_{IL}	X	V_{IH}	X
$\overline{CS2}$ (27)	V_{IL}	X	X	V_{IH}	X	V_{IL}	X	X	V_{IH}
V_{PP} (1)	+5 V	+5 V			+5 V	+25 V	+25 V		
V_{CC}* (26/28)	+5 V	+5 V			+5 V	+5 V	+5 V		
Q (11 to 13, 15 to 19)	Q	HI-Z			HI-Z	D	HI-Z		

X=Don't care.

* Do not use the internal jumper of 26-28 to conduct PC board currents.

read/output disable

When the outputs of two or more TMS 2564's are paralled on the same bus, the output of any particular device in the circuit can be read with no interference from the competing outputs of the other devices. To read the output of the TMS 2564, the low-level signal is applied to the PD/$\overline{PGM}$ and $\overline{CS}$ pins. All other devices in the circuit should have their outputs disabled by applying a high-level signal to one of these pins. Output data is accessed at pins Q1 to Q8. Data can be accessed in 450 ns = $t_{a(A)}$.

power down

Active power dissipation can be cut by over 80 percent by applying a high TTL signal to the PD/$\overline{PGM}$ pin. In this mode all outputs are in a high-impedance state.

erasure

Before programming, the TMS 2564 is erased by exposing the chip through the transparent lid to high intensity ultraviolet (wavelength 2537 angstroms). The recommended minimum exposure dose (= UV intensity X exposure time) is fifteen watt-seconds per square centimeter. A typical 12 milliwatt per square centimeter, filterless UV lamp will erase the device in about 21 minutes. The lamp should be located about 2.5 centimeters above the chip during erasure. After erasure, all bits are in the high state.

start programming

After erasure (all bits in logic high state), logic "0's" are programmed into the desired locations. A low can be erased only by ultraviolet light. The programming mode is achieved when Vpp is 25 V. Data is presented in parallel (8 bits) on pins Q1 to Q8. Once addresses and data are stable, a 50 millisecond low TTL pulse should be applied to the $\overline{PGM}$ pin at each address location to be programmed. Maximum pulse width is 55 milliseconds. Locations can be programmed in any order. More than one TMS 2564 can be programmed when the devices are connected in parallel. During programming both chip select signals should be held low unless program inhibit is desired.

inhibit programming

When two or more TMS 2564's are connected in parallel, data can be programmed into all devices or only chosen devices. TMS 2564's not intended to be programmed should have a high level applied to PD/$\overline{PGM}$ or $\overline{CS1}$ or $\overline{CS2}$.

logic symbol†

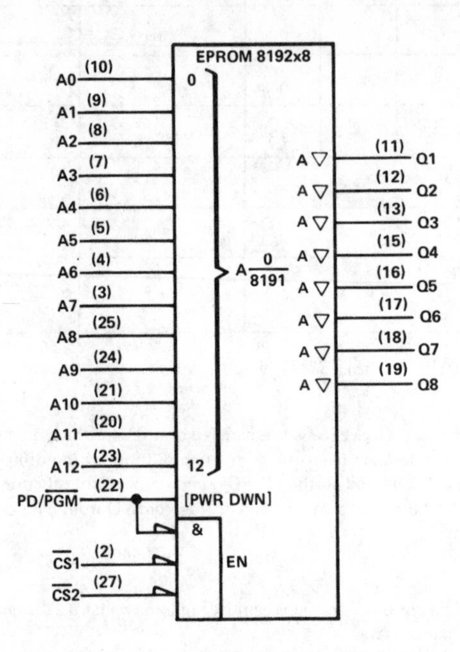

† This symbol is in accordance with IEEE Std 91/ANSI Y32.14 and recent decisions by IEEE and IEC. See explanation on page 289.

absolute maximum ratings over operating free-air temperature range (unless otherwise noted)*

Supply voltage, V_{CC} (see Note 1)	−0.3 to 6 V
Supply voltage, V_{PP} (see Note 1)	−0.3 to 28 V
All input voltages (see Note 1)	−0.3 to 6 V
Output voltage (operating with respect to V_{SS})	−0.3 to 6 V
Operating free-air temperature range	0°C to 70°C
Storage temperature range	−55°C to 125°C

NOTE 1: Under absolute maximum ratings, voltage values are with respect to the most-negative supply voltage, V_{SS} (substrate).

* Stresses beyond those listed under "Absolute Maximum Ratings" may cause permanent damage to the device. This is a stress rating only and functional operation of the device at these or any other conditions beyond those indicated in the "Recommended Operating Conditions" section of this specification is not implied. Exposure to absolute-maximum-rated conditions for extended periods may affect device reliability.

TEXAS INSTRUMENTS
INCORPORATED
POST OFFICE BOX 225012 • DALLAS, TEXAS 75265

recommended operating conditions

PARAMETER	MIN	NOM	MAX	UNIT
Supply voltage, V_{CC} (see Note 2)	4.75	5	5.25	V
Supply voltage, V_{PP} (see Note 3)		V_{CC}		V
Supply voltage, V_{SS}		0		V
High-level input voltage, V_{IH}	2		V_{CC}+1	V
Low-level input voltage, V_{IL}	−0.1†		0.8	V
Read cycle time, $t_{c(rd)}$	450			ns
Operating free-air temperature, T_A	0		70	°C

NOTES: 2. V_{CC} must be applied before or at the same time as V_{PP} and removed after or at the same time as V_{PP}. The device must not be inserted into or removed from the board when V_{PP} or V_{CC} is applied so that the device is not damaged.

3. V_{PP} can be connected to V_{CC} directly (except in the program mode). V_{CC} supply current in this case would be $I_{CC} + I_{PP}$. During programming, V_{PP} must be maintained at 25 V (± 1V).

† The algebraic convention, where the more negative limit is designated as minimum, is used in this data sheet for logic voltage levels and time intervals.

electrical characteristics over full ranges of recommended operating conditions

PARAMETER		TEST CONDITIONS	TMS 2564			UNIT
			MIN	TYP†	MAX	
V_{OH}	High-level output voltage*	I_{OH} = −400 μA	2.4			V
V_{OL}	Low-level output voltage*	I_{OL} = 2.1 mA			0.45	V
I_I	Input current (leakage)	V_I = 5.25 V			10	μA
I_O	Output current (leakage)	V_O = 5.25 V			10	μA
I_{PP1}	V_{PP} supply current	V_{PP} = 5.25 V PD/$\overline{PGM}$ = V_{IL}			18	mA
I_{PP2}	V_{PP} supply current (during program pulse)	PD/$\overline{PGM}$ = V_{IL}			30	mA
I_{CC1}	V_{CC} supply current (standby)	PD/$\overline{PGM}$ = V_{IH}		15	30	mA
I_{CC2}	V_{CC} supply current (active)	PD/$\overline{PGM}$ = V_{IL}		80	160	mA

† Typical values are at T_A = 25°C and nominal voltages.
* AC and DC tests are made at 10% and 90% points using a 50% pattern.

capacitance over recommended supply voltage and operating free-air temperature range f = 1 MHz*

PARAMETER	TEST CONDITIONS	TYP†	MAX	UNIT
C_i Input capacitance	V_I = 0 V, f = 1 MHz	4	6	pF
C_o Output capacitance	V_O = 0 V, f = 1 MHz	8	12	pF

† All typical values are T_A = 25°C and nominal voltage.
* This parameter is tested on sample basis only.

switching characteristics over full ranges of recommended operating conditions (see note 4)

PARAMETER		TEST CONDITIONS (SEE NOTES 4 AND 5)	MIN	TYP†	MAX	UNIT
$t_{a(A)}$	Access time from address	C_L = 100 pF, 1 Series 74 TTL load, $t_r \leq 20$ ns, $t_f \leq 20$ ns See Figure 1		280	450	ns
$t_{a(S)}$	Access time from $\overline{CS}$1 and $\overline{CS}$2 (whichever occurs last)				120	ns
$t_{a(PR)}$	Access time from PD/$\overline{PGM}$			280	450	ns
$t_{v(A)}$	Output data valid after address change		0			ns
$t_{dis(S)}$	Output disable time from chip select during read only (whichever occurs last) ‡		0		100	ns
$t_{dis(PR)}$	Output disable time from PD/$\overline{PGM}$ during standby ‡		0		100	ns

† All typical values are at T_A = 25°C and nominal voltages.
‡ Value calculated from 0.5 volt delta to measured output level.

recommended timing requirements for programming T_A = 25°C (see note 4)

PARAMETER		MIN	TYP†	MAX	UNIT
$t_{w(PR)}$	Pulse width, program pulse	45	50	55	ms
$t_{r(PR)}$	Rise time, program pulse	5			ns
$t_{f(PR)}$	Fall time, program pulse	5			ns
$t_{su(A)}$	Address setup time	2			μs
$t_{su(D)}$	Data setup time	2			μs
$t_{su(VPP)}$	Setup time from V_{PP}	0			ns
$t_{h(A)}$	Address hold time	2			μs
$t_{h(D)}$	Data hold time	2			μs
$t_{h(PR)}$	Program pulse hold time	0			ns
$t_{h(VPP)}$	V_{PP} hold time	0			ns

† Typical values are at nominal voltages.

NOTES: 4. For all switching characteristics and timing measurements, input pulse levels are 0.65 V to 2.2 V and V_{PP} = 25 V ± 1 V during programming. AC and DC timing measurements are made at 90% points using a 50% pattern.

5. Common test conditions apply for t_{dis} except during programming. For $t_{a(A)}$, $t_{a(S)}$, and t_{dis}, PD/$\overline{PGM}$ = V_{IL}.

PARAMETER MEASUREMENT INFORMATION

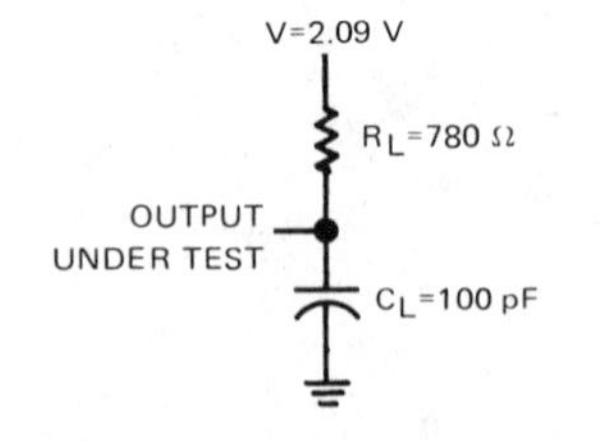

FIGURE 1 – TYPICAL OUTPUT LOAD CIRCUIT

TEXAS INSTRUMENTS
INCORPORATED
POST OFFICE BOX 225012 • DALLAS, TEXAS 75265

read cycle timing

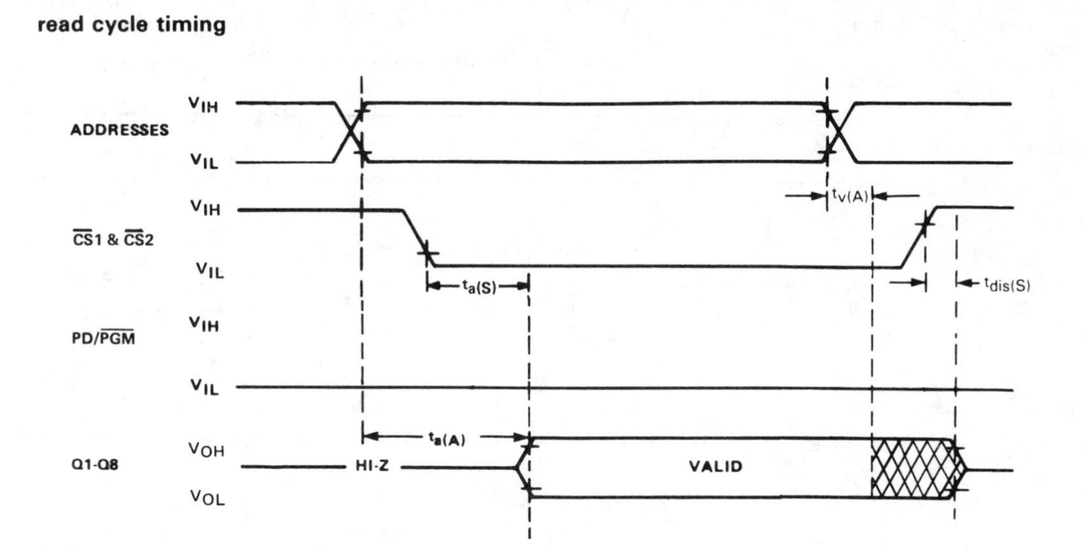

standby mode

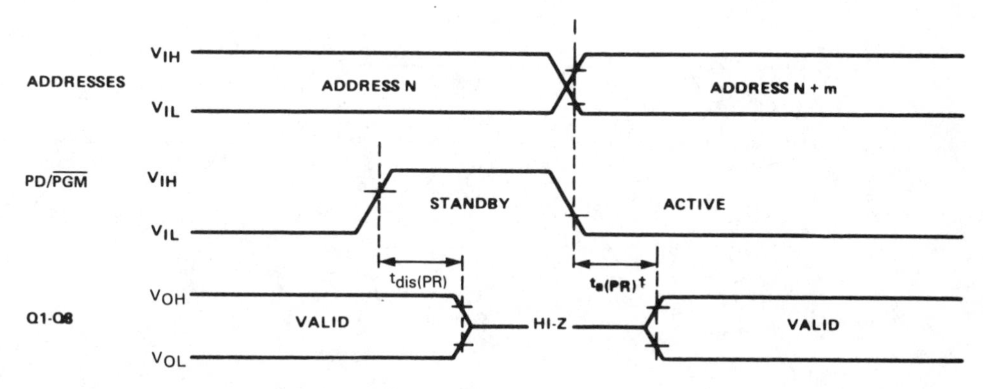

† $t_{a(PR)}$ referenced to PD/$\overline{PGM}$ or the address, whichever occurs last.
$\overline{CS}$1 and $\overline{CS}$2 in Don't Care State in Standby Mode.

program cycle timing

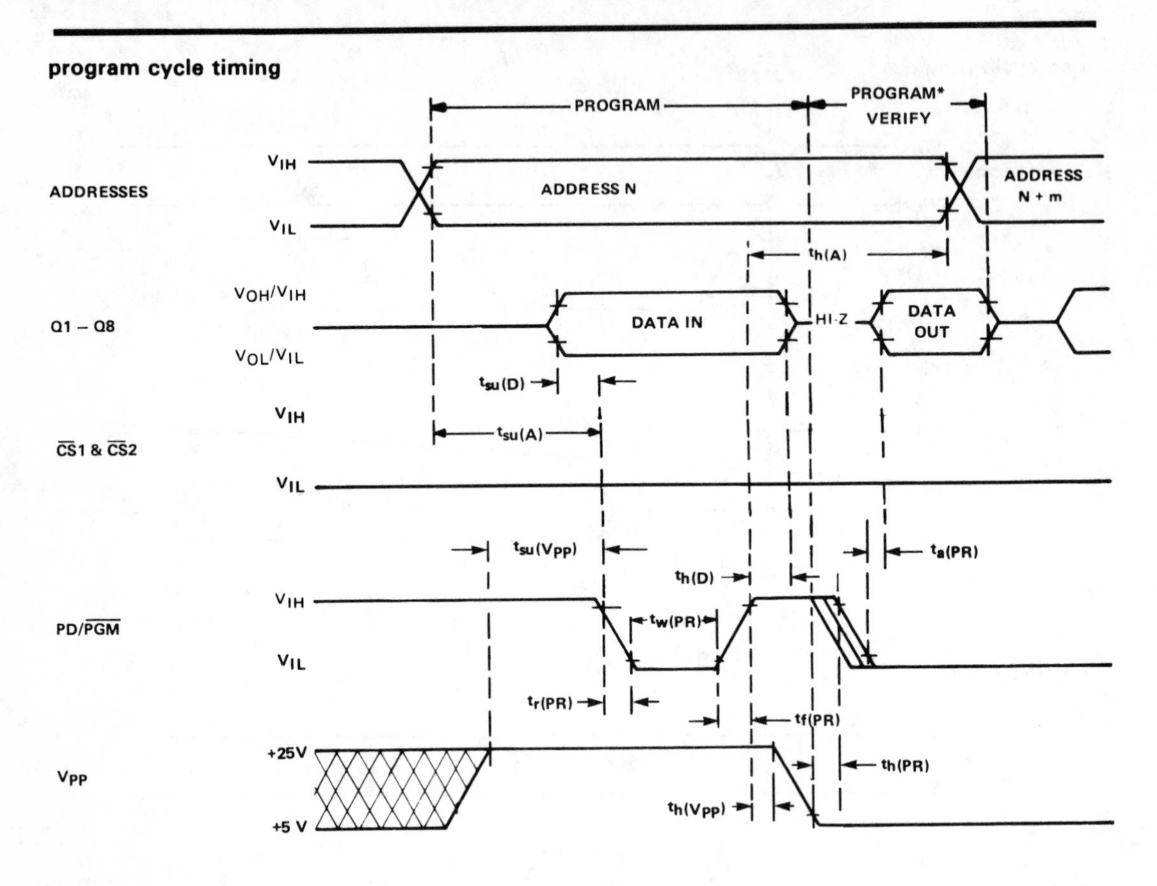

* Equivalent to read mode.

TEXAS INSTRUMENTS
INCORPORATED
POST OFFICE BOX 225012 • DALLAS, TEXAS 75265

typical device characteristics (read mode)

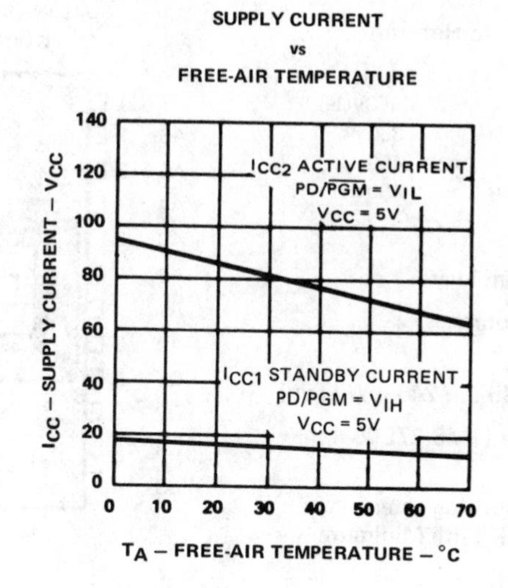

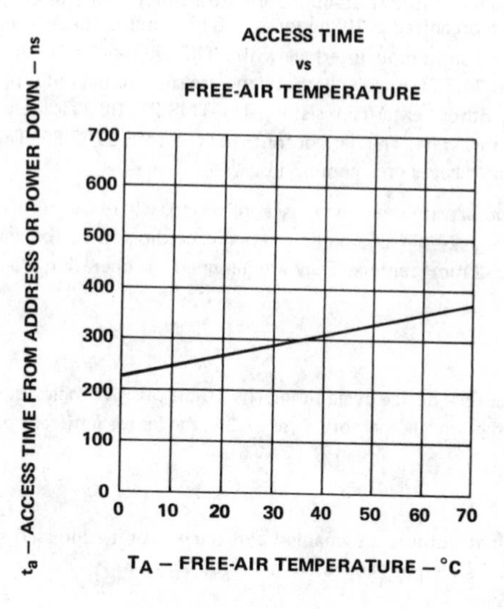

TEXAS INSTRUMENTS
INCORPORATED
POST OFFICE BOX 225012 • DALLAS, TEXAS 75265

MOS
LSI

TMS 2708-35 JL, TMS 2708-45 JL, TMS 27L08-45 JL 1024-WORD BY 8-BIT ERASABLE PROGRAMMABLE READ-ONLY MEMORIES

DECEMBER 1979—REVISED MAY 1982

- 1024 X 8 Organization
- All Inputs and Outputs Fully TTL Compatible
- Static Operation (No Clocks, No Refresh)
- Performance Ranges:

	Max Access	Min Cycle
TMS 2708-35	350 ns	350 ns
TMS 2708-45	450 ns	450 ns
TMS 27L08-45	450 ns	450 ns

- 3-State Outputs for OR-Ties
- N-Channel Silicon-Gate Technology
- 8-Bit Output for Use in Microprocessor-Based Systems
- Low Power on TMS 27L08-45 . . . 245 mW (Typ)
- 10% Power Supply Tolerance (TMS 27L08-45 Only)
- Plug-Compatible Pin-Outs Allowing Interchangeability/Upgrade to 16K With Minimum Board Change

24-PIN CERPAK
DUAL-IN-LINE PACKAGE
(TOP VIEW)

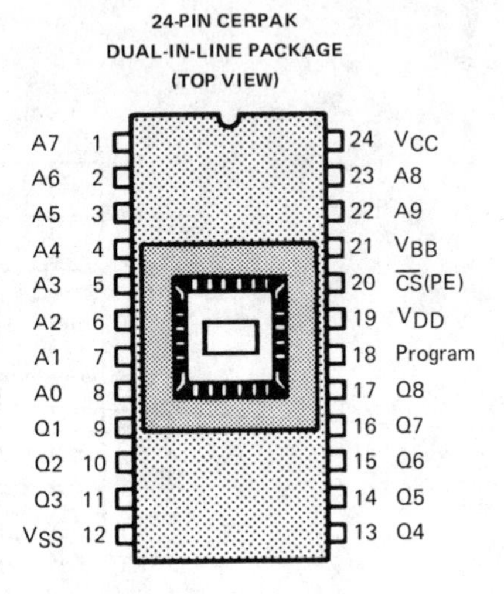

description

The TMS 2708-35, TMS 2708-45, and TMS 27L08-45 JL are ultra-violet light-erasable, electrically programmable read only memories. They have 8,192 bits organized as 1024 words of 8-bit length. The devices are fabricated using N-channel silicon-gate technology for high speed and simple interface with MOS and bipolar circuits. All inputs (including program data inputs) can be driven by Series 74 TTL circuits without the use of external pull-up resistors. Each output can drive one Series 74 or 74LS TTL circuit without external resistors. The TMS 27L08 guarantees 200 mV dc noise immunity in the high state and 250 mV in the low state. The data outputs for the TMS 2708-35, TMS 2708-45, and TMS 27L08-45 are three-state for OR-tying multiple devices on a common bus.

These EPROMs are designed for high-density fixed-memory applications where fast turn arounds and/or program changes are required. They are supplied in a 24-pin dual-in-line ceramic cerdip (JL suffix) package designed for insertion in mounting-hole rows on 600-mil (15.2 mm) centers. They are designed for operation from 0°C to 70°C.

operation (read mode)

address (A0-A9)

The address-valid interval determines the device cycle time. The 10-bit positive-logic address is decoded on-chip to select one of the 1024 words of 8-bit length in the memory array. A0 is the least-significant bit and A9 is the most-significant bit of the word address.

chip select, program enable [$\overline{CS}$ (PE)]

When the chip select is low, all eight outputs are enabled and the eight-bit addressed word can be read. When the chip select is high, all eight outputs are in a high-impedance state.

TEXAS INSTRUMENTS
INCORPORATED
POST OFFICE BOX 225012 • DALLAS, TEXAS 75265

data out (Q1-Q8)

The chip must be selected before the eight-bit output word can be read. Data will remain valid until the address is changed or the chip is deselected. When deselected, the three-state outputs are in a high-impedance state. The outputs will drive TTL circuits without external components.

program

The program pin must be held below V_{CC} in the read mode.

operation (program mode)

erase

Before programming, the TMS 2708-35, TMS 2708-45, or TMS 27L08-45 is erased by exposing the chip through the transparent lid to high-intensity ultraviolet light (wavelength 2537 angstroms). The recommended minimum exposure dose (= UV intensity × exposure time) is fifteen watt-seconds per square centimeter. Thus, a typical 12 milliwatt per square centimeter, filterless UV lamp will erase the device in a minimum of 21 minutes. The lamp should be located about 2.5 centimeters above the chip during erasure. After erasure, all bits are in the high state.

5

programming

Programming consists of successively depositing a small amount of charge to a selected memory cell that is to be changed from the erased high state to the low state. A low can be changed to a high only by erasure. Programming is normally accomplished on a PROM or EPROM Programmer, an example of which is TI's Universal PROM Programming Module in conjunction with the 990 prototyping system. Programming must be done at room temperature (25°C) only.

to start programming (see program cycle timing diagram)

First bring the $\overline{CS}$ (PE) pin to +12 V to disable the outputs and convert them to inputs. This pin is held high for the duration of the programming sequence. The first word to be programmed is addressed (it is customary to begin with the "0" address) and the data to be stored is placed on the Q1-Q8 program inputs. Then a +25 V program pulse is applied to the program pin. After 0.1 to 1.0 milliseconds the program pin is brought back to 0 V. After at least one microsecond the word address is sequentially changed to the next location, the new data is set up and the program pulse is applied.

Programming continues in this manner until all words have been programmed. This constitutes one of N program loops. The entire sequence is then repeated N times with $N \times t_{w(PR)} \geqslant 100$ ms. Thus, if $t_{w(PR)} = 1$ ms; then N = 100, the minimum number of program loops required to program the EPROM.

to stop programming

After cycling through the N program loops, the last program pulse is brought to 0 V, then Program Enable [$\overline{CS}$ (PE)] is brought to V_{IL} which takes the device out of the program mode. The data supplied by the programmer must be removed before the address is changed since the program inputs are now data outputs and change of address could cause a voltage conflict on the output buffer. Q1-Q8 outputs are invalid up to 10 microseconds after the program enable pin is brought from $V_{IH(PE)}$ to V_{IL}.

logic symbol†

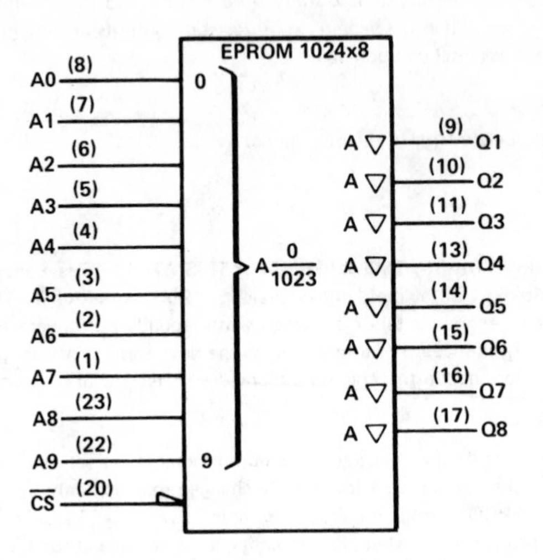

† This symbol is in accordance with IEEE Std 91/ANSI Y32.14 and recent decisions by IEEE and IEC. See explanation on page 289.

absolute maximum ratings over operating free-air temperature range (unless otherwise noted)*

Supply voltage, V_{CC} (see Note 1)	−0.3 to 15 V
Supply voltage, V_{DD} (see Note 1)	−0.3 to 20 V
Supply voltage, V_{SS} (see note 1)	−0.3 to 15 V
All input voltage (except program) (see Note 1)	−0.3 to 20 V
Program input (see Note 1)	−0.3 to 35 V
Output voltage (operating, with respect to V_{SS})	−2 to 7V
Operating free-air temperature range	0 °C to 70 °C
Storage temperature range	−55 °C to 125 °C

NOTE 1: Under absolute maximum ratings, voltage values are with respect to the most-negative supply voltage, V_{BB} (substrate), unless otherwise noted. Throughout the remainder of this data sheet, voltage values are with respect to V_{SS}.

* Stresses beyond those listed under "Absolute Maximum Ratings" may cause permanent damage to the device. This is a stress rating only and functional operation of the device at these or any other conditions beyond those indicated in the "Recommended Operating Conditions" section of this specification is not implied. Exposure to absolute-maximum-rated conditions for extended periods may affect device reliability.

recommended operating conditions

PARAMETER	TMS2708-35, TMS2708-45 MIN	NOM	MAX	TMS27L08-45 MIN	NOM	MAX	UNIT
Supply voltage, V_{BB}	−4.75	−5	−5.25	−4.5	−5	−5.5	V
Supply voltage, V_{CC}	4.75	5	5.25	4.5	5	5.5	V
Supply voltage, V_{DD}	11.4	12	12.6	10.8	12	13.2	V
Supply voltage, V_{SS}		0			0		V
High-level input voltage, V_{IH} (except program and program enable)	2.4		$V_{CC}+1$	2.2		$V_{CC}+1$	V
High-level program enable input voltage, $V_{IH(PE)}$	11.4	12	12.6	10.8	12	13.2	V
High-level program input voltage, $V_{IH(PR)}$	25	26	27	25	26	27	V
Low-level input voltage, V_{IL} (except program)	V_{SS}		0.65	V_{SS}		0.65	V
Low-level program input voltage, $V_{IL(PR)}$ Note: $V_{IL(PR)}$ max ≤ $V_{IH(PR)}$ −25 V	V_{SS}		1	V_{SS}		1	V
High-level program pulse input current (sink), $I_{IH(PR)}$			40			40	mA
Low-level program pusle input current (source), $I_{IL(PR)}$			3			3	mA
Operating free-air temperature, T_A	0		70	0		70	°C

electrical characteristics over full ranges of recommended operating conditions (unless otherwise noted)

PARAMETER		TEST CONDITIONS	TMS 2708-35, TMS 2708-45 MIN	TYP†	MAX	TMS 27L08-45 MIN	TYP†	MAX	UNIT
V_{OH}	High-level output voltage	I_{OH} = −100 μA	3.7			3.7			V
		I_{OH} = −1 mA	2.4			2.4			
V_{OL}	Low-level output voltage	I_{OL} = 1.6 mA			0.45			0.40	V
I_I	Input current (leakage)	V_I = 0 V to 5.25 V		1	10		1	10	μA
I_O	Output current (leakage)	$\overline{CS}$(PE) = 5 V, V_O = 0.4 V to 5.25 V		1	10		1	10	μA
I_{BB}	Supply current from V_{BB}	All inputs high, $\overline{CS}$(PE) = 5 V, T_A = 0°C (worst case)		30	45		9	18	mA
I_{CC}	Supply current from V_{CC}			6	10		.9	6	mA
I_{DD}	Supply current from V_{DD}			50	65		20	34	mA
$P_{D(AV)}$	Power Dissipation	T_A = 70°C			800			350	mW
		T_A = 0°C CS = 0V					245	475	
		T_A = 0°C CS = +5 V					290	580	

†All typical values are at T_A = 25°C and nominal voltages.

capacitance over recommended supply voltage range and operating free-air temperature range, f = 1 MHz

PARAMETER		TYP†	MAX	UNIT
C_i	Input capacitance	4	6	pF
C_o	Output capacitance	8	12	pF

†All typical values are at T_A = 25°C and nominal voltages.

switching characteristics over recommended supply voltage range and operating free-air temperature range

PARAMETER		TEST CONDITIONS	TMS2708-35		TMS2708 TMS27L08		UNIT
			MIN	MAX	MIN	MAX	
$t_{a(ad)}$	Access time from address	C_L = 100 pF, 1 Series 74 TTL load, $t_{f(CS)}$, $t_{f(ad)}$ = 20 ns		300		450	ns
$t_{a(CS)}$	Access time from $\overline{CS}$			120		120	ns
$t_{v(A)}$	Output data valid after address change		0		0		ns
t_{dis}	Output disable time†		0	120	0	120	ns
$t_{c(rd)}$	Read cycle time		300		450		ns

† Value calculated from 0.5 volt delta to measured output level.

T_A = 25°C program characteristics over recommended supply voltage range

PARAMETER		MIN	MAX	UNIT
$t_{w(PR)}$	Pulse width, program pulse	0.1	1	ms
t_T	Transition times (except program pulse)		20	ns
$t_{T(PR)}$	Transition times, program pulse	50	2000	ns
$t_{su(ad)}$	Address setup time	10		μs
$t_{su(da)}$	Data setup time	10		μs
$t_{su(PE)}$	Program enable setup time	10		μs
$t_{h(ad)}$	Address hold time	1000		ns
$t_{h(ad,da R)}$	Address hold time after program input data stopped	0		ns
$t_{h(da)}$	Data hold time	1000		ns
$t_{h(PE)}$	Program enable hold time	500		ns
$t_{CL,adX}$	Delay time, $\overline{CS}$(PE) low to address change	0		ns

PARAMETER MEASUREMENT INFORMATION

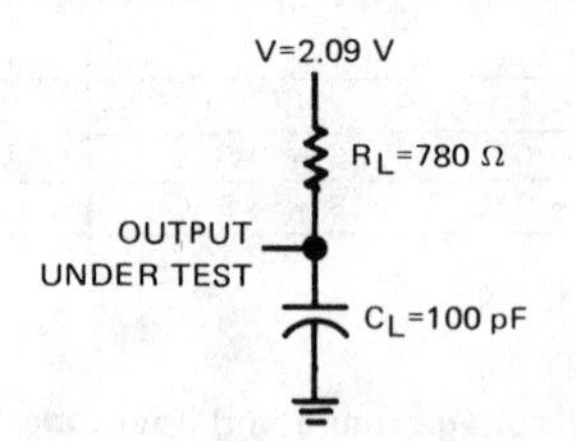

FIGURE 1 – TYPICAL OUTPUT LOAD CIRCUIT

read cycle timing

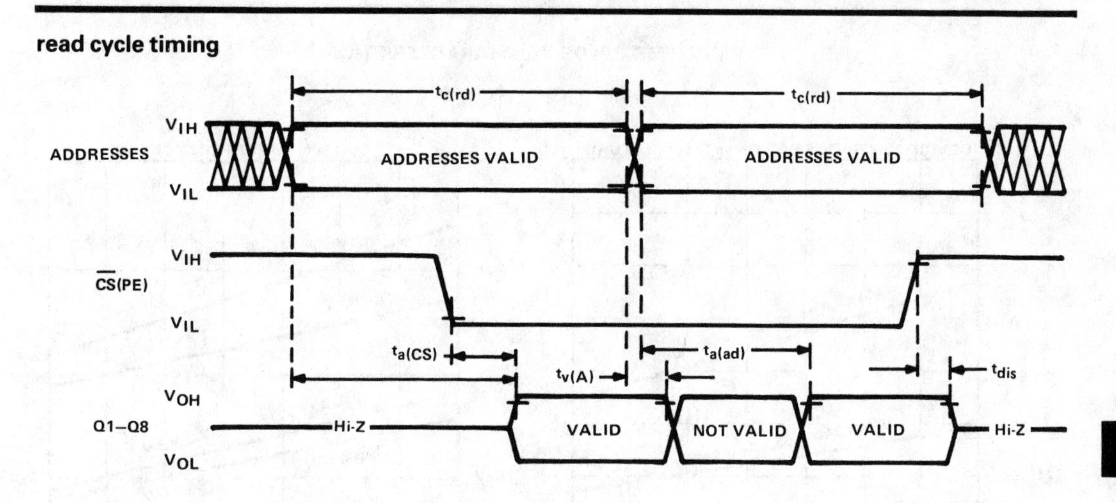

program cycle timing

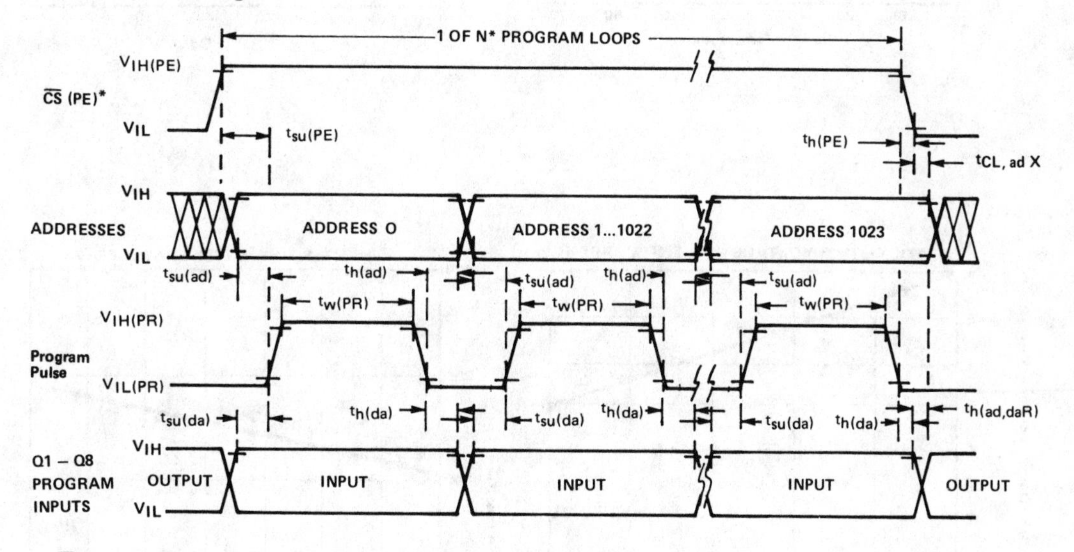

*$\overline{CS}$ (PE) is at +12 V through N program loops where N ⩾ 100 ms/tw (PR).

NOTE: Q1-Q8 outputs are invalid up to 10 μsec after programming [$\overline{CS}$(PE) goes low].

All timing reference points in this data sheet (inputs and outputs) are 90% points.

TYPICAL TMS 27L08-45 CHARACTERISTICS

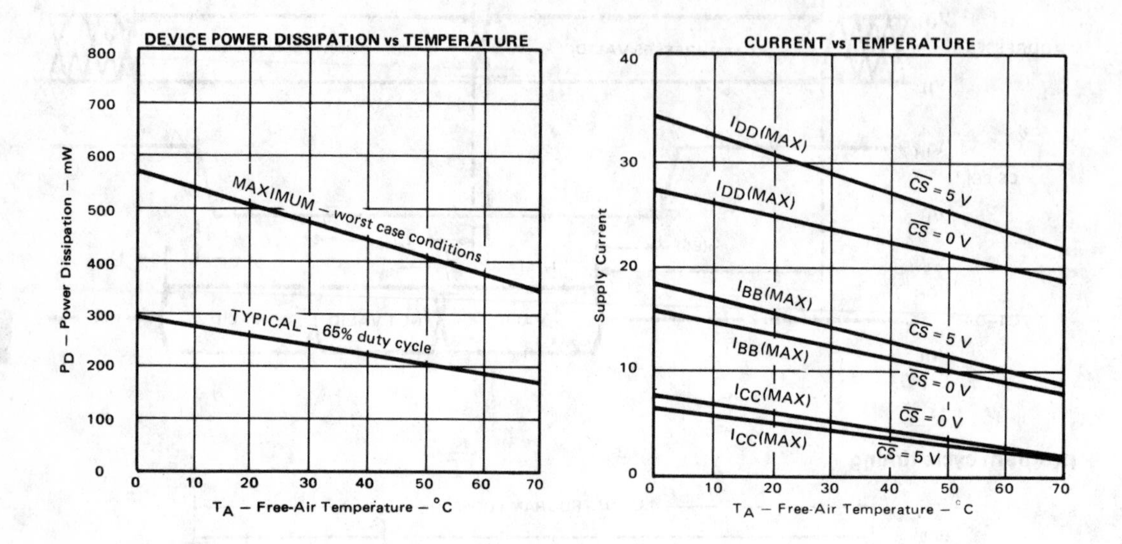

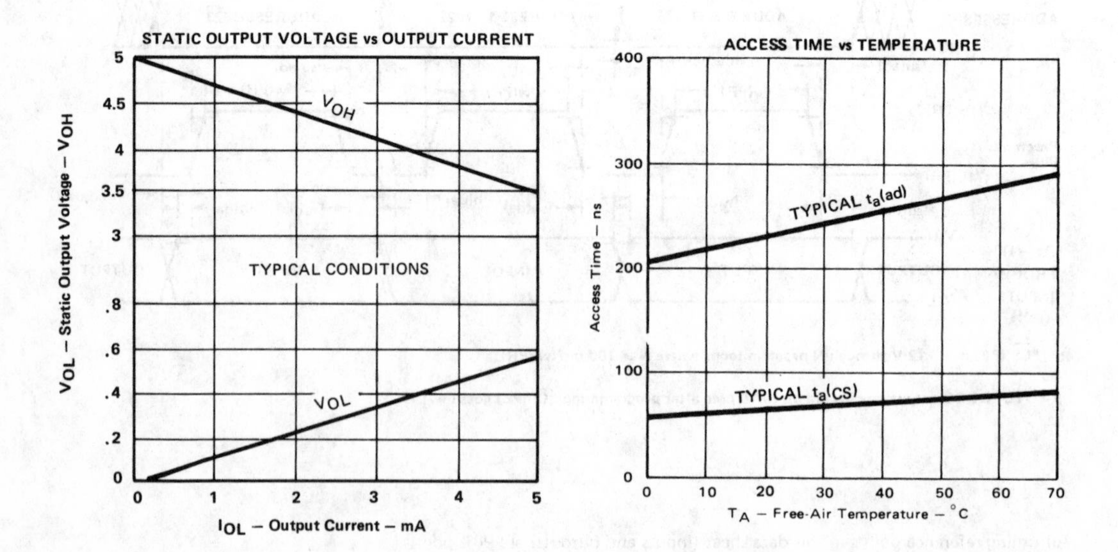

TEXAS INSTRUMENTS
INCORPORATED
POST OFFICE BOX 225012 • DALLAS, TEXAS 75265

TMS 2716-30 JL AND TMS 2716-45 JL 2048-WORD BY 8-BIT ERASABLE PROGRAMMABLE READ-ONLY MEMORIES

DECEMBER 1979—REVISED MAY 1982

- **2048 X 8 Organization**
- **All Inputs and Outputs Fully TTL Compatible**
- **Static Operation (No Clocks, No Refresh)**
- **Performance Ranges:**

	ACCESS TIME (MAX)	CYCLE TIME (MIN)
TMS 2716-30	300 ns	300 ns
TMS 2716-45	450 ns	450 ns

- **3-State Outputs for OR-Ties**
- **N-Channel Silicon-Gate Technology**
- **8-Bit Output for Use in Microprocessor-Based Systems**
- **Low Power . . . 315 mW (Typical)**

24-PIN CERPAK DUAL-IN-LINE PACKAGE (TOP VIEW)

A7	1	24	V_{CC}(PE)
A6	2	23	A8
A5	3	22	A9
A4	4	21	V_{BB}
A3	5	20	A10
A2	6	19	V_{DD}
A1	7	18	$\overline{CS}$(program)
A0	8	17	Q8
Q1	9	16	Q7
Q2	10	15	Q6
Q3	11	14	Q5
V_{SS}	12	13	Q4

5

description

The TMS 2716 is an ultra-violet light-erasable, electrically programmable read only memory. It has 16,384 bits organized as 2048 words of 8-bit length. The device is fabricated using N-channel silicon-gate technology for high-speed and simple interface with MOS and bipolar circuits. All inputs (including program data inputs) can be driven by Series 74 circuits without the use of external pull-up resistors and each output can drive one Series 74 or 74LS TTL circuit without external resistors. The TMS 2716 guarantees 250 mV dc noise immunity in the low state. Data outputs are three-state for OR-tying multiple devices on a common bus. The TMS 2716 is plug-in compatible with the TMS 2708 and the TMS 27L08. Pin compatible mask programmed ROMs are available for large volume requirements.

This EPROM is designed for high-density fixed-memory applications where fast turn arounds and/or program changes are required. It is supplied in a 24-pin dual-in-line cerpak (JL suffix) package designed for insertion in mounting-hole rows on 600-mil (15.2 mm) centers. It is designed for operation from 0 °C to 70 °C.

operation (read mode)

address (A0-A10)

The address-valid interval determines the device cycle time. The 11-bit positive-logic address is decoded on-chip to select one of 2048 words of 8-bit length in the memory array. A0 is the least-significant bit and A10 most-significant bit of the word address.

chip select, program [$\overline{CS}$ (Program)]

When the chip select is low, all eight outputs are enabled and the eight-bit addressed word can be read. When the chip select is high, all eight outputs are in a high-impedance state.

program

In the program mode, the chip select feature does not function as pin 18 inputs only the program pulse. The program mode is selected by the V_{CC}(PE) pin. Either 0 V or + 12 V on this pin will cause the TMS 2716 to assume program cycle.

data out (Q1-Q8)

The chip must be selected before the eight-bit output word can be read. Data will remain valid until the address is changed or the chip is deselected. When deselected, the three-state outputs are in a high-impedance state. The outputs will drive TTL circuits without external components.

TEXAS INSTRUMENTS
INCORPORATED
POST OFFICE BOX 225012 • DALLAS, TEXAS 75265

operation (program mode)

erase

Before programming, the TMS 2716 is erased by exposing the chip through the transparent lid to high intensity ultraviolet light (wavelength 2537 angstroms). The recommended minimum exposure dose (= UV intensity × exposure time) is fifteen watt-seconds per square centimeter. Thus, a typical 12 milliwatt per square centimeter, filterless UV lamp will erase the device in a minimum of 21 minutes. The lamp should be located about 2.5 centimeters above the chip during erasure. After erasure, all bits are in the high state.

programming

Programming consists of successively depositing a small amount of charge to a selected memory cell that is to be changed from the erased high state to the low state. A low can be changed to a high only by erasure. Programming is normally accomplished on a PROM or EPROM Programmer, an example of which is TI's Universal PROM Programming Module in conjunction with the 990 prototyping system. Programming must be done at room temperature (25 °C) only.

to start programming (see program cycle timing diagram)

First bring the V_{CC}(PE) pin to +12 V or 0 V to disable the outputs and convert them to inputs. This pin is held high for the duration of the programming sequence. The first word to be programmed is addressed (it is customary to begin with the "0" address) and the data to be stored is placed on the Q1-Q8 program inputs. Then a +26 V program pulse is applied to the program pin. After 0.1 to 1.0 milliseconds the program pin is brought back to 0 V. After at least one microsecond the word address is sequentially changed to the next location, the new data is set up and the program pulse is applied.

Programming continues in this manner until all words have been programmed. This constitutes one of N program loops. The entire sequence is then repeated N times with $N \times t_{w(PR)} \geq 100$ ms. Thus, if $t_{w(PR)} = 1$ ms; then N = 100, the minimum number of program loops required to program the EPROM.

to stop programming

After cycling through the N program loops, the last program pulse is brought to 0 V, then Program Enable V_{CC}(PE) is brought back to ±5 volts which takes the device out of the program mode. The data supplied by the programmer must be removed before the address is changed since the program inputs are now data outputs and a change of address could cause a voltage conflict on the output buffer. Q1-Q8 outputs are invalid up to 10 microseconds after the program enable pin is brought from $V_{IH(PE)}$ to $V_{IL(PE)}$.

logic symbol†

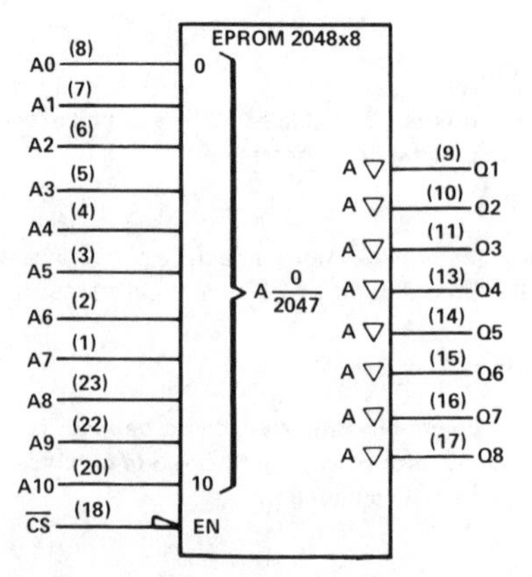

† This symbol is in accordance with IEEE Std 91/ANSI Y32.14 and recent decisions by IEEE and IEC. See explanation on page 289.

TEXAS INSTRUMENTS
INCORPORATED
POST OFFICE BOX 225012 • DALLAS, TEXAS 75265

absolute maximum ratings over operating free-air temperature range (unless otherwise noted)*

Supply voltage, V_{CC} (see Note 1) . . . . −0.3 to 15 V
Supply voltage, V_{DD} (see Note 1) . . . . −0.3 to 20 V
Supply voltage, V_{SS} (see Note 1) . . . . −0.3 to 15 V
All input voltage (except program) (see Note 1) . . . . −0.3 to 20 V
Program input (see Note 1) . . . . −0.3 to 35 V
Output voltage (operating, with respect to V_{SS}) . . . . −2 to 7 V
Operating free-air temperture range . . . . 0°C to 70°C
Storage temperature range . . . . −55°C to 125°C

* Stresses beyond those listed under "Absolute Maximum Ratings" may cause permanent damage to the device. This is a stress rating only and functional operating of the device at these or any other conditions beyond those indicated in the "Recommended Operting Conditions" section of this specification is not implied. Exposure to absolute-maximum-rated conditions for extended periods may affect device reliability.

NOTE 1: Under absolute maximum ratings, voltage values are with respect to the most-negative supply voltage, V_{BB} (substrate), unless otherwise noted. Throughout the remainder of this data sheet, voltage values are with respect to V_{SS}.

recommended operating conditions

PARAMETER	MIN	NOM	MAX	UNIT
Supply voltage, V_{BB}	−4.75	−5	−5.25	V
Supply voltage, V_{CC}	4.75	5	5.25	V
Supply voltage, V_{DD}	11.4	12	12.6	V
Supply voltage, V_{SS}		0		V
High-level input voltage, V_{IH} (except program and program enable)	2.4		$V_{CC}+1$	V
High-level program enable input voltage, $V_{IH(PE)}$	11.4	12	12.6	V
High-level program input voltage, $V_{IH(PR)}$	25	26	27	V
Low-level input voltage, V_{IL} (except program)	V_{SS}		0.65	V
Low-level program input voltage, $V_{IL(PR)}$ Note: $V_{IL(PR)}$ max $\leqslant V_{IH(PR)} - 25$ V	V_{SS}		1	V
High-level program pulse input current (sink), $I_{IH(PR)}$			40	mA
Low-level program pulse input current (source), $I_{IL(PR)}$			3	mA
Operating free-air temperature, T_A	0		70	°C

electrical characteristics over full ranges of recommended operating conditions (unless otherwise noted)

PARAMETER		TEST CONDITIONS	MIN	TYP†	MAX	UNIT
V_{OH}	High-level output voltage	$I_{OH} = -100\ \mu A$	3.7			V
		$I_{OH} = -1$ mA	2.4			
V_{OL}	Low-level output voltage	$I_{OL} = 1.6$ mA			0.45	V
I_I	Input current (leakage)	$V_I = 0$ V to 5.25 V		1	10	μA
I_O	Output current (leakage)	$\overline{CS}$ (Program) = 5 V, $V_O = 0.4$ V to 5.25 V		1	10	μA
I_{BB}	Supply current from V_{BB}	All inputs high,		10	20	mA
I_{CC}	Supply current from V_{CC}	$\overline{CS}$ (Program) = 5 V,		1	8	mA
I_{DD}	Supply current from V_{DD}	$T_A = 0°C$ (worst case)		26	45	mA
I_{PE}	Supply current from PE on V_{CC} Pin	$V_{PE} = V_{DD}$		2	4	mA
$P_{D(AV)}$	Power Dissipation	$T_A = 70°C$			540	mW
		$T_A = 0°C$ $\overline{CS} = 0V$		315	595	
		$T_A = 0°C$ $\overline{CS} = +5$ V		375	720	

† All typical values are at $T_A = 25°C$ and nominal voltages.

capacitance over recommended supply voltage range and operating free-air temperature range, f = 1 MHz

PARAMETER		TYP†	MAX	UNIT
C_i	Input capacitance [except CS (Program)]	4	6	pF
$C_{i(CS)}$	$\overline{CS}$ (Program) input capacitance	20	30	pF
C_o	Output capacitance	8	12	pF

† All typical values are at T_A = 25°C and nominal voltages.

switching characteristics over recommended supply voltage range and operating free-air temperature range

PARAMETER		TEST CONDITIONS	TMS2716-30		TMS2716-45		UNIT
			MIN	MAX	MIN	MAX	
$t_{a(ad)}$	Access time from address	C_L = 100 pF 1 Series 74 TTL Load $t_{f(CS)}$, $t_{f(ad)}$ = 20 ns See Figure 1		300		450	ns
$t_{a(CS)}$	Access time from $\overline{CS}$			120		120	ns
$t_{v(A)}$	Output data valid after address change		0		0		ns
t_{dis}	Output disable time†		0	120	0	120	ns
$t_{c(rd)}$	Read cycle time		300		450		ns

† Value calculated from 0.5 volt delta to measured output level.

T_A = 25°C program characteristics over recommended supply voltage range

PARAMETER		MIN	MAX	UNIT
$t_{w(PR)}$	Pulse width, program pulse	0.1	1	ms
t_T	Transition times (except program pulse)		20	ns
$t_{T(PR)}$	Transition times, program pulse	30	2000	ns
$t_{su(ad)}$	Address setup time	10		μs
$t_{su(da)}$	Data setup time	10		μs
$t_{su(PE)}$	Program enable setup time	10		μs
$t_{h(ad)}$	Address hold time	1000		ns
$t_{h(ad,da R)}$	Address hold time after program input data stopped	0		ns
$t_{h(da)}$	Data hold time	1000		ns
$t_{h(PE)}$	Program enable hold time	500		ns
$t_{CL,adX}$	Delay time, $\overline{CS}$ (Program) low to address change	0		ns

PARAMETER MEASUREMENT INFORMATION

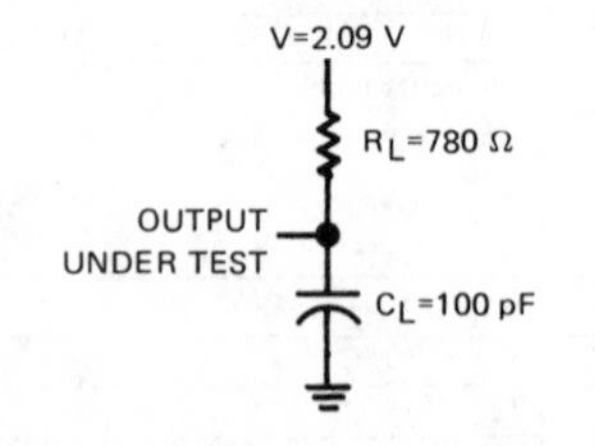

FIGURE 1 – TYPICAL OUTPUT LOAD CIRCUIT

TEXAS INSTRUMENTS
INCORPORATED
POST OFFICE BOX 225012 • DALLAS, TEXAS 75265

read cycle timing

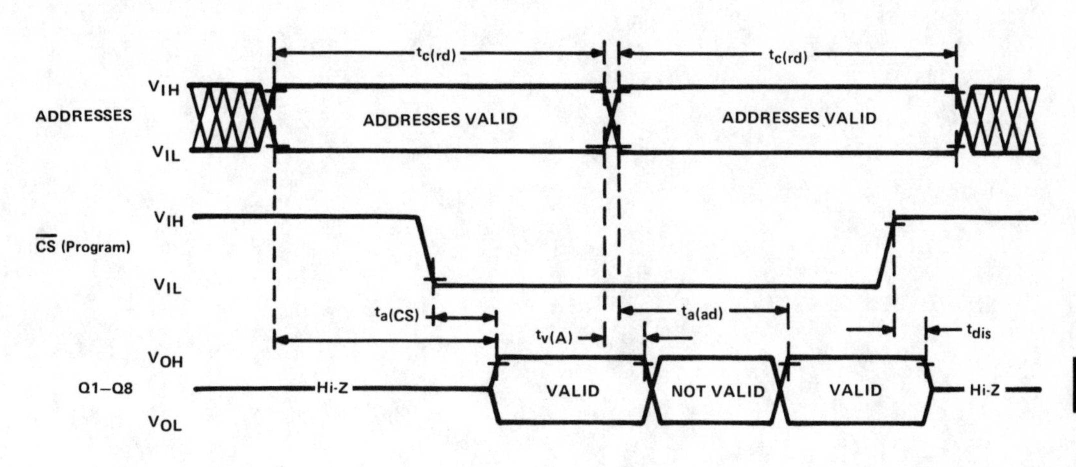

program cycle timing

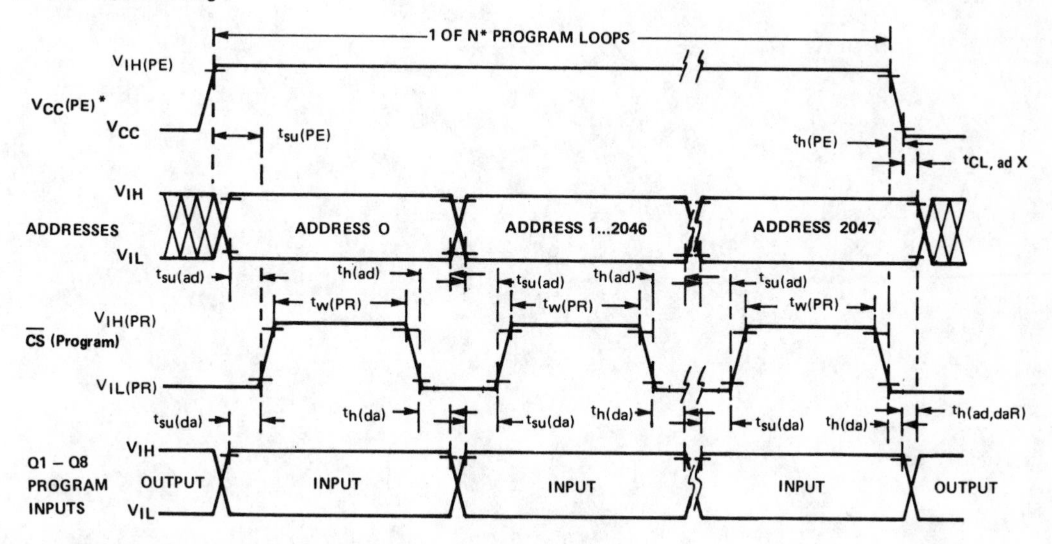

*V_{CC} (PE) is at 0 V or +12 V through N program loops where N $\geqslant$ 100 ms/tw (PR).

NOTE: Q1-Q8 outputs are invalid up to 10 μsec after programing (V_{CC} (PE) goes low).

ROM
Data Sheets

TMS 4732 JL, NL
4096-WORD BY 8-BIT READ-ONLY MEMORY

MAY 1977 – REVISED MAY 1982

- 4096 x 8 Organization
- All Inputs and Outputs TTL-Compatible
- Fully Static (No Clocks, No Refresh)
- Single 5 V Power Supply
- Maximum Access Time from Address . . . 300 ns
- Minimum Access Time from Power Down . . . 300 ns
- Typical Power Dissipation . . . 275 mW
- 3-State Outputs for OR-Ties
- Pin-Compatible with TMS 2532 EPROM
- Two Output Enable Controls for Chip Select Flexibility

24-PIN CERAMIC AND PLASTIC DUAL-IN-LINE PACKAGES (TOP VIEW)

A7	1	24	VCC
A6	2	23	A8
A5	3	22	A9
A4	4	21	S2/$\overline{S}$2
A3	5	20	$\overline{S}$1/S1/$\overline{E}$
A2	6	19	A10
A1	7	18	A11
A0	8	17	Q8
Q1	9	16	Q7
Q2	10	15	Q6
Q3	11	14	Q5
VSS	12	13	Q4

description

The TMS 4732 is a 32,768-bit read-only memory organized as 4096 words of 8-bit length. This makes the TMS 4732 ideal for microprocessor based systems. The device is fabricated using N-channel silicon-gate technology for high speed and simple interface with bipolar circuits.

All inputs can be driven directly by Series 74 TTL circuits without the use of any external pull-up resistor. Each output can drive one Series 74 or 74S load without external resistors. The data outputs are three-state for OR-tieing multiple devices on a common bus. Two chip select controls allow data to read. These controls are programmable, providing additional system decode flexibility. The data is always available, it is not dependent on external clocking of the control pins.

The TMS 4732 is designed for high-density fixed-memory applications such as logic function generation and microprogramming. The part is pin compatible with the TMS 2532 4096 x 8 EPROM, which aids in prototyping and code verification.

This ROM is supplied in 24-pin dual-in-line plastic (NL suffix) or ceramic (JL suffix) packages designed for insertion in mounting-hold rows on 600-mil centers. The device is designed for operation from 0°C to 70°C.

operation

address (A0–A11)

The address-valid interval determines the device cycle time. The 12-bit positive-logic address is decoded on-chip to select one of 4096 words of 8-bit length in the memory array. A0 is the least-significant bit and A11 the most-significant bit of the word address.

chip select/output enable (pins 20 and 21)

Each of these pins can be programmed during mask fabrication to be active with either a high or a low level input. When both signals are active, all eight outputs are enabled and the eight-bit addressed word can be read. When either signal is not active, all eight outputs are in a high-impedance state.

power down ($\overline{E}$)

A mask programmable option is to utilize pin 20 in a power-down mode. In this mode, pin 20 is clocked. When it is high, the chip is put into a standby mode. This reduces I_{CC1}, which in the active state is 80 mA maximum, to a standby I_{CC2} of 20 mA maximum.

data out (Q1–Q8)

The eight outputs must be enabled by both pins 20 and 21 before the output word can be read. Data will remain valid until the address is changed or the outputs are disabled (chip deselected). When disabled, the three-state outputs are in a high-impedance state. Q1 is considered the least-significant bit, Q8 the most-significant bit.

The outputs will drive two Series 54/74 TTL circuits without external components.

logic symbol†

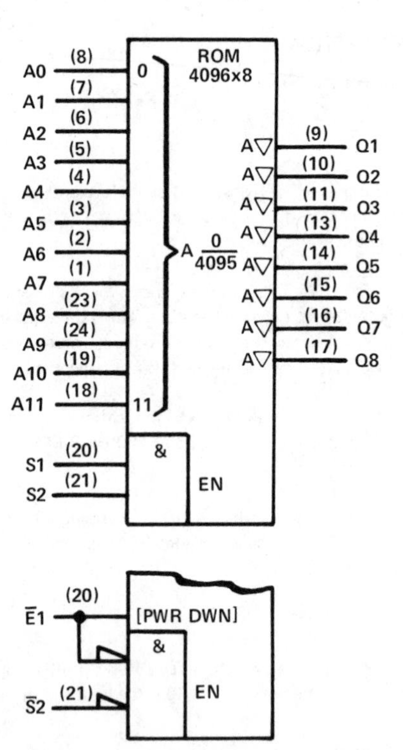

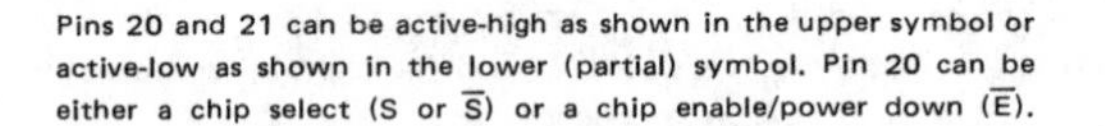
Pins 20 and 21 can be active-high as shown in the upper symbol or active-low as shown in the lower (partial) symbol. Pin 20 can be either a chip select (S or $\overline{S}$) or a chip enable/power down ($\overline{E}$).

† This symbol is in accordance with IEEE Std 91/ANSI Y32.14 and recent decisions by IEEE and IEC. See explanation on page 289.

functional block diagram

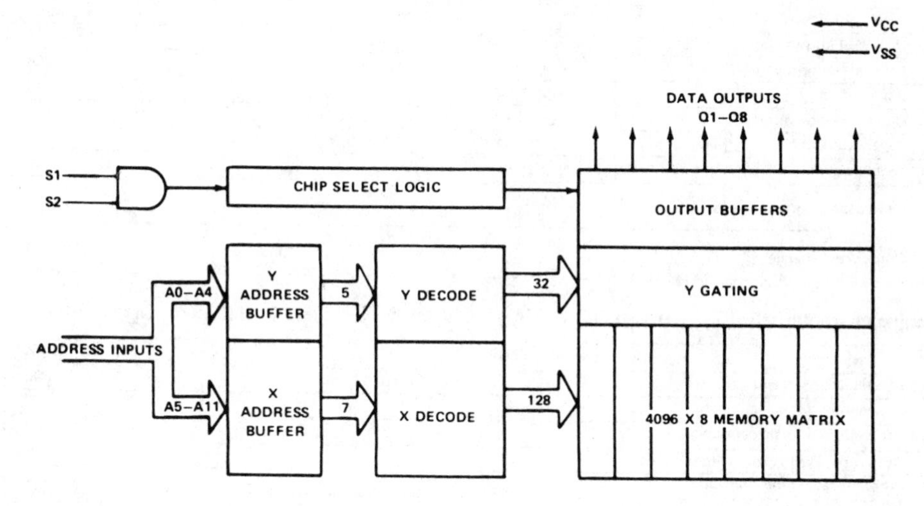

6

absolute maximum ratings

Supply voltage to ground potential (see Note 1)	−0.5 to 7 V
Applied output voltage (see Note 1)	−0.5 to 7 V
Applied input voltage (see Note 1)	0.5 to 7 V
Power dissipation	500 mW
Ambient operating temperature	0°C to 70°C
Storage temperature	−55°C to 150°C

Note 1: Voltage values are with respect to V_{SS}.

recommended operating conditions

PARAMETER	MIN	NOM	MAX	UNIT
Supply voltage, V_{CC}	4.5	5	5.5	V
High-level input voltage, V_{IH}	2	2.4	V_{CC}+1	V
Low-level input voltage, V_{IL}	0.5		.8	V
Operating free-air temperature, T_A	0		70	°C

electrical characteristics, T_A = 0°C to 70°C, V_{DD} = 5 V ±10% (unless otherwise noted)

PARAMETER		TEST CONDITIONS		MIN	MAX	UNIT
V_{OH}	High-level output voltage	V_{CC} = 4.5 V,	I_{OH} = –400 μA	2.4		V
V_{OL}	Low-level output voltage	V_{CC} = 4.5 V,	I_{OL} = 3.2 mA		0.4	V
I_I	Input current	V_{CC} = 5.5 V,	$0_V \leq V_{IN} \leq 5.5$ V		10	μA
I_O	Output leakage current	V_O = 0.4 V to V_{CC},	Chip deselected		±10	μA
I_{CC1}	Supply current from V_{CC} (active)	V_{CC} = 5.5 V,	$V_I = V_{CC}$ Output not loaded		80	mA
I_{CC2}	Supply current from V_{CC} (power down)	V_{CC} = 5.5 V			20	mA
C_i	Input capacitance	V_O = 0 V, f = 1 MHz	T_A = 25°C,		6	pF
C_o	Output capacitance	V_O = 0 V, f = 1 MHz	T_A = 25°C,		12	pF

switching characteristics, T_A = 0°C to 70°C, V_{CC} = 5 V ±10%, 2 series 74 TTL loads, C_L = 100 pF*

PARAMETER		MIN	MAX	UNIT
$t_{a(AD)}$	Access time from address		300	ns
$t_{a(S)}$	Access time from chip select		120	ns
$t_{a(PD)}$	Access time from power down		300	ns
$t_{v(A)}$	Output data valid after address change	20		ns
t_{dis}	Output disable time from chip select		100	ns

* All AC measurements are made at 10% and 90% points.

read cycle timing

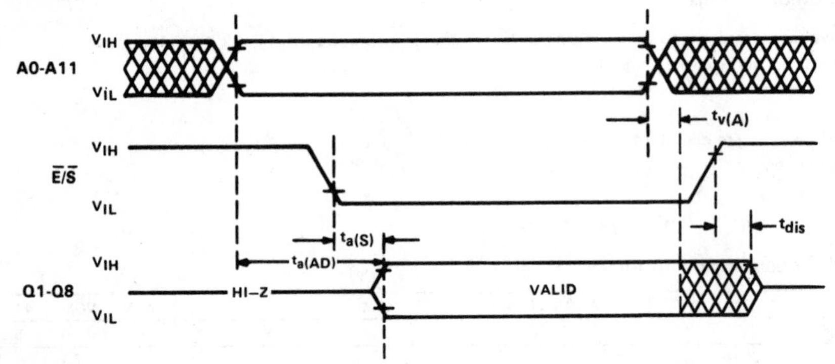

standby mode

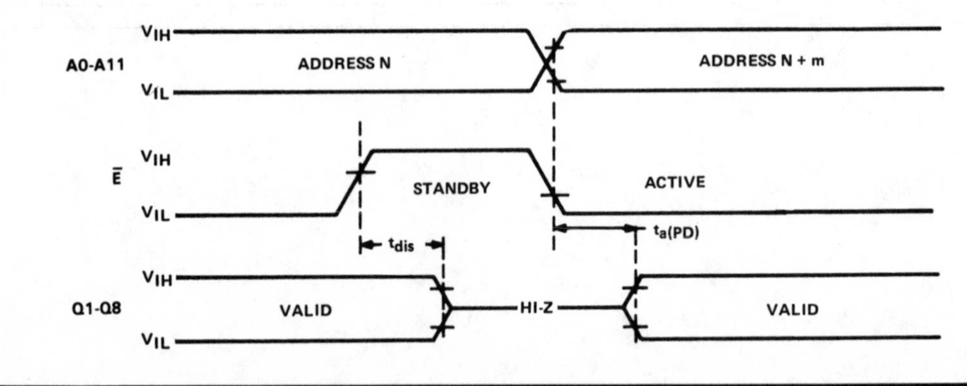

TEXAS INSTRUMENTS
INCORPORATED
POST OFFICE BOX 225012 • DALLAS, TEXAS 75265

PROGRAMMING DATA

PROGRAMMING REQUIREMENTS: The TMS 4732JL, NL is a fixed program memory in which the programming is performed by TI at the factory during the manufacturing cycle to the specific customer inputs supplied in the format below. The device is organized as 4096 8-bit words with address locations numbered 0 to 4095. The 8-bit words can be coded as a 2-digit hexadecimal number between 00 and FF. All data words and addresses in the following format are coded in hexadecimal numbers. In coding all binary words must be in positive logic before conversion to hexadecimal. Q1 is considered the least significant bit and Q8 the most significant bit. For addresses, A0 is least significant bit and A11 is the most significant.

The input media containing the programming data can be in the form of cards or EPROMs.

Either 16K, 32K, or 64K EPROMs can be used or any combination of them.

The following is a description of how the cards must be formatted, should they be used instead of EPROMs.

INPUT CARD FORMAT

Each code deck submitted by customer shall consist of the following:

1. Title Card
2. Comment Cards
3. Start of Data Card
4. Data Cards

The cards shall be standard 80 column cards with the information in the following format:

TITLE CARD

Card Column	Information
1 – 5	The word 'TITLE' shall be punched in these columns.
6	Blank
7,8	The letters 'ZA' shall be punched in these columns.
9 – 14	Leave blank. A special device code number will be assigned by Texas Instruments. (left justified)
15	Blank
16 – 30	Customer's Part Number, if required. (left justified)
31	Blank
32	Customer's Part Number to be included as part of device symbolization. Options: Y = Yes N = No
33 – 36	Blank
37	Type of Package Options: C = ceramic P = plastic
38	Blank
39 – 40	Customer Defined Option for Device Mode. Options: PD = power down mode CS = chip select mode

6

TEXAS INSTRUMENTS
INCORPORATED
POST OFFICE BOX 225012 • DALLAS, TEXAS 75265

Card Column	Information
41	Logic Level for device pin 20. Options: Blank = power down mode 1 = chip select mode, outputs enabled with high level. 0 = chip select mode, outputs enabled with low level.
42	Logic Level for device pin 21. Options: Blank = NC 1 = chip select mode, outputs enabled with high level. 0 = chip select mode, outputs enabled with low level.
43	Blank.
44	Blank
45 – 49	Texas Instruments Device Series (4732A, 4732AI, etc.) (left justified)

COMMENT CARDS

Any number of comment cards may be used for specifying the customer's name, individual to contact, telephone number, address, any special instructions, etc. The format for these cards is as follows: The letter 'C' (for comment) must be punched in column 1, columns 2–4 must be blank, and comments can be punched in columns 5–80.

START OF DATA CARD

This card is to identify that the next card will be the beginning of customer's code. Format is as follows: Columns 1–4 must have '&ROM' punched in them. The remainder of card is blank.

DATA CARDS

There will be 128 data cards supplied for each customer code. Each card will contain (in hexidecimal) the data for 32 memory locations. Each data card shall be in the following format:

Card Column	Hexadecimal Information
1 – 3	Hexidecimal address of first word on the card, four bits in length.
4	Blank.
5 – 68	Data. Each 8-bit data byte is represented by two ASCII characters to represent a hexidecimal value of '00' to 'FF'.
69 – 70	Checksum. The checksum is the negative of the sum of all 8-bit bytes in the record from columns 1 to 68, evaluate modulo 256 (carry from high order bit ignored). For purposes of calculating the checksum, the value of column 4 is defined as zero.) Adding together, modulo 256, all 8-bit bytes from column 1 to 68 (column 4=0), then adding the checksum, results in zero.

EXAMPLE JCL DECK TO RUN GATE PLACEMENT

```
//   CIC JOB CARD
//   EXEC GATEPLM, DEV=TM4732A, DOMTAPE=volume serial number
Input Cards
//   The input card to PFP is: //PFP DD DSN=&&PFPIN, DISP=OLD
```

MOS
LSI

TMS 4764 JL, NL
8192-WORD BY 8-BIT READ-ONLY MEMORY

JUNE 1981 — REVISED MAY 1982

- 8192 X 8 Organization
- Fully Static (No Clocks, No Refresh)
- All Inputs and Outputs TTL Compatible
- Single 5 V Power Supply
- Optional Power-down or Chip-Select
- Maximum Access Time from Address . . . 300 ns
- Maximum Access Time from Power down . . . 300 ns
- Typical Active Power Dissipation . . . 275 mW
- Typical Standby Power Dissipation . . . 65 mW

24-PIN CERAMIC AND PLASTIC DUAL-IN-LINE PACKAGES (TOP VIEW)

A7	1	24	VCC
A6	2	23	A8
A5	3	22	A9
A4	4	21	A12
A3	5	20	S/$\overline{S}$/$\overline{E}$
A2	6	19	A10
A1	7	18	A11
A0	8	17	Q8
Q1	9	16	Q7
Q2	10	15	Q6
Q3	11	14	Q5
VSS	12	13	Q4

6

description

The TMS 4764 is a 65,536-bit read-only memory organized as 8192 words of 8-bit length. This makes the TMS 4764 ideal for microprocessor based systems. The device is fabricated using N-channel silicon-gate technology for high speed and simple interface with bipolar circuits.

All inputs can be driven directly by Series 74 TTL circuits without the use of any external pull-up resistor. Each output can drive two Series 74 or 74S loads without external resistors. The data outputs are three-state for OR-tieing multiple devices on a common bus. Pin 20 is programmable, providing additional system flexibility. The data is always available, it is not dependent on external clocking of pin 20.

The TMS 4764 is designed for high-density fixed-memory applications such as logic function generation and microprogramming. It is pin compatible with TI's full line of ROMs and EPROMs.

This ROM is supplied in 24-pin dual-in-line plastic (NL suffix) or ceramic (JL suffix) packages designed for insertion in mounting-hold rows on 600-mil centers. The device is designed for operation from 0°C to 70°C.

operation

address (A0—A12)

The address-valid interval determines the device cycle time. The 13-bit positive-logic address is decoded on-chip to select one of 8192 words of 8-bit length in the memory array. A0 is the least-significant bit and A12 the most-significant bit of the word address.

chip select (S or $\overline{S}$)

Pin 20 can be programmed during mask fabrication to be active with eighter a high- or a low-level input. When the signal is active, all eight outputs are enabled and the eight-bit addressed word can be read. When the signal is not active, all eight outputs are in a high-impedance state.

power down ($\overline{E}$)

A mask programmable option is to utilize pin 20 in a power-down mode. In this mode, pin 20 is clocked. When it is high, the chip is put into a standby mode. This reduces I_{CC1}, which in the active state is 80 mA maximum, to a standby I_{CC2} of 20 mA.

TEXAS INSTRUMENTS
INCORPORATED
POST OFFICE BOX 225012 • DALLAS, TEXAS 75265

data out (Q1–Q8)

The eight outputs must be enabled by pin 20 before the output word can be read. Data will remain valid until the address is changed or the outputs are disabled (chip deselected). When disabled, the three-state outputs are in a high-impedance state. Q1 is considered the least-significant bit, Q8 the most-significant bit.

The outputs will drive two Series 54/74 TTL circuits without external components.

logic symbol†

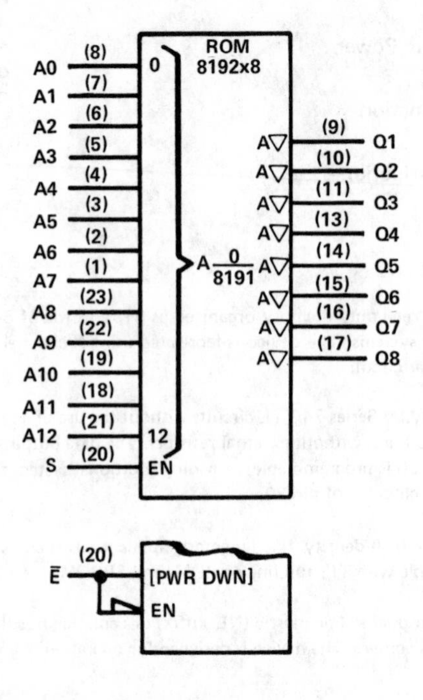

Pin 20 can be active-high as shown in the upper symbol or active-low as shown in the lower (partial) symbol. It can be either a chip select (S or S̄) or a chip enable/power down (Ē).

† This symbol is in accordance with IEEE Std 91/ANSI Y32.14 and recent decisions by IEEE and IEC. See explanation on page 289.

TEXAS INSTRUMENTS
INCORPORATED
POST OFFICE BOX 225012 • DALLAS, TEXAS 75265

functional block diagram

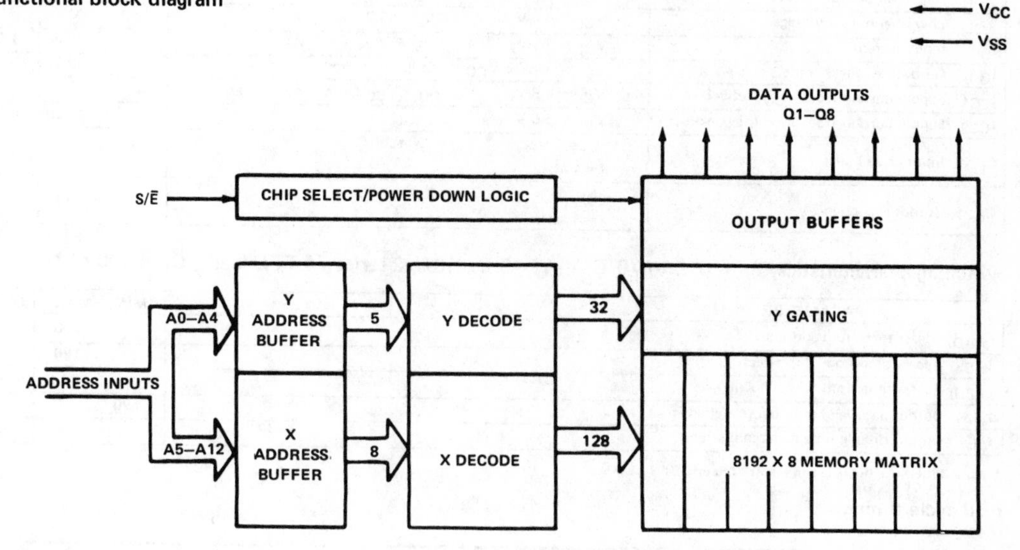

absolute maximum ratings

Supply voltage to ground potential (see Note 1) . . . . −0.5 to 7 V
Applied output voltage (see Note 1) . . . . −0.5 to 7 V
Applied input voltage (see Note 1) . . . . 0.5 to 7 V
Power dissipation . . . . 500 mW
Operating free-air temperature . . . . 0°C to 70°C
Storage temperature . . . . −55°C to 150°C

Note 1: Voltage values are with respect to V_{SS}.

recommended operating conditions

PARAMETER	MIN	NOM	MAX	UNIT
Supply voltage, V_{CC}	4.5	5	5.5	V
High-level input voltage, V_{IH}	2		V_{CC}+1	V
Low-level input voltage, V_{IL}	−0.5		0.8	V
Operating free-air temperature, T_A	0		70	°C

TEXAS INSTRUMENTS
INCORPORATED
POST OFFICE BOX 225012 • DALLAS, TEXAS 75265

electrical characteristics, $T_A = 0°C$ to $70°C$, $V_{DD} = 5\ V \pm 10\%$ (unless otherwise noted)

PARAMETER		TEST CONDITIONS		MIN	MAX	UNIT
V_{OH}	High-level output voltage	$V_{CC} = 4.5\ V$,	$I_{OH} = -400\ \mu A$	2.4		V
V_{OL}	Low-level output voltage	$V_{CC} = 4.5\ V$,	$I_{OL} = 3.2\ mA$		0.4	V
I_I	Input current	$V_{CC} = 5.5\ V$,	$0_V \leq V_{IN} \leq 5.5\ V$		10	μA
I_O	Output leakage current	$V_O = 0.4\ V$ to V_{CC},	Chip deselected		±10	μA
I_{CC1}	Supply current from V_{CC} (active)	$V_{CC} = 5.5\ V$,	$V_I = V_{CC}$ Output not loaded		80	mA
I_{CC2}	Supply current from V_{CC} (power down)	$V_{CC} = 5.5\ V$			20	mA
C_i	Input capacitance	$V_O = 0\ V$, f = 1 MHz	$T_A = 25°C$,		6	pF
C_o	Output capacitance	$V_O = 0\ V$, f = 1 MHz	$T_A = 25°C$,		12	pF

switching characteristics, $T_A = 0°C$ to $70°C$, $V_{CC} = 5\ V \pm 10\%$, 2 series 74 TTL loads, $C_L = 100\ pF$*

PARAMETER		MIN	MAX	UNIT
$t_{a(AD)}$	Access time from address		300	ns
$t_{a(S)}$	Access time from chip select		120	ns
$t_{a(PD)}$	Access time from power down		300	ns
$t_{v(A)}$	Output data valid after address change	20		ns
t_{dis}	Output disable time from chip select		100	ns

* All AC measurements are made at 10% and 90% points

read cycle timing

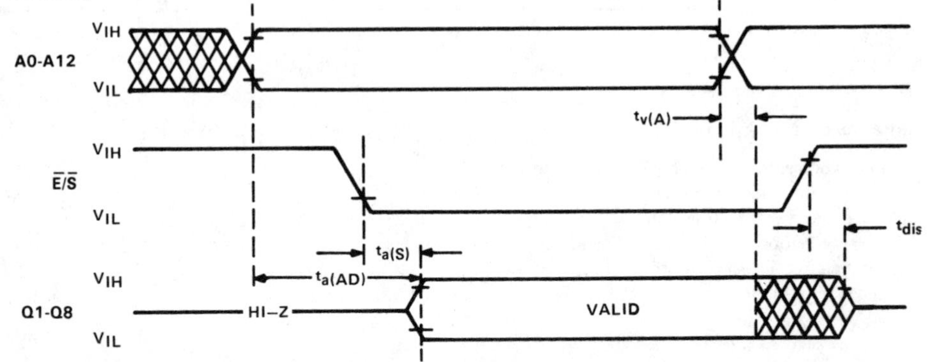

standby mode

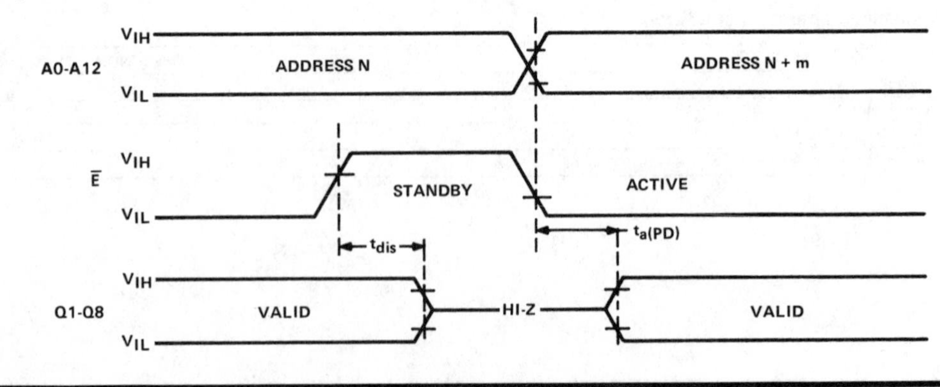

TEXAS INSTRUMENTS
INCORPORATED
POST OFFICE BOX 225012 • DALLAS, TEXAS 75265

PROGRAMMING DATA

PROGRAMMING REQUIREMENTS: The TMS 4764JL, NL is a fixed program memory in which the programming is performed by TI at the factory during manufacturing cycle to the specific customer inputs supplied in the format below. The device is organized as 8192 8-bit words with address locations numbered 0 to 8191. The 8-bit words can be coded as a 2-digit hexadecimal number between 00 and FF. All data words and addresses in the following format are coded in hexadecimal numbers. In coding all binary words must be in positive logic before conversion to hexadecimal. Q1 is considered the least significant bit and Q8 the most significant bit. For addresses, A0 is least significant bit and A12 is the most significant.

The input media containing the programming data can be in the form of cards or EPROMs.

Either 16K, 32K, or 64K EPROMs can be used, or any combination of them.

The following is a description of how the cards must be formatted, should they be used instead of EPROMs.

PROGRAMMING INSTRUCTIONS – 64K ROM

Each code deck submitted by customer shall consist of the following:

1. Title Card
2. Comment Cards
3. Start of Data Card
4. Data Cards

The cards shall be standard 80 column cards with the information in the following format:

TITLE CARD

Card Column	Information
1 – 5	The word 'TITLE' shall be punched in these columns.
6	Blank
7,8	The letters 'ZA' shall be punched in these columns.
9 – 14	Leave blank. A special device code number will be assigned by Texas Instruments. (left justified)
15	Blank
16 – 30	Customer's Part Number, if required. (left justified)
31	Blank
32	Customer's Part Number to be included as part of device symbolization. Options: Y = Yes N = No
33	Blank
34 – 35	# of Pins on Device Package Options: 24 = 24-pin package 28 = 28-pin package
36	Blank

6

Card Column	Information
37	Type of Package Options: C = ceramic P = plastic
38	Blank
39 – 40	Customer Defined Option for Device Mode. Options: PD = power down mode CS = chip select mode
41	Logic Level for pin 20 on 24-pin package or pin 23 on 28-pin package. Options: Blank = power down mode 1 = chip select mode, outputs enabled with high level. 0 = chip select mode, outputs enabled with low level.
42	Logic Level for pin 27 on 28-pin package. Options: Blank = for 24-pin package (see columns 34 – 35) 1 = chip select mode, outputs enabled with high level. 0 = chip select mode, outputs enabled with low level.
43	Logic Level for pin 2 on 28-pin package. Options: Blank = for 24-pin package (see columns 34–35) 1 = chip select mode, outputs enabled with high level. 0 = chip select mode, outputs enabled with low level.
44	Blank
45 – 49	Texas Instruments Device Series (ie. 4764, 4864, etc.) (left justified)

COMMENT CARDS

Any number of comment cards may be used for specifying the customer's name, individual to contact, telephone number, address, any special instructions, etc. The format for these cards is as follows: The letter 'C' (for comment) must be punched in column 1, columns 2-4 must be blank, and comments can be punched in columns 5-80.

START OF DATA CARD

This card is to identify that the next card will be the beginning of customer's code. Format is as follows: Columns 1-4 must have '&ROM' punched in them. The remainder of card is blank.

TEXAS INSTRUMENTS
INCORPORATED
POST OFFICE BOX 225012 • DALLAS, TEXAS 75265

DATA CARDS

There will be 256 data cards supplied for each customer code. Each card will contain (in hexadecimal) the data for 32 memory locations. Each data card shall be in the following format:

Card Column	Information
1 – 4	Hexadecimal address of first word on the card, four bits in length.
5,6	Blank
7 – 70	Data. Each 8-bit data byte is represented by two ASCII characters to represent a hexadecimal value of '00' to 'FF'.
71,72	Checksum. The checksum is the negative of the sum of all 8-bit bytes in the record from columns 1 to 70, evaluate modulo 256 (carry from high order bit ignored). For purposes of calculating the checksum, the value of columns 5 and 6 are defined to be zero. Adding together, modulo 256, all 8-bit bytes from columns 1 to 70 (columns 5 and 6 = 0), then adding the checksum, results in zero.
73 – 76	Blank
77 – 80	Card sequence number, in decimal. (right justified).

Memory Systems Data Sheets

TMM10010 SERIES
ADD-IN MEMORY MODULES

OCTOBER 1981

- High Density: 128K and 256K Bytes X 16/18 Bits on Dual Wide Board
- Implements All Parity Detection, Control and Logging Functions (Optional)
- Low Power Consumption, Single +5 Volt Supply
- Hardware/Software Compatible with LSI-11 Systems, Q-Bus Plus*
- Control and Status Register Stores Parity Error Information
- 18 to 22 Address Lines (1/4 to 4 Megabytes), User Selectable with 4K Word Granularity
- Internal, Transparent Refresh
- Compatible with Existing Memory and Parity Diagnostic Programs

description

The TMM10010 Series add-in memory modules are offered in four high-density versions. Each is fully compatible with the Q-Bus Plus* (LSI-11/23).

All modules use a single 5-volt power supply to ensure low power consumption and high performance.

The TMM10010 memories undergo 100% testing at the component level, as well as the board level.

Battery backup is jumper selectable.

Internal distributed refresh cycles are transparent to the external system bus.

*Trademark of Digital Equipment Corporation.

operating modes

The TMM10010 has two basic modes of operation, memory and I/O. Memory operations involve data transfers to and from memory and the refreshing of memory. I/O operations use data transfers to and from the Control and Status Register.

specifications

memory capacity

MODEL	CAPACITY	BITS/WORD
TMM10010-01	128K Bytes (131,072)*	16 + 2 parity bits
TMM10010-02	128K Bytes (131,072)	16
TMM10010-05	256K Bytes (262,144)	16 + 2 parity bits
TMM10010-06	256K Bytes (262,144)	16

*K = 1024

memory access and cycle times

OPERATION	ACCESS TIME (ns)		CYCLE TIME (ns)	
	TYP	MAX	TYP	MAX
DATI (Memory Read)	175	195	360	395
DATO(B) (Memory Write)	75	95	360	395

power requirements (supply voltage = 5 V ±5%)

MODE	MODEL	POWER (WATTS)		CURRENT (AMPS)	
		TYP	MAX	TYP	MAX
Operating	−01, −05	13.5	16.2	2.7	3.1
	−02, −06	11.4	13.7	2.3	2.6
Standby	−01, −05	11.8	14.2	2.4	2.7
	−02, −06	10.8	12.0	2.2	2.3
Battery Backup	All	3.7	4.4	0.7	0.8

environmental conditions

MODE	TEMPERATURE	RELATIVE HUMIDITY
Operating	5 °C to 60 °C	10% to 95% (Noncondensing)
Storage	−40 °C to +85 °C	

dimensions

8.43 in. (21.07 cm) × 5.187 in. (12.967 cm), single width, double height.

TEXAS INSTRUMENTS
INCORPORATED
POST OFFICE BOX 225012 • DALLAS, TEXAS 75265

pin list

PIN	FUNCTION	PIN	FUNCTION
AA1	NOT USED	BA1	BDCOK H
AB1	NOT USED	BB1	NOT USED
AC1	BDAL <16> L	BC1	BDAL <18> L
AD1	BDAL <17> L	BD1	BDAL <19> L
AE1	NOT USED	BE1	BDAL <20> L
AF1	NOT USED	BF1	BDAL <21> L
AH1	NOT USED	BH1	NOT USED
AJ1	GROUND	BJ1	GROUND
AK1	NOT USED	BK1	REFR OSC OUT
AL1	NOT USED	BL1	REFR REQ IN
AM1	GROUND	BM1	GROUND
AN1	NOT USED	BN1	NOT USED
AP1	NOT USED	BP1	NOT USED
AR1	BREF L	BR1	NOT USED
AS1	NOT USED	BS1	NOT USED
AT1	GROUND	BT1	NOT USED
AU1	NOT USED	BU1	NOT USED
AV1	+5 V BATTERY	BV1	+5 V POWER
AA2	+5 V POWER	BA2	+5 V POWER
AB2	NOT USED	BB2	NOT USED
AC2	GROUND	BC2	GROUND
AD2	NOT USED	BD2	NOT USED
AE2	BDOUT L	BE2	BDAL <2> L
AF2	BRPLY L	BF2	BDAL <3> L
AH2	BDIN L	BH2	BDAL <4> L
AJ2	BSYNC L	BJ2	BDAL <5> L
AK2	BWTBT L	BK2	BDAL <6> L
AL2	NOT USED	BL2	BDAL <7> L
AM2	BIAKI L	BM2	BDAL <8> L
AN2	BIAKO L	BN2	BDAL <9> L
AP2	BBS 7 L	BP2	BDAL <10> L
AR2	BDMGI L	BR2	BDAL <11> L
AS2	BDMGO L	BS2	BDAL <12> L
AT2	BINIT L	BT2	BDAL <13> L
AU2	BDAL <0> L	BU2	BDAL <14> L
AV2	BDAL <1> L	BV2	BDAL <15> L

options (switch/jumper selectable)

Address Space	18-22 bits (1/4 to 4 Megabytes)
Starting Address	Any 4K word boundary
I/O Page Size	1024, 2048, 3584 or 4096 words
Parity (Controller)	All functions implemented on board
Battery Backup	User selectable

TMM20000 SERIES
ADD-IN MEMORY FOR PDP-11 COMPUTERS

JUNE 1980, REVISED MAY 1981

- **Fully Compatible with PDP-11 UNIBUS* (Modified or Extended) Systems**
- **High Density . . . 1 Megabyte . . . 512K (16 Bit) Words Plus 6 Bits for Error Detection and Correction**
- **Single-Bit Error Correction, Double Bit Error Detection Greatly Enhance System Reliability**
- **Twenty-two Address Lines Allow for Expansion to 2M Words**
- **Single +5-Volt Supply, Low Power Consumption**
- **Error Logging Capability Isolates Failures to Individual Memory Components**

description

The TMM20000 high-density, high-speed add-in modules are completely compatible with the DEC* PDP-11 family of UNIBUS* computers. System dependability is markedly improved using the single-bit error correction, double-bit error detection feature.

The TMM20000 Series modules' 22 address lines permit expansion to 2M words. These modules operate from a single +5-volt power supply with low power consumption.

* Trademark of Digital Equipment Corporation.

TEXAS INSTRUMENTS
INCORPORATED
POST OFFICE BOX 225012 • DALLAS, TEXAS 75265

error detection and correction (EDAC)

On-board circuitry performs single-bit error correction and double-bit error detection. Control (CSR) and error (ESR) status registers, with single- and double-bit error display, can capture in real time all information necessary to pinpoint any single-bit failure to the exact location of the failing memory component. All error detection and correction operations are transparent to the operating system.

Battery backup is jumper selectable.

system reliability

The TMM20000 modules undergo 100% testing, not only at board level but at memory component level as well. All modules and memory components are burned-in for improved reliability.

specifications

memory capacity

MODEL	WORDS	BITS/WORD
TMM20000-01	256K words X 22 bits (262,144 words)	16 data bits plus 6 error-detection bits
TMM20000-02	128K words X 22 bits (131,072 words)	16 data bits plus 6 error-detection bits
TMM20000-04	512K words X 22 bits (524,288 words)	16 data bits plus 6 error-detection bits

access and cycle times (see Note 1)

OPERATION		STATUS REGISTERS ACCESS TIME (ns)		MEMORY REGISTERS ACCESS TIME (ns)		MEMORY REGISTERS CYCLE TIME (ns)	
		TYP	MAX	TYP	MAX	TYP	MAX
DATI (Memory Read)	No error detected (see Note 2)		125	400	430	620	665
	Error detected (see Note 2)		125		540		965
DATO (Memory Write)			125	40	60	620	665
DATOB (Memory Write/Byte)			125	40	60	920	965

power requirements

MODE	CONDITIONS	128K WORDS POWER (W)		256K WORDS POWER (W)		512K WORDS POWER (W)		FUNCTION
		TYP	MAX	TYP	MAX	TYP	MAX	
Operating	+5 V (±5%)	16	20	17	20	17	20	Continuous read/write cycles
Standby		15	18	16	19	16	19	Memory ready and refresh every 15 μs
Battery backup	Standby	9	12	10	12	10	12	Memory array and refresh support cycles operating in standby modes

environmental conditions

MODE	TEMPERATURE	RELATIVE HUMIDITY
Operating	5°C to 50°C	10% to 95% (Noncondensing)
Storage	−40°C to +85°C	

dimensions

15.68 in. (39,2 cm) X 8.47 in. (21.0 cm), single width, hex height.

NOTES: 1. Access and cycle times are measured from receipt of Bus MSYN to transmission of Bus SSYN.
2. Access and cycle times are extended when an error is detected and corrected.

pin list (UNIBUS pin assignments)

PIN	STANDARD	MODIFIED	EXTENDED
AA1	INIT L		
AA2	NOT USED		
AB1	NOT USED		
AB2	NOT USED		
AC1	D00 L		
AC2	GROUND		
AD1	D02 L		
AD2	D01 L		
AE1	D04 L		
AE2	D03 L		
AF1	D06 L		
AF2	D05 L		
AH1	D08 L		
AH2	D07 L		
AJ1	D10 L		
AJ2	D09 L		
AK1	D12 L		
AK2	D11 L		
AL1	D14 L		
AL2	D13 L		
AM1	NOT USED		
AM2	D15 L		
AN1	NOT USED	NOT USED	A21 L
AN2	PB L		
AP1	NOT USED	NOT USED	A20 L
AP2	NOT USED		
AR1	NOT USED		
AR2	NOT USED		
AS1	NOT USED		
AS2	NOT USED		
AT1	GROUND		
AT2	NOT USED		
AU1	NOT USED		
AU2	NOT USED		
AV1	NOT USED		
AV2	NOT USED		
BA1	NOT USED		
BA2	+5 V		
BB1	NOT USED		
BB2	NOT USED		
BC1	NOT USED		
BC2	GROUND		
BD1	NOT USED	+5 BATT	+5 BATT
BD2	NOT USED		
BE1	NOT USED	NOT USED	A19 L
BE2	NOT USED	NOT USED	A18 L
BF1	NOT USED		
BF2	DCLO L		
BH1	A01 L		
BH2	A00 L		
BJ1	A03 L		
BJ2	A02 L		
BK1	A05 L		
BK2	A04 L		
BL1	A07 L		
BL2	A06 L		
BM1	A09 L		
BM2	A08 L		
BN1	A11 L		
BN2	A10 L		
BP1	A13 L		
BP2	A12 L		
BR1	A15 L		
BR2	A14 L		
BS1	A17 L		
BS2	A16 L		
BT1	GROUND		
BT2	C1 L		
BU1	SSYN L		
BU2	C0 L		
BV1	MSYN L		
BV2	NOT USED		
CA1	NPG IN/OUT		
CA2	+5 V		
CB1	NPG IN/OUT		
CT1	GROUND		
DA2	+5 V		
DK2	BG7 IN/OUT		
DL2	BG7 IN/OUT		
DM2	BG6 IN/OUT		
DN2	BG6 IN/OUT		
DP2	BG5 IN/OUT		
DR2	BG5 IN/OUT		
DS2	BG4 IN/OUT		
DT1	GROUND		
DT2	BG4 IN/OUT		
EA2	+5 V		
ET1	GROUND		
FA2	+5 V		
FT1	GROUND		

selectable options

- Modified or extended UNIBUS: switch select
- I/O page size: 2K, 4K, or 8K words (switch select)
- Starting address (switch select): any 16K word boundary
- Total address space (switch select):
  modified UNIBUS – 128K words (256K bytes)
  extended UNIBUS – 2048K words (4M bytes)
- Control and status register (CSR) address location (switch select):
  1 of 16
- Error status register (ESR): on or off
- Battery backup: enable or disable (jumper)

TMM 30000 SERIES
ADD-IN MEMORY FOR VAX 11/780* COMPUTERS

DECEMBER 1980

- Fully Compatible with VAX 11/780* Computer, Including Battery Backup Option
- Very-High Density . . . 1/2 Megabyte, 1 Megabyte on a Single Board
- Extensive Testing and Burn-In Ensure High Reliability
- Single +5-Volt Supply, Low-Power Consumption
- Replaces Two to Four VAX M8210* Memory Boards; Greatly Improving System Reliability Due to Reduced TTL and High Reliability 64K Technology
- +5-Volt LED and +5-Volt Battery LED
- Board Select LED
- On-Line/Off-Line Switch
- Full One Year Warranty

description

The TMM30000 high-density, high-speed add-in modules are hardware/software-compatible with the DEC VAX 11/780* computer.

TI high-temperature burn-in and module test procedures enhance system dependability.

The TMM30000 Series modules require memory sizing boards to replace the equivalent number of M8210 memory boards, e.g.; TMM30000-01 needs three and the TMM30000-03 needs one. These memory sizing boards are furnished with the TMM30000 Series Modules.

LEDs indicate when +5-volt power or +5-volt battery power is applied, or when +5-volt power is applied and the battery backup is not used. The board select LED indicates only when a memory bank is accessed.

*Trademark of Digital Equipment Corporation

TEXAS INSTRUMENTS
INCORPORATED
POST OFFICE BOX 225012 • DALLAS, TEXAS 75265

specifications

memory capacity

MODEL	WORDS	BITS/WORD	MEMORY SIZING BOARDS REQUIRED
TMM30000-01	131,072	72 (64 + 8EDAC)	3
TMM30000-03	65,536	72 (64 + 8EDAC)	1

timing (typical)

Access time: 250 ns

Cycle time: 530 ns

(System cycle time determined by memory controller)

power (typical)

MODEL	WATTS (Operating)*	WATTS (Standby)**
TMM30000-01	20.5	16.5
TMM30000-03	15.5	13.5

*Back-to-back memory cycles, distributed refresh, +5 volts at 25°C
**Distributed refresh, +5 volts at 25°C

environmental conditions

MODE	TEMPERATURE	RELATIVE HUMIDITY
Operating	0°C to 50°C	10% to 95% (Noncondensing)
Storage	−40°C to 70°C	

dimensions

11.95 in. (19.87 cm) X 15.688 in. (39.22 cm)

VAX 11/780 BACKPLANE DESIGNATION (TYPICAL INSTALLATION)

(Each array module slot is assigned 1/4 megabyte of MOS memory)

1 Megabyte Boundaries

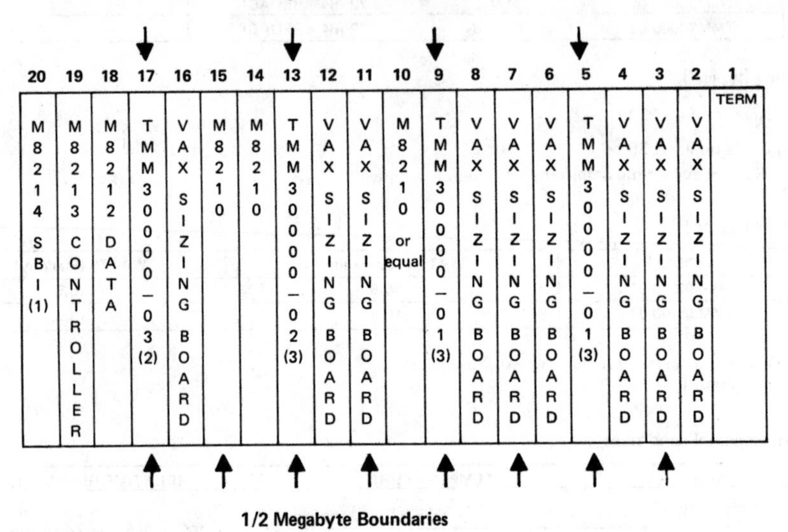

NOTES: 1. Synchronous backplane interconnect interface
2. Place on 1/2-megabyte boundaries. (No more than one set per backplane.)
3. Place on 1-megabyte boundaries

TEXAS INSTRUMENTS
INCORPORATED
POST OFFICE BOX 225012 • DALLAS, TEXAS 75265

TMM40010A SERIES MEMORY MODULES FOR MULTIBUS* SYSTEMS

NOVEMBER 1981, REVISED FEBRUARY 1982

- **Very High Density . . . 64K Bytes to 512K Bytes on One Board**
- **Completely Compatible with Intel Multibus* Protocol**
- **Single-Bit Error Correction and Double-Bit Detection with Inhibit Capability**
- **Lower and Upper Memory Addresses Are Independently Selectable with 4K-Byte Granularity**
- **Single +5-Volt Supply for Low Power Consumption**
- **Control-Status and Error-Status Registers (Accessible via Two I/O Ports)**
- **Battery Backup and ROM Overlay Capability**
- **LEDs Indicate Type of Error and Exact Location of Failing DRAM**
- **Switch Selectable 20 or 24-Bit Address Bus**
- **Holes Provided for Pull-Up SIP Resistor for Those Desiring 21, 22, or 23 Address Lines**

description

The TMM40010A Series memory modules are available in four high-density versions all of which are compatible with the Intel Multibus*. All modules employ three types of error logging: visual, processor polling, and nonvectored interrupt.

The TMM40010A supports word transfer and high, swap, and low byte transfers.

The power-fail sense flip-flop on the power supply may be read and reset via the control status register.

*Trademark of Intel Corporation.

specifications

memory capacity

MODEL NO.	CAPACITY
TMM40010A-01	256K (262,144) Bytes
TMM40010A-02	128K (131,072) Bytes
TMM40010A-04	512K (524,288) Bytes
TMM40010A-07	64K (65,536) Bytes

operating voltage: 5 V ±5%

typical power (supply voltage = 5 V)

MODEL NO.	OPERATING (WATTS)	REFRESH ONLY (WATTS)	BATTERY BACKUP (WATTS)
TMM40010A-01	14	12.7	7.5
TMM40010A-02	12	11.8	6.5
TMM40010A-04	14	12.7	7.5
TMM40010A-07	12.1	11.3	4.8

system timing

	PARAMETER		MIN	TYP	MAX	UNIT
$t_{a(RD)}$	Time from MRDC active† to valid read data			325	355	ns
$t_{a(WT)}$	Time from MWTC active† to write data latched			90	130	ns
$t_{c(RD,WT)}$	Cycle time, read or write			700	730	ns
$t_{c(REFRESH)}$	Cycle time, refresh			750	850	ns
$t_{XACK(RD)}$	Time from MRDC active† to XACK active			390	420	ns
$t_{XACK(WT)}$	Time from MWTC active† to XACK active		92	132	158	ns
$t_{AACK(RD)}$	Time from MRDC active† to AACK active	Tap 2		285		ns
		Tap 3		250		
		Tap 4		215		
		Tap 5		110		
$t_{AACK(WT)}$	Time from MWTC active† to AACK active		92	132	158	ns
$t_{XACK(I/O)}$	Time from IORC or IOWC active to XACK active		26	45	79	ns
t_{AACK}	Time from IORC or IOWC active to AACK active		26	45	79	ns

† No refresh

NOTE: Advanced acknowledge (AACK/) may be used to provide acknowledges of 74 ns, 110 ns, 147 ns, and 257 ns in advance of valid read data (worst case).

environmental conditions

MODE	TEMPERATURE	RELATIVE HUMIDITY
Operating	0 °C to 70 °C‡	10% to 95% (Noncondensing)
Storage	−40 °C to 80 °C	

‡ With adequate air flow

dimensions

12.0 in. (30,48 cm) × 6.75 in. (17,15 cm)

additional features

- 256 byte or 4K byte I/O may be selected
- Interrupts strappable to any nonvectored interrupt line
- Save first error/last error selectable via Control Status Register
- LEDs may be cleared in software by writing to Error Status Register
- Normally asynchronous transparent refresh may be synchronized with the processor, or boards may be synchronized with each other in multiboard systems.
- Corrected data is written back into memory when a single-bit error is detected.

pin assignments

CONNECTOR P1

	COMPONENT SIDE			CIRCUIT SIDE		
	PIN	MNEMONIC	DESCRIPTION	PIN	MNEMONIC	DESCRIPTION
Power Supplies	1	GND	Signal ground	2	GND	Signal ground
	3	+5 V	+5 V	4	+5 V	+5 V
	5	+5 V	+5 V	6	+5 V	+5 V
	7	+12 V	+12 V	8	+12 V	+12 V
	9	−5 V	−5 V	10	−5 V	−5 V
	11	GND	Signal ground	12	GND	Signal ground
Bus Controls	13	BCLK/	Bus clock	14	INIT/	Initialize
	15	BPRN/	Bus priority in	16	BPRO/	Bus priority out
	17	BUSY/	Bus busy	18	BREQ/	Bus request
	19	MRDC/	Memory read command	20	MWTC/	Memory write command
	21	IORC/	I/O read command	22	IOWC/	I/O write command
	23	XACK/	Transfer acknowledge	24	INH1/	Inhibit 1 disable RAM
Bus Controls and Address	25	AACK/	Advanced acknowledge	26	INH2/	Inhibit 2 disable PROM or ROM
	27	BHEN/	Byte high enable	28	AD10/	Address bus
	29	CBRQ/	Common bus request	30	AD11/	
	31	CCLK/	Constant clock	32	AD12/	
	33	INTA/	Interrupt acknowledge	34	AD13/	
Interrupts	35	INT6/	Parallel interrupt requests	36	INT7/	Parallel interrupt requests
	37	INT4/		38	INT5/	
	39	INT2/		40	INT3/	
	41	INT0/		42	INT1/	
Address	43	ADRE/	Address bus	44	ADRF/	Address bus
	45	ADRC/		46	ADRD/	
	47	ADRA/		48	ADRB/	
	49	ADR8/		50	ADR9/	
	51	ADR6/		52	ADR7/	
	53	ADR4/		54	ADR5/	
	55	ADR2/		56	ADR3/	
	57	ADR0/		58	ADR1/	
Data	59	DATE/	Data bus	60	DATF/	Data bus
	61	DATC/		62	DATD/	
	63	DATA/		64	DATB/	
	65	DAT8/		66	DAT9/	
	67	DAT6/		68	DAT7/	
	69	DAT4/		70	DAT5/	
	71	DAT2/		72	DAT3/	
	73	DAT0/		74	DAT1/	

TEXAS INSTRUMENTS
INCORPORATED
POST OFFICE BOX 225012 • DALLAS, TEXAS 75265

CONNECTOR P1 (Continued)

	COMPONENT SIDE			CIRCUIT SIDE		
	PIN	MNEMONIC	DESCRIPTION	PIN	MNEMONIC	DESCRIPTION
Power Supplies	75	GND	Signal GND	76	GND	Signal ground
	77	Reserved		78	Reserved	
	79	−12 V	−12 V	80	−12 V	−12 V
	81	+5 V	+5 V	82	+5 V	+5 V
	83	+5 V	+5 V	84	+5 V	+5 V
	85	GND	Signal GND	86	GND	Signal ground

pin assignments (continued)

CONNECTOR P2

COMPONENT SIDE			CIRCUIT SIDE		
PIN	MNEMONIC	DESCRIPTION	PIN	MNEMONIC	DESCRIPTION
1	GND	Signal ground	2	GND	Signal ground
3	5VB	+5-volt battery	4	+5VB	+5-volt battery
5		Reserved	6	VCCPP	+5-volt pulsed power
7	−5VB	−5-volt battery	8	−5VB	−5-volt battery
9		Reserved	10		Reserved
11	12VB	+12-volt battery	12	12VB	+12-volt battery
13	PFSR/	Power failure sense reset	14		Reserved
15	−12VB	−12-volt battery	16	−12VB	−12-volt battery
17	PFSN/	Power failure sense	18	ACLO	AC low
19	PFIN	Power failure interrupt	20	MPRO/	Memory protect
21	GND	Signal ground	22	GND	Signal ground
23	+15 V	+15 V	24	+15V	+15 V
25	−15 V	−15 V	26	−15V	−15 V
27	PAR1/	Parity 1	28	HALT/	Bus master halt
29	PAR2/	Parity 2	30	WAIT	Bus master wait state
31	Reserved		32	ALE	Bus master ALE
33			34		Reserved
35			36		
37			38	AUX RESET/	Reset switch
39	REFR1/	Refresh option	40	REFR2/	Refresh test
41	Reserved		42		Reserved
43			44		
45			46		
47			48		
49			50		
51			52		
53			54		
55	ADR16/	Address bus	56	ADR17/	Address bus
57	ADR14/	Address bus	58	ADR15/	Address bus
59			60		

/indicates board interprets electrical low on the bus as a Logical One.

Applications Information

8

Applications Brief

JUNE 1980

64K DYNAMIC RAM ARCHITECTURE

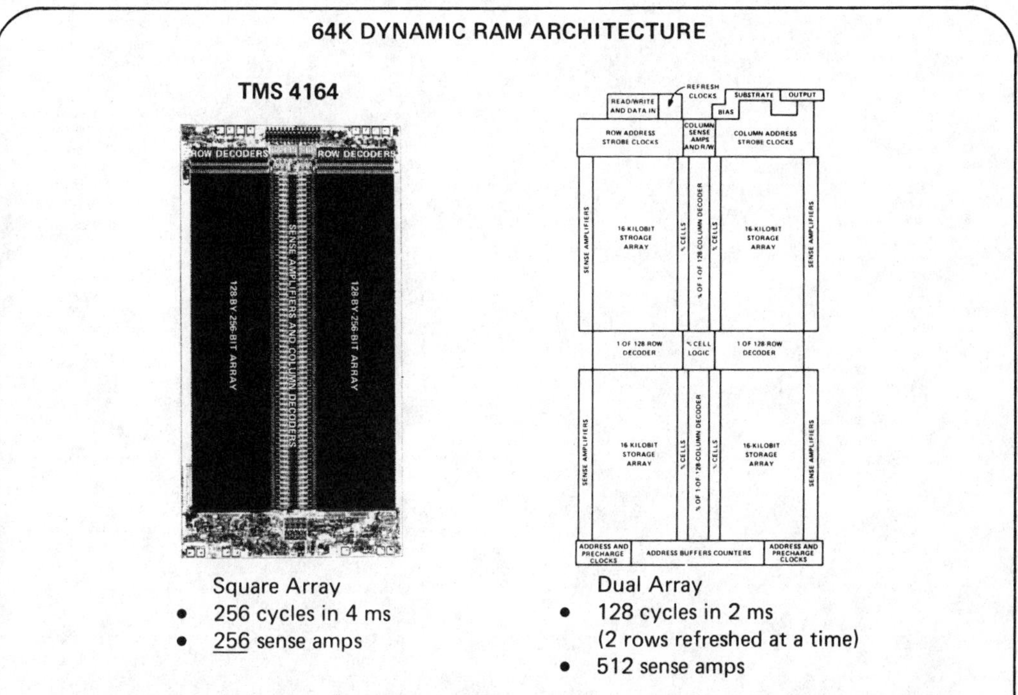

Square Array
- 256 cycles in 4 ms
- 256 sense amps

Dual Array
- 128 cycles in 2 ms (2 rows refreshed at a time)
- 512 sense amps

Of the two basic 64K Dynamic RAM architectures illustrated above, TI chose the historically successful industry standard organization for its TMS 4164. This 256-column by 256-row square array results in the following attributes:

- Superior Performance – The 256-cycle approach requires only half the number of sense amplifiers as the 128-cycle approach. Fewer sense amplifiers means lower power dissipation (125 mW typical) and lower junction temperatures (parameters which limit cycle time for many 16K Dynamic RAMs). The TMS 4164-15 is thus able to operate at a dramatically improved 280-ns cycle time versus 375 ns for 16K circuits.

- Improved Reliability – Another advantage resulting from the utilization of only 256 sense amplifiers is the increased chip area which can be devoted to the memory array (array area/bar size = 0.52). This allows each cell to store greater charge and, coupled with the reduced on-chip temperatures, ensures a decreased refresh rate (4 ms per cell versus 2 ms). In addition, current transients and system noise levels are reduced by as much as 2 to 1.

- Low Cost – With only 256 sense amplifiers and reduced on-chip routing, the TMS 4164 organization optimizes bar size and yields a cost-effective, reliable, and producible memory component.

Applications Brief

JULY 1980

64K DYNAMIC RAM REFRESH ANALYSIS
SYSTEM DESIGN CONSIDERATIONS

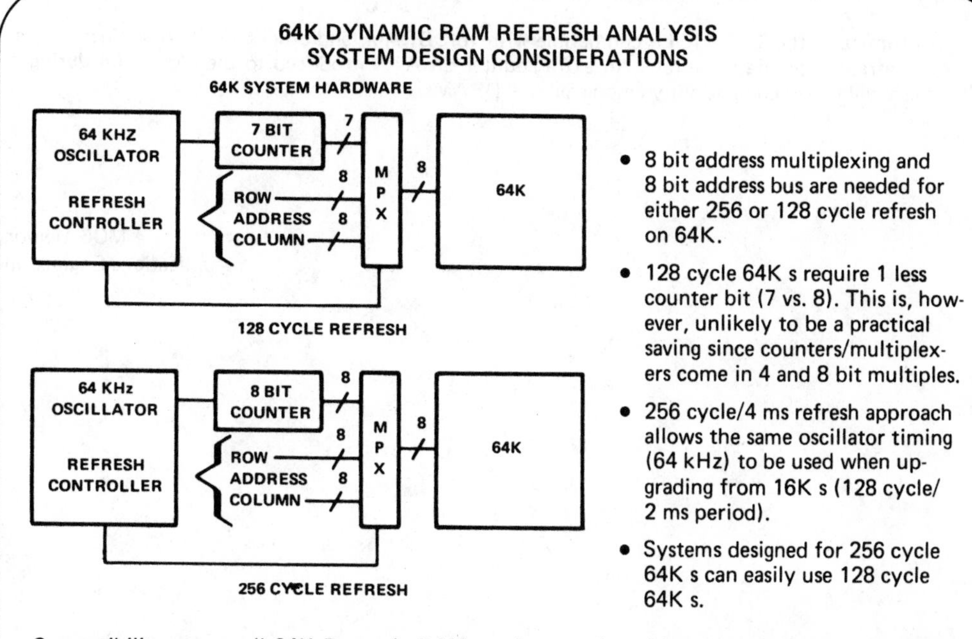

- 8 bit address multiplexing and 8 bit address bus are needed for either 256 or 128 cycle refresh on 64K.
- 128 cycle 64K s require 1 less counter bit (7 vs. 8). This is, however, unlikely to be a practical saving since counters/multiplexers come in 4 and 8 bit multiples.
- 256 cycle/4 ms refresh approach allows the same oscillator timing (64 kHz) to be used when upgrading from 16K s (128 cycle/ 2 ms period).
- Systems designed for 256 cycle 64K s can easily use 128 cycle 64K s.

Compatibility among all 64K Dynamic RAM vendors can be achieved by designing to TI's 4164 64K x 1 Dynamic RAM. The TMS 4164 requires all 256 rows to be refreshed within 4 ms. Competitive 64K DRAMs which are not able to achieve the 256 cycle, 4 ms refresh rate require twice the number of sense amplifiers as the TMS 4164 and half the number of refresh addresses. A 64K DRAM which requires the 128 cycle, 2 ms refresh treats the 256 cycle, 4 ms refresh as two refresh events in 2 ms each.

Simply: 256 cycle in 4 ms = 2 (128 cycle in 2 ms)

The extra address bit, A7, during refresh is treated by these vendors as a don't care situation.

The TMS 4164 has the same refresh rate as the 4116, 16K x 1 Dynamic RAM, which requires 128 rows to be refreshed in 2 ms. Most 4116 based systems already contain the extra refresh counter bit required for upgrading to the 64K. Those implemented with the 74LS393, 8-bit counter already do.

For a given cycle time, say 280 ns, the 256 cycle 4 ms refresh architecture of the TMS 4164 requires the same refresh overhead as the 128 cycle 2 ms approach as can be seen by the following calculations:

$$\text{Refresh overhead} = \frac{\text{refresh cycles in given time}}{\text{available cycles in given time}} = \frac{\text{256 cycles}}{\text{4 ms/280 ns per cycle}} = \frac{\text{128 cycles}}{\text{2 ms/280 ns per cycle}} = 1.8\%$$

8

However, the TMS 4164 provides the user with the following advantages:

- Half the number of sense amplifiers, small chip size, low cost.
- Lower power yielding lower temperature and increased reliability.
- More chip area devoted to memory array allowing greater detectable cell charge and improved performance.

In summary, the TMS 4164 is compatible with 16K DRAMs and other 64K DRAMs since they are all refreshed at the same rate. An extra counter bit A7, introduced to the TMS 4164 during refresh will insure compatibility among all 64K DRAMs.

MOS Memory
Applications Engineering

Applications Brief

AUGUST 1981

256-CYCLE REFRESH CONVERSION

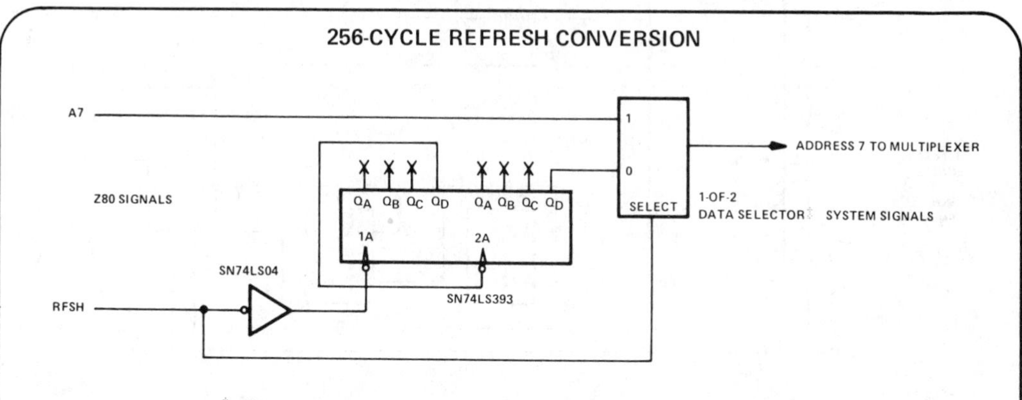

‡ This may be implemented in discrete logic, with an SN74LS157 or other scheme.

CIRCUIT TO CONVERT Z80 128-CYCLE REFRESH TO 256-CYCLE REFRESH REQUIRED BY TMS 4164

Adding the circuit above to Z80-based systems increases availability and competitive pricing among those vendors who have announced 64K dynamic RAMs[1] including

* TEXAS INSTRUMENTS
— Fairchild
— Inmos
— National
— Signetics

Z80 users who are considering an upgrade in dynamic memory from 16K to 64K have much to gain by modifying the Z80 128-cycle refresh. The circuit shown above converts the Z80 128-cycle refresh to the 256-cycle refresh required by TI's TMS 4164 64K dynamic RAM. Designing in the 256-cycle refresh results in broader choice of 64K dynamic memory available to the designer. Designing with only 128-cycle capability severely limits the potential sources for 64K dynamic RAMs.

Adding the circuit shown above to the Z80-based system also allows the designer to take advantage of the TMS 4164's low cost and low power dissipation that result from using only half as many sense amplifiers as the 128-cycle approach.

[1] Electronics, May 22, 1980

8

EXPANSION OF 3242 FOR 256-CYCLE REFRESH

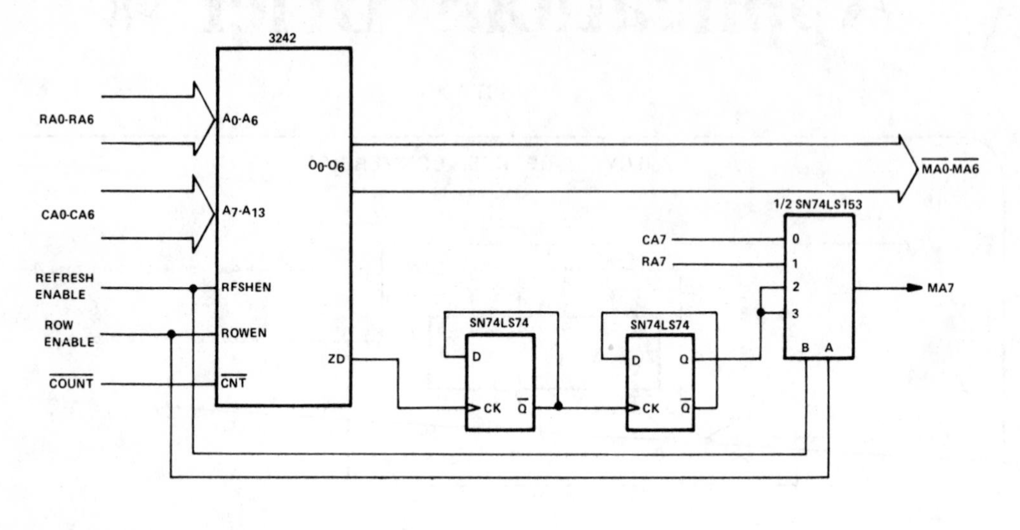

RA0-RA7 - ROW ADDRESS

CA0-CA7 - COLUMN ADDRESSES

$\overline{MA0}$-$\overline{MA6}$ - MEMORY ADDRESSES (INVERTED)

MA7 - MEMORY ADDRESS (NON-INVERTED)

Adding the above circuit to those systems which utilize a 3242 allows the use of 256-cycle-refresh parts. The circuit on the following page may also be incorporated into designs using the 3242A where the zero-detect output is not available.

These designs are presented to demonstrate possible implementation, however, particular system requirements may suggest alternate circuits to optimize component layout.

NOTE: The SN74LS153 may be replaced with an SN74LS352 to obtain inverted signal for all memory addresses.

EXPANSION OF 3242A FOR 256-CYCLE REFRESH

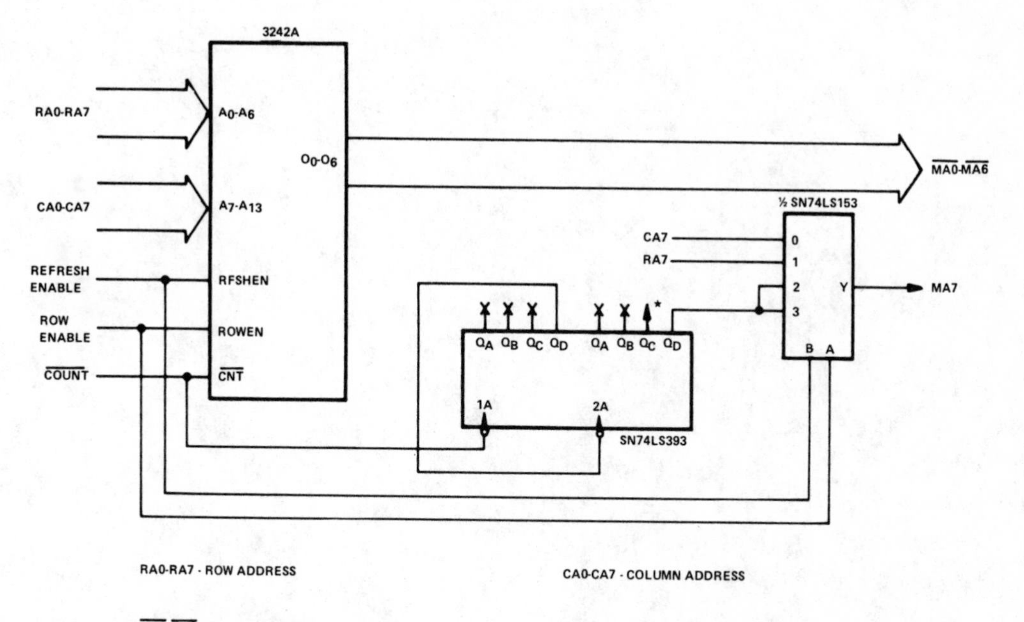

RA0-RA7 - ROW ADDRESS

CA0-CA7 - COLUMN ADDRESS

$\overline{MA0}$-$\overline{MA6}$ - MEMORY ADDRESS (INVERTED)

MA7 - MEMORY ADDRESS (NON-INVERTED)

★ This output may be used to implement 256-cycle refresh for 3232 devices in conjunction with the other half of the SN74LS153.

In summary, these circuits show how to convert a system to a design with maximum flexibility that can use either 128-cycle or 256-cycle 64K dynamic RAMs. Meeting this requirement allows the use of any 64K RAM which will ultimately result in the lowest cost and most reliable system.

MOS Memory
Applications Engineering

Applications Brief

TMS 4116 VS. TMS 4164 DATA SHEET DIFFERENCES

SPECIFICATION	SYMBOL	−15		−20		−25	
		4116	4164	4116	4164	4116	4164
Page Mode Cycle Time	$t_{c(P)}$	170	160	225	225	275	275
Read or Write Cycle Time	$t_{c(rd)}$ $t_{c(W)}$	375	260	375	330	410	410
Read-Modify-Write Cycle Time	$t_{c(rdW)}$	375	285	375	345	515	455
Pulse Width, $\overline{CAS}$ High	$t_{w(CH)}$	60	50	80	80	100	100
Pulse Width, $\overline{RAS}$ Low	$t_{w(RL)}$	150	150	200	200	250	250
Column Address Setup Time	$t_{su(CA)}$	−10	−5	−10	−5	−10	−5
Delay Time, $\overline{CAS}$ High to $\overline{RAS}$ Low	t_{CHRL}	−20	0	−20	0	−20	0
Delay Time, $\overline{RAS}$ Low to $\overline{W}$ Low (RMW) Cycle)	t_{RLWL}	120	110	160	130	200	190
Delay Time, $\overline{W}$ Low to $\overline{CAS}$ Low (Early W)	t_{WLCL}	−20	−5	−20	−5	−20	−5
Delay Time, $\overline{CAS}$ Low to $\overline{W}$ Low (RMW Cycle)	$t_{(CLWL)}$	70	60	95	65	125	105
Refresh Period	t_{rf}	2 ms	4 ms	2 ms	4 ms	2 ms	4 ms
Data Hold Time After $\overline{CAS}$ Low	$t_{h(CLD)}$	45	60	55	80	75	110
Data Hold Time After $\overline{RAS}$ Low	$t_{h(RLD)}$	95	145	120	180	160	210
Write Hold Time After $\overline{CAS}$ Low	$t_{h(CLW)}$	45	60	55	80	75	110
Write Hold Time After $\overline{RAS}$ Low	$t_{h(RLW)}$	95	110	120	145	160	195
Column Address Hold Time After $\overline{RAS}$ Low	$t_{h(RLCA)}$	95	95	120	140	160	190

All times are in nanoseconds unless otherwise noted.

Here's what these specification differences mean when upgrading to the TMS 4164.

Memory cycle times are reduced in all modes: page mode cycle time as low as 160 ns, read or write cycle time to 280 ns, and read-modify-write cycle time as low as 280 ns.

The TMS 4164 has a 2 cycle, 4 ms refresh which results in the same refresh rate as the 4116's 128 cycle; 2 ms refresh. With the same refresh rate, the faster cycle time of the 64K is the key to lower refresh overhead. The result is refresh overhead reduced from 2.4% (16K) to 1.8% (64K).

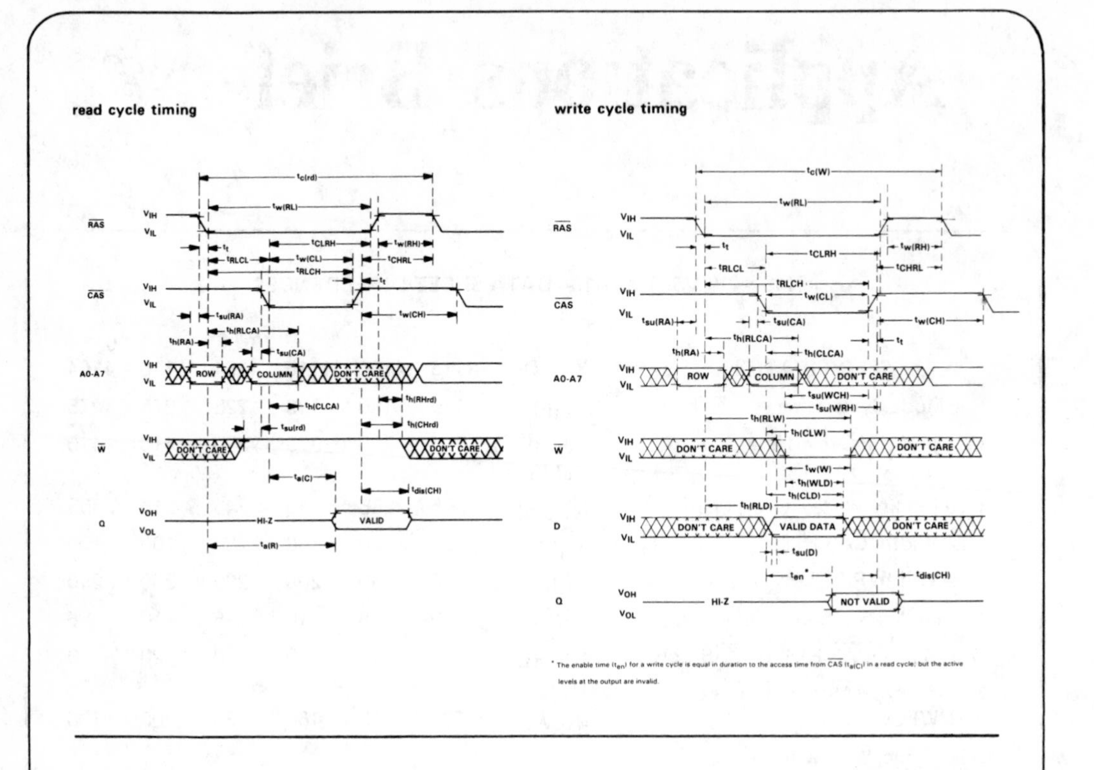

Column address setup time is longer on all speed range parts and column address hold time is increased on −20 and −25 parts. This means column address must be presented sooner and held longer when using the TMS 4164.

Data In, Write Command, and RAS low have longer hold times with the 4164. The removal of these signals is late in the cycle and usually a "cleanup" operation. For this reason, the longer hold times do not affect system performance as the access time from $\overline{\text{RAS}}$ remains the same as the 4116.

Other differences include a shorter delay time from $\overline{\text{RAS}}$ and $\overline{\text{CAS}}$ to $\overline{\text{W}}$ low on the read-modify-write cycle and a shorter delay time from $\overline{\text{W}}$ low to $\overline{\text{CAS}}$ low on an early write cycle. These differences are expected with the shorter cycle times of the 4164.

Overall the TMS 4164 specifications show significant improvements over the TMS 4116. These improvements mean upgrading to the 4164 not only yields a denser memory layout but also significant speed advantage.

MOS Memory
Applications Engineering

Applications Brief

AUGUST 1981

TTL ADDRESS DRIVERS AND LINE TERMINATION IN MOS MEMORY ARRAYS

State-of-the-art MOS memory logic levels are compatible with the voltage levels of TTL logic families. MOS inputs require very little current, so that a standard TTL output is capable of driving up to 32 memory devices on a bus line. However, the extremely high speeds of the TTL devices (from 2 to 3 nanoseconds transition time) can induce ringing due to transmission-line effects even on printed circuit lines only 7 inches long.

A printed circuit trace 0.015 inches wide on 0.062-inch double-sided board can be represented by a transmission line with distributed capacitance (C_D) and inductance (L_D) of 15 picofarads and 0.2 microhenries per foot. From this, we can calculate the characteristic impedance of the transmission line.

$$Zo = \sqrt{\frac{L}{C}} = \sqrt{\frac{0.20 \times 10^{-6}}{15 \times 10^{-12}}}$$

$$= 115\ \Omega$$

Adding an MOS memory input (C_I = 5 picofarads) every half inch, as shown in Figure 1 below, will increase the distributed capacitance effectively to 135 picofarads per foot.

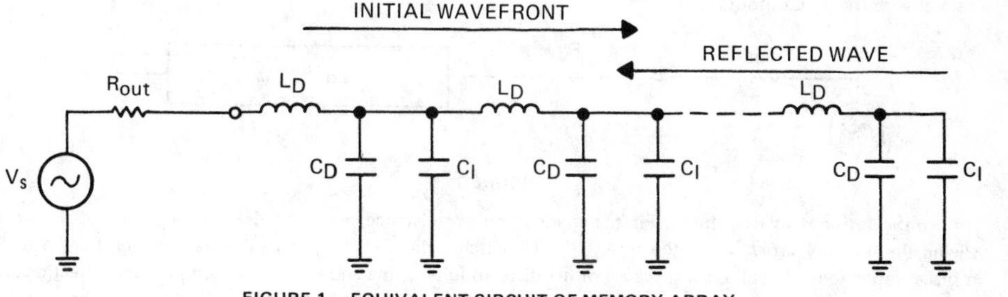

FIGURE 1 – EQUIVALENT CIRCUIT OF MEMORY ARRAY

Characteristic impedance of the transmission line can then be calculated in the following manner:

$$Zo = \sqrt{\frac{0.20 \times 10^{-6}}{135 \times 10^{-12}}} = 39\ \Omega$$

The equivalent impedance of the line with 1 memory input every half inch is only 39 Ω.

8

Whenever a discontinuity occurs in a transmission line, a portion of the signal traveling down the line will be reflected. These reflections add to the original waveform and cause distortion of the rising and falling edges in the form of plateaus, undershoot, and ripple.

One method that could be used to reduce this distortion is parallel termination. This involves tying a resistor between a point at the end of the array and another voltage source as shown in Figure 2.

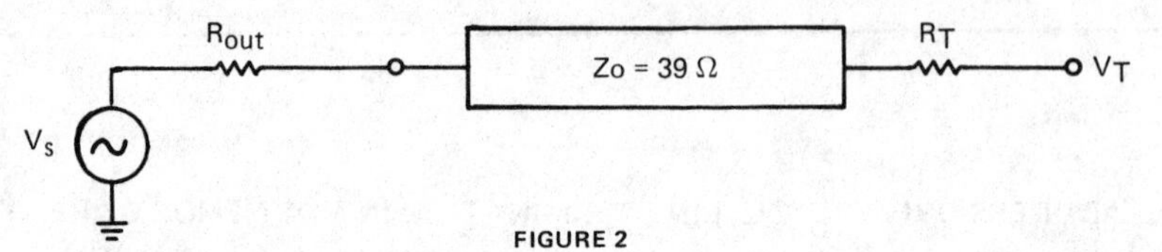

FIGURE 2

Typically, +5 V would be used for V_T because it is always available in systems using TTL devices. The problem with this type of termination is the amount of current required to pull the line low. For $R_T = 39\ \Omega$, the current calculations for a low level would be:

$$I_{OL} = \frac{5\ V - 0.4\ V}{39\ \Omega} = 118\ mA$$

This obviously eliminates this configuration for such a low value of terminating resistor. Even if $V_T = 2\ V$, the current values are still not acceptable for TTL drivers.

$$I_{OL} = \frac{2\ V - 0.4\ V}{39\ \Omega} = 41\ mA$$

$$I_{OH} = \frac{2\ V - 2.4\ V}{39\ \Omega} = 10\ mA$$

The alternative in this case is a series terminator at the source, as shown in Figure 3. This places a resistor directly in series with the TTL output.

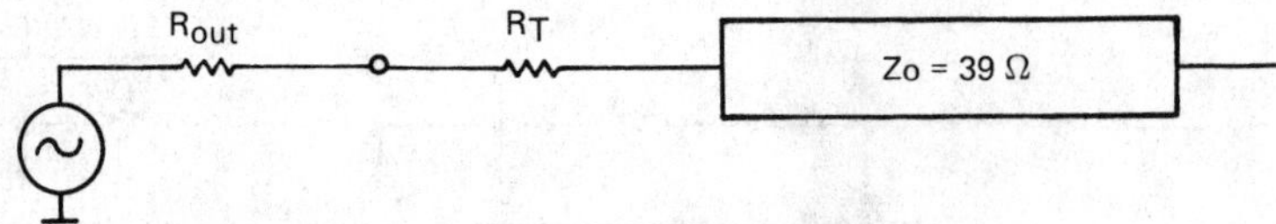

FIGURE 3

This configuration does two things. First, the added series resistance slows the rise and fall times of the signals driving the memory array. Also, the series resistor matches the TTL output to the transmission line. With R_T in place, reflections will still occur because of the discontinuity at the end of the line, but primary reflections will be dissipated by the terminating resistor so secondary reflections will not occur. The effect of the primary reflections can be minimized by the proper selection of R_T. The resistor in series does not add to the dc current requirement to drive the line.

A perfect match is not possible, however, because of the different output impedances of the driver at high and low levels. R_{out} is approximately 10 Ω in the low state and approximately 80 Ω in the high state. Considering the different output impedances and the increased rise and fall times caused by the added resistance, the best value of R_T is chosen by trial and error.

Finally, consideration must be given to the routing of the lines through the array. If the driver is driving more than one row so that several parallel branches are formed, care should be taken to keep each of the paths to the same length. The arrangement shown in Figure 4 is one way to achieve this in a multiple row layout.

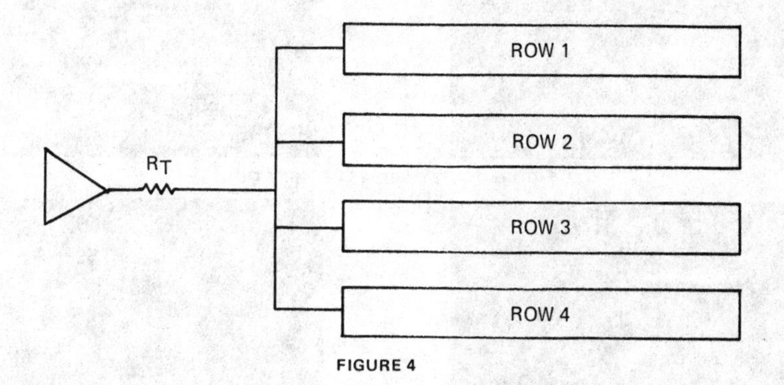

FIGURE 4

As an example, a memory system with 32 TMS4116's arranged as 4 rows by 8 devices was driven with a SN74S240 driver and a series terminator. The responses for no termination; 15 Ω, 22 Ω, 33 Ω, 47 Ω, and 68 Ω are shown in Figures 5 through 10. The high-to-low transitions showed undershoot and ringing that decreased as the terminator increased. The best high-to-low transition; (considering undershoot, ringing, and edge time) was achieved at 47 Ω. The low-to-high response looked good even with no termination. This can be explained best by the fact that the TTL output impedance in the high level is 80 Ω. As the termination increased, the low-to-high transition time increased.

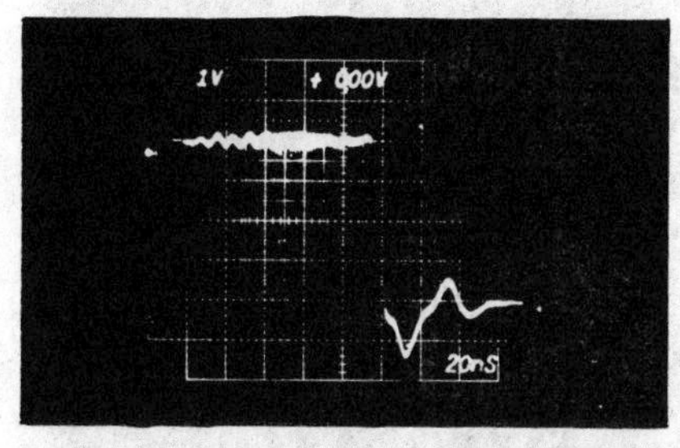

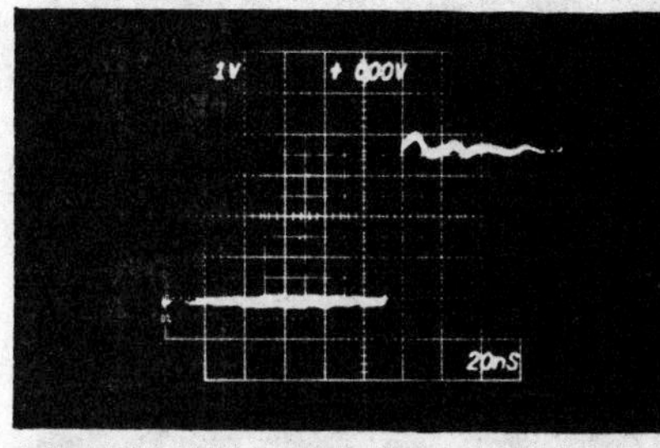

FIGURE 5 – NO TERMINATION

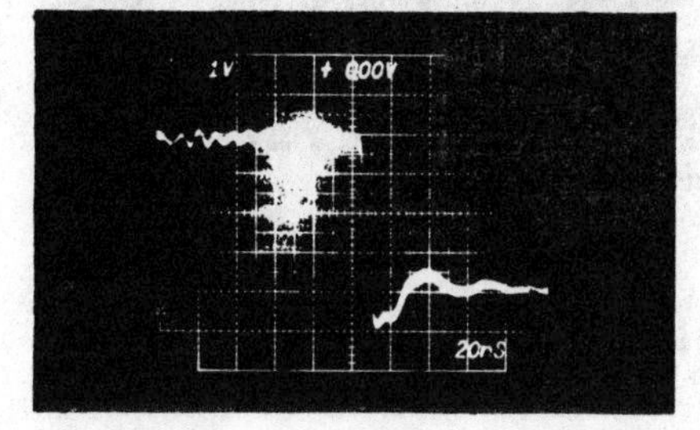

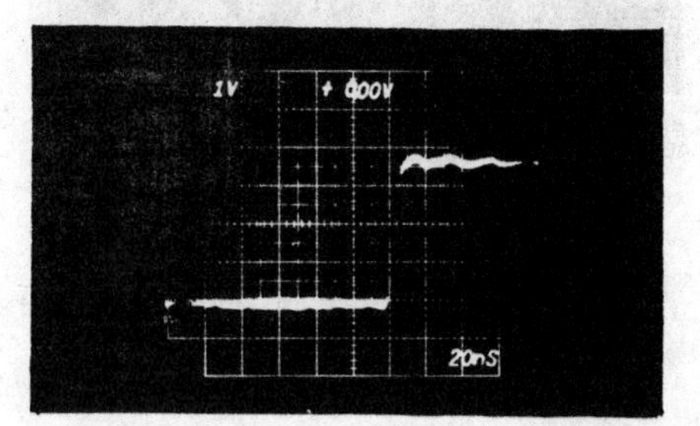

FIGURE 6 – 15-Ω SERIES TERMINATION

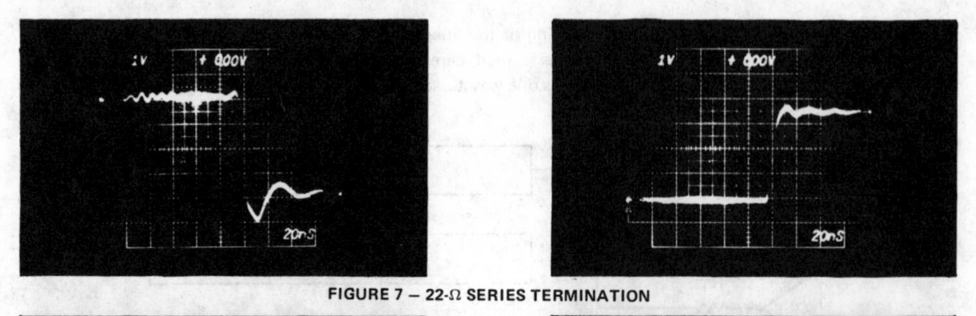

FIGURE 7 – 22-Ω SERIES TERMINATION

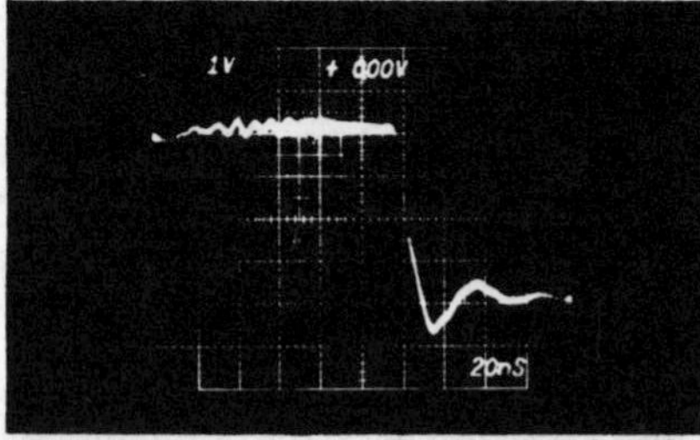

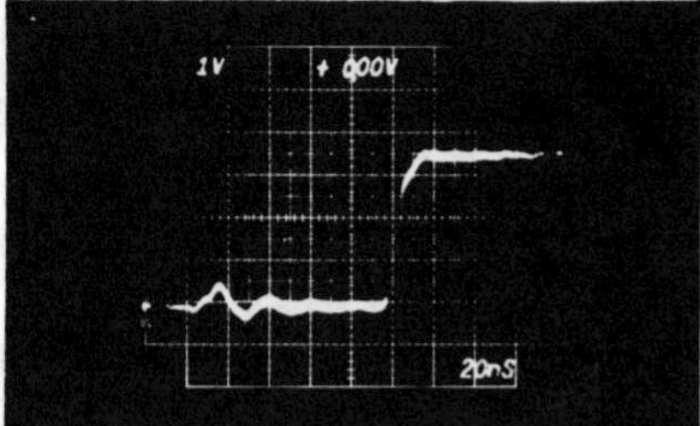

FIGURE 8 – 33-Ω SERIES TERMINATION

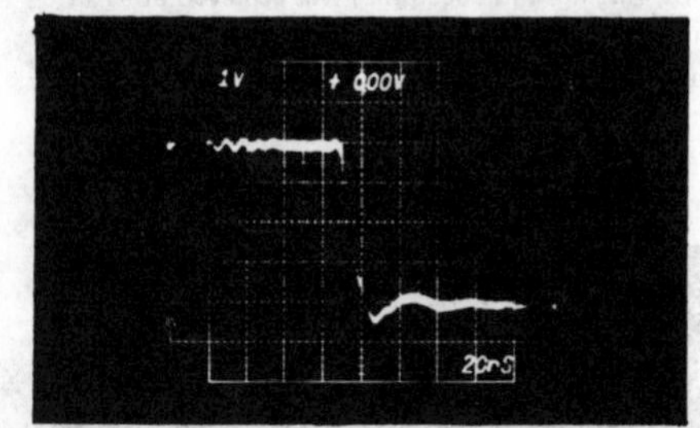

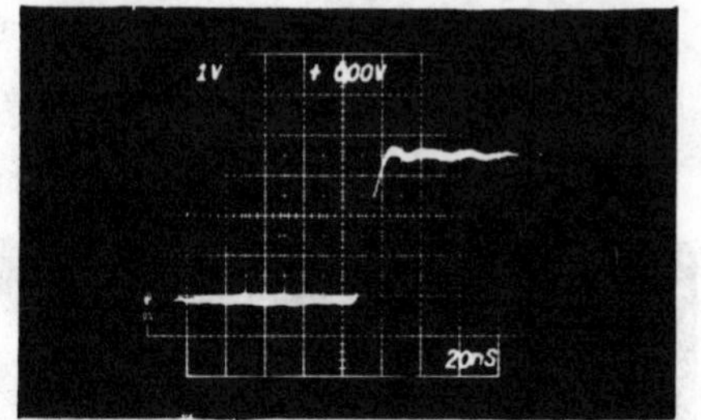

FIGURE 9 – 47-Ω SERIES TERMINATION

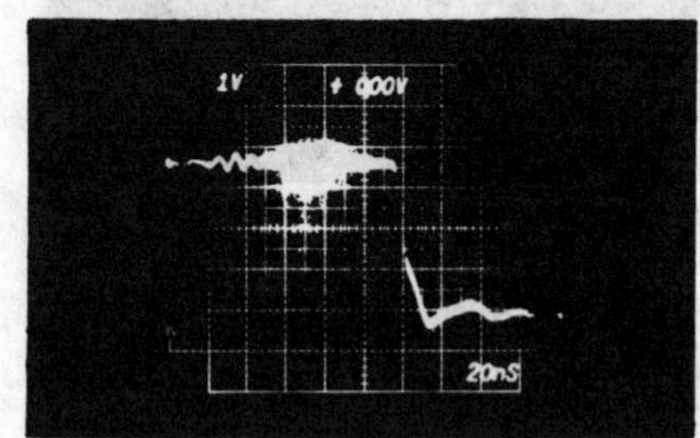

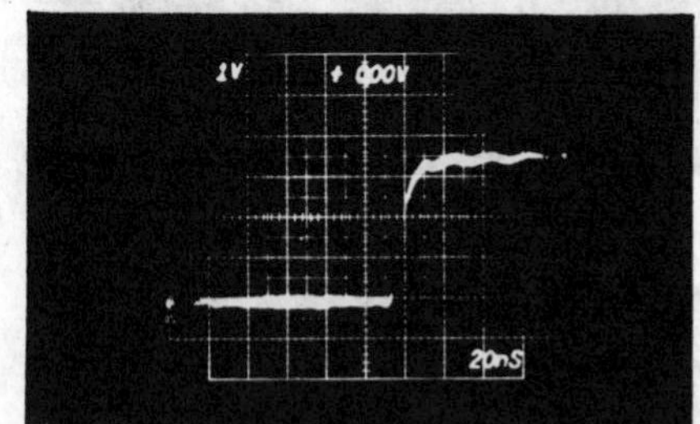

FIGURE 10 – 68-Ω SERIES TERMINATION

The photographs in this article show how the series terminating resistor affects both the rising and falling edges in a typical system. Since every layout will vary because of type of memory used, number of devices in the array, and actual memory layout, the method of trial and error should be followed for each memory design.

MOS Memory
Application Engineering

Applications Brief

TMS 4164 SYSTEM POWER REQUIREMENTS

The TMS 4164 dissipates considerably less power than the TMS 4116. On a per-bit basis, the average operating power of the 64K is less than one-eighth of the 16K. To take advantage of this low power dissipation, it is necessary to calculate the maximum average current a memory system will consume. Worst-case power supply requirements can be determined through the use of the following equations:

$$I_{DD} = N\,[R \times I_{DD3} + (1\text{-}R)\,(A \times I_{DD1} + (1\text{-}A)\,I_{DD2})]$$

where I_{DD} = maximum average current of system
N = number of devices in system
R = refresh overhead
A = relative time memory is active
I_{DD3} = average refresh current (Table 1 or Figure 1)
I_{DD2} = average standby current (Table 1)
I_{DD1} = average active current (Table 1 or Figure 1)

Parameter A is calculated by dividing the word size by the number of devices in the system. This assumes only one word of memory can be accessed at one time.

$$A = \frac{\#\ \text{devices/word}}{\#\ \text{devices/system}}$$

Parameter R is the ratio of the time required to refresh the memory to the time the memory is available to be accessed. This can be calculated by multiplying the cycle time and refresh rate.

$$R = \text{cycle time} \times \text{refresh rate}$$

As an example, examine a system with 64 4164's organized as 512 kilobytes of memory. Assume cycle time of 350 ns and minimum refresh rate (64 kHz).

$$N = 64$$

$$A = \frac{8}{64} = .125$$

$$R = 350 \times 10^{-9} \times 64 \times 10^{3} = .022$$

$$I_{DD} = 37 \times 10^{-3}\ A$$

$$I_{DD2} = 5 \times 10^{-3}\ A$$

$$I_{DD3} = 32 \times 10^{-3}\ A$$

8

TABLE 1 – I_{DD} ELECTRICAL CHARACTERISTICS FOR TMS 4164–20

PARAMETER		TEST CONDITIONS	MIN	TYP	MAX	UNIT
I_{DD1}	Average operating current during read or write cycle	t_c = minimum cycle		24	34	mA
I_{DD2}	Standby current	After 1 memory cycle, $\overline{RAS}$ and $\overline{CAS}$ high		3.5	5	mA
I_{DD3}	Average refresh current	t_c = minimum cycle $\overline{RAS}$ low, $\overline{CAS}$ high		19	26	mA

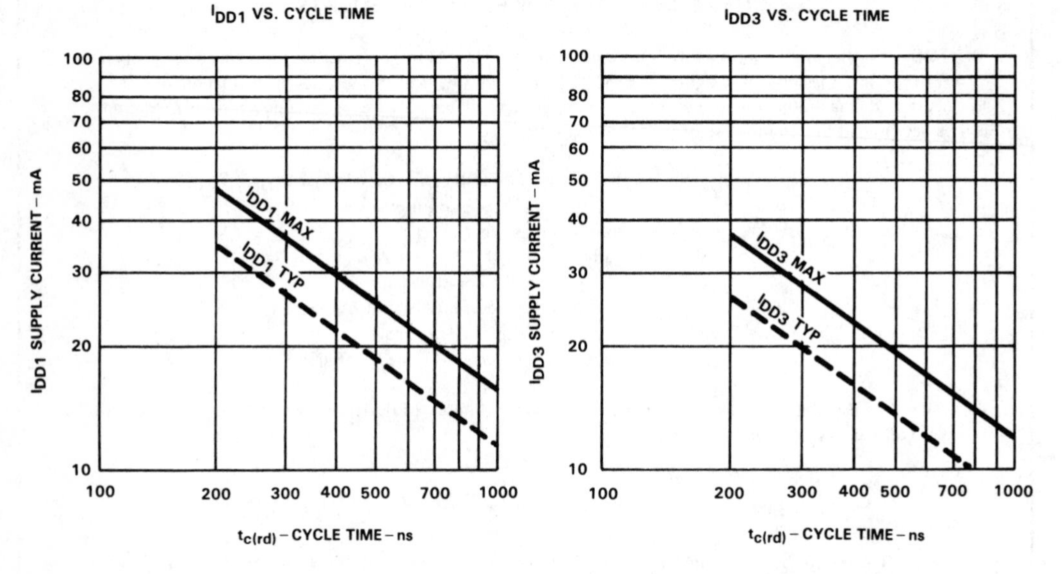

FIGURE 1 – I_{DD} vs CYCLE TIME

Substituting into original equation

$$
\begin{aligned}
I_{DD} &= 64\ [.022\ (26 \times 10^{-3}) + (1 - .022)\ [(.125)\ (34 \times 10^{-3}) + (1 - .125)\ (5 \times 10^{-3})]] \\
&= 64\ [5.72 \times 10^{-4} + .978\ [4.25 \times 10^{-3} + 4.375 \times 10^{-3}]] \\
&= 64\ [5.72 \times 10^{-4} + 8.435 \times 10^{-3}] \\
&= 5.77 \times 10^{-1} A = 577 \text{ mA maximum average current}
\end{aligned}
$$

To determine the maximum average current with the system on standby, let A = 0

$$
\begin{aligned}
I_{DD} &= 64\ [.022\ (26 \times 10^{-3}) + (1 - .022)\ [0 + 5 \times 10^{-3}]] \\
&= 64\ [5.72 \times 10^{-4} + 4.89 \times 10^{-3}] \\
&= 3.50 \times 10^{-1} = 350 \text{ mA maximum average current standby mode.}
\end{aligned}
$$

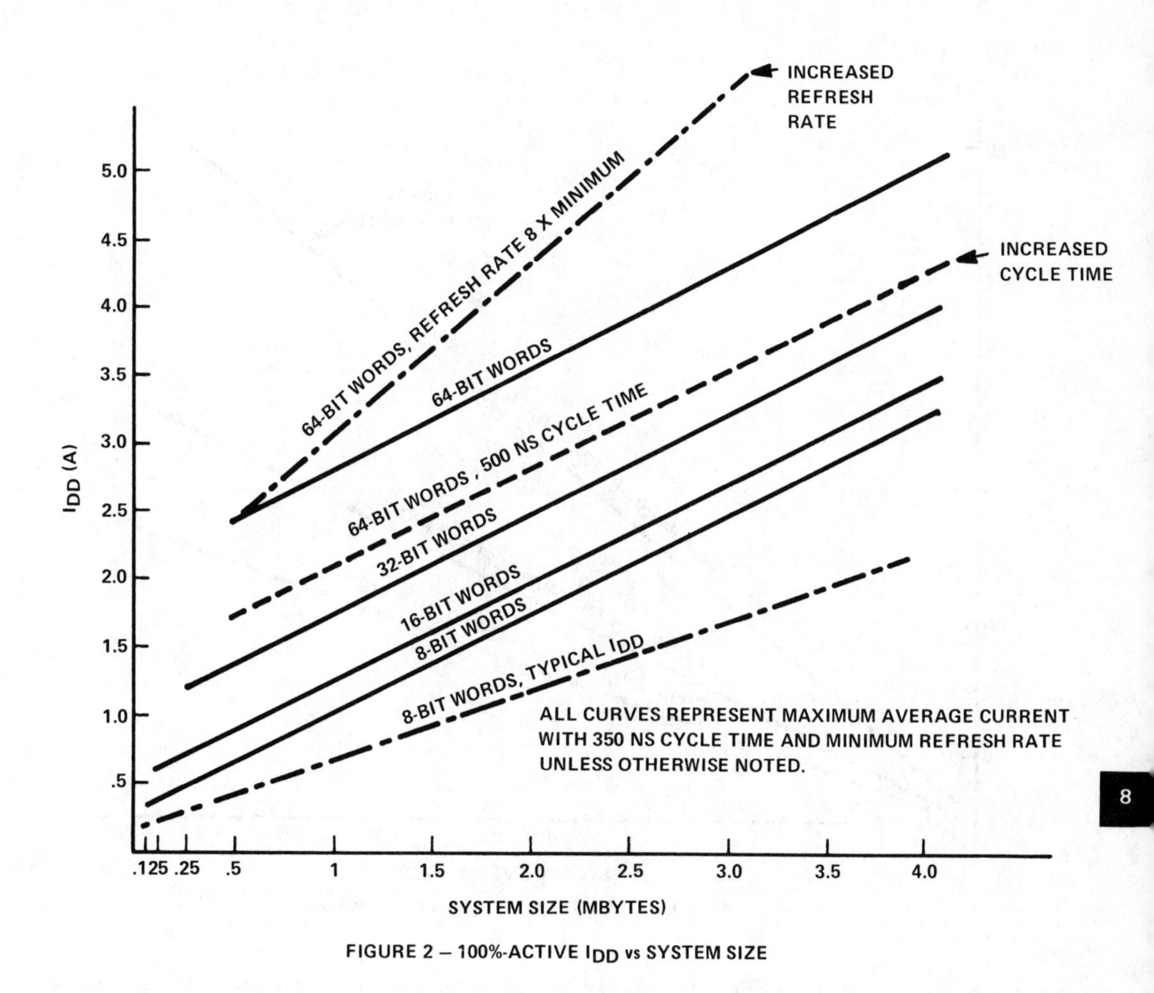

FIGURE 2 – 100%-ACTIVE I_{DD} vs SYSTEM SIZE

The graph in Figure 2 was plotted using the equation for maximum average current. This graph shows the maximum amount of current a memory system could draw with the memory organized in different word sizes. The graph also shows the effect of increasing cycle and refresh time on system power requirements. Finally, the graph shows the difference between maximum average current and typical average current.

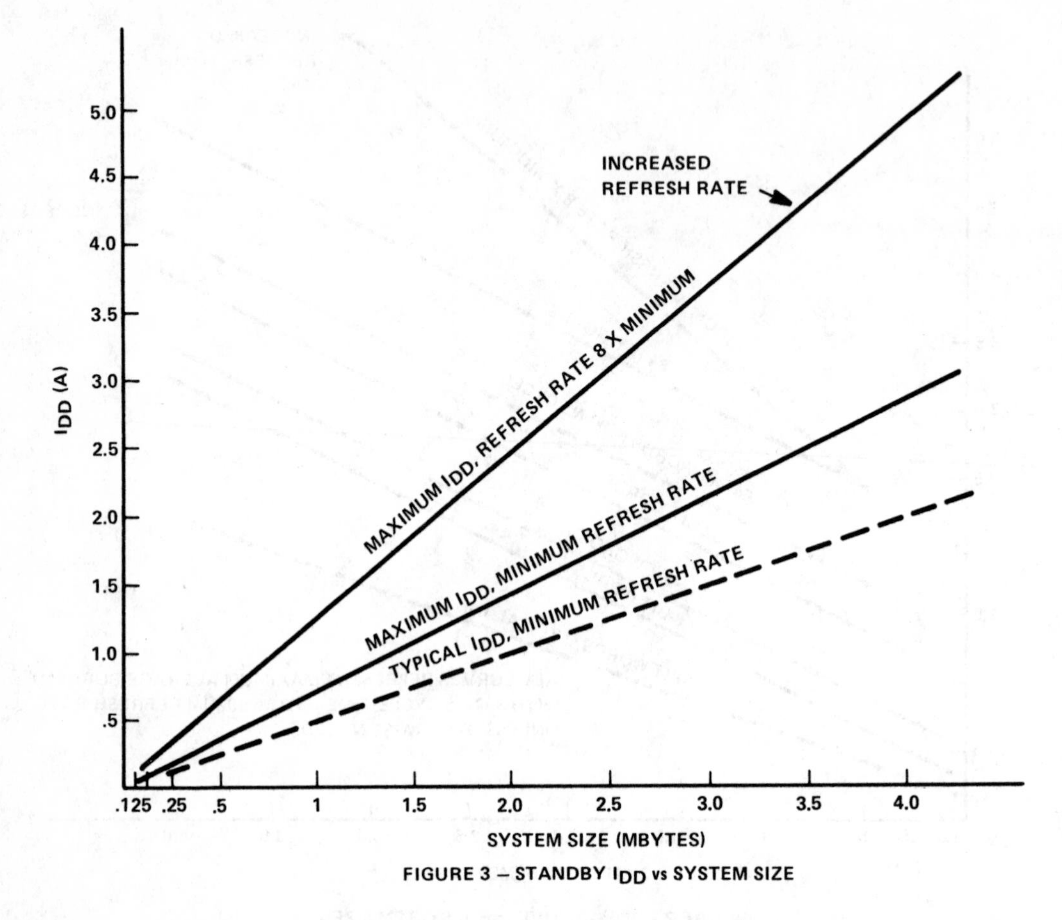

FIGURE 3 – STANDBY I_{DD} vs SYSTEM SIZE

Figure 3 shows the current necessary to operate various system sizes in the standby mode. The current required at refresh rate 8X higher is also plotted to show the effect of refresh rate on system power requirements. The dotted line shows the calculations of typical average current using minimum refresh time.

These calculations make it possible to properly match power supply ratings to memory system requirements. They are also necessary in designing a battery-backup system for which worst-case standby current requirements would be needed. These calculations emphasize the low power requirements of the TMS 4164 in a system environment.

MOS Memory
Applications Engineering

Applications Brief

APRIL 1981

TMS 4164 PRINTED CIRCUIT BOARD LAYOUT

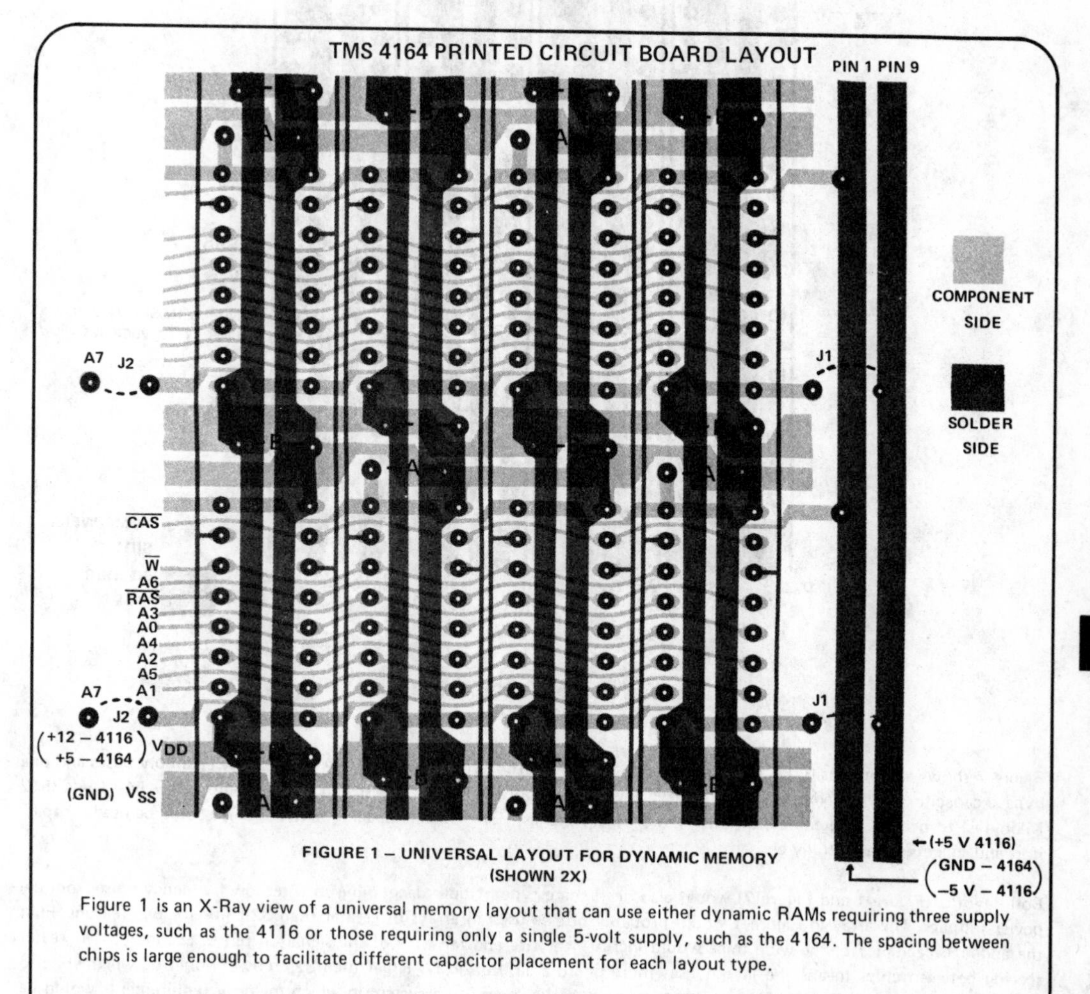

FIGURE 1 – UNIVERSAL LAYOUT FOR DYNAMIC MEMORY (SHOWN 2X)

Figure 1 is an X-Ray view of a universal memory layout that can use either dynamic RAMs requiring three supply voltages, such as the 4116 or those requiring only a single 5-volt supply, such as the 4164. The spacing between chips is large enough to facilitate different capacitor placement for each layout type.

To use this layout for 4116's, jumpers would be placed at J1 to connect pin 9 to +5 volts. The bus that is connected to pin 1 should be at –5 volts for 4116's. Capacitors would be placed between –5 volts and ground in the spaces provided on every other chip (A). Remaining chips would have capacitors placed between the +12-volt line and ground (B). Bypass capacitors for the +5-volt line are not as critical so they can be placed around the periphery of the array.

A different jumper setting and capacitor placement must be used for a 4164 array. Jumpers J2 would connect pin 9 to A7 and J1 would be left open. Pin 1 no longer needs to be –5 volts since pin 1 on the 4164 is not connected internally. This line could be grounded to reduce crosstalk between $\overline{CAS}$ and the power supply line. Capacitors can now be placed between +5 volts and ground (B) on every chip to yield a quiet memory layout.

8

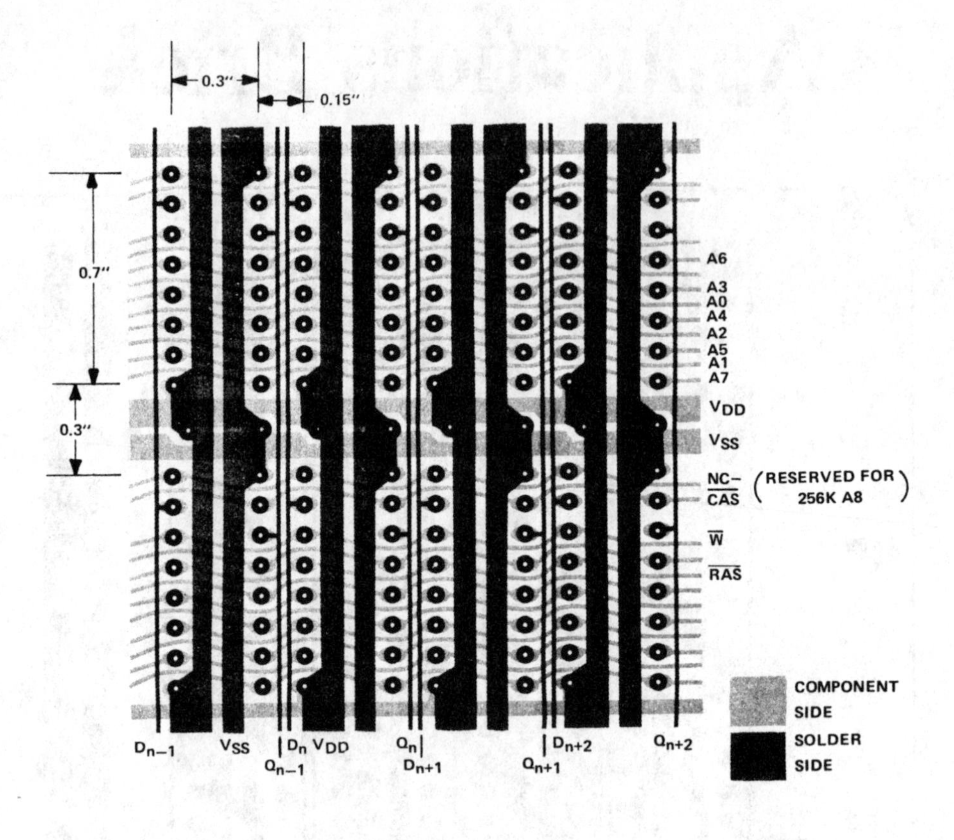

FIGURE 2 – PC BOARD LAYOUT FOR TMS 4164
(SHOWN 2X)

Figure 2 shows a high density PC board layout for the TMS 4164. This layout features gridded power supply buses and one bypass capacitor per dynamic RAM to achieve low-noise supply voltages. The spacing used would allow 64-4164's (512 Kilobytes) to occupy an area of 3.6" x 8" (28.8 square inches). Even smaller spacing between chips could be used if capacitors and sockets were carefully chosen.

Both layouts (Figure 1 and Figure 2) would also need some type of bulk decoupling to filter low frequency noise from the power supplies. The array in Figure 1 would probably not be used in arrays of greater than 32 chips for two reasons. First the added capacitance of the wide trace would increase the rise and fall time of the signal on pin 9. Secondly, the greater spacing between chips means the layout uses more board area. For arrays larger than 32 chips it would be worthwhile to design a separate board for 4116's and 4164's. For a smaller board in a system in which memory requirements would be expected to increase, the universal layout could save the step of redesigning the memory board.

MOS Memory
Applications Engineering

Applications Brief

TMS 4164
INTERNAL TOPOLOGY

For complete testing and characterization of the TMS 4164 with respect to cell pattern sensitivity, it is necessary to know its true address bit significance. The sixteen address bits required to access the 65,536 cell locations are multiplexed onto eight inputs as eight row (entered by falling edge of $\overline{RAS}$) and eight column (entered by falling edge of $\overline{CAS}$) addresses.

The pinout of the TMS 4164 (Figure 1A) shows these address lines as A0 – A7. The pinout uses this particular address arrangement to maintain compatibility to earlier (4K and 16K) Dynamic RAM memories, although the TMS 4164 uses a different binary weighting on these lines internally. In a system there is no particular advantage to this order of addressing since the device requires no special sequencing to read or write a given memory location.

The bit map of the TMS 4164 array can be obtained using the true address bit significance as shown below for both row and column addressing.

DESIRED ROW OR COLUMN ADDRESS		WEIGHT	TMS 4164 PIN NAME	PIN #
(MSB)	A7	2^7	A7	9
	A6	2^6	A0	5
	A5	2^5	A2	6
	A4	2^4	A1	7
	A3	2^3	A5	10
	A2	2^2	A4	11
	A1	2^1	A3	12
(LSB)	A0	2^0	A6	13

8

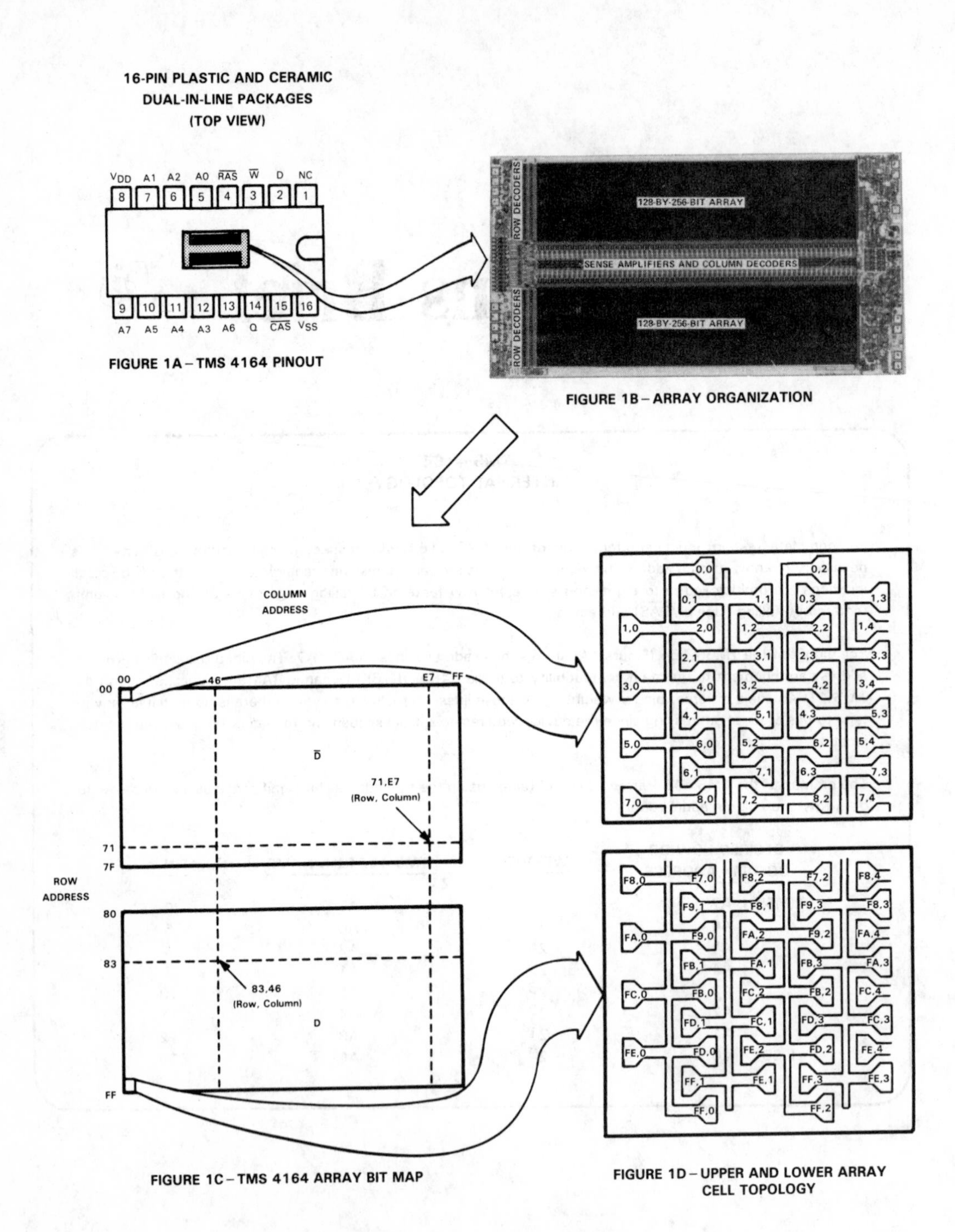

FIGURE 1A – TMS 4164 PINOUT

FIGURE 1B – ARRAY ORGANIZATION

FIGURE 1C – TMS 4164 ARRAY BIT MAP

FIGURE 1D – UPPER AND LOWER ARRAY CELL TOPOLOGY

Figure 1A shows the chip pinout, Figure 1B is a closeup of the array, Figure 1C shows the bit map for the rows and columns, and Figure 1D is a closeup of the cell topology in the array.

Internally the cells are arranged so as to maximize the cell size within the available area. This layout is shown in Figure 1D. The neighboring cells surrounding any particular cell are considered here for their degree of influence on that cell. Each cell has two nearest neighbor cells located in an adjacent column. These have a greater degree of influence upon the cell than do the near neighbors. The near and nearest neighbors for a specific cell can be obtained using the algorithm given below and Figure 1D.

Let (R, C) represent any cell location where R = row address and C = column address.

If row and column addresses are either both even or both odd:

	Row Address $\leq 7F_H$	Row Address $\geq 80_H$
Nearest Neighbors	R−2 C+1	R−2 C−1
	R+0 C+1	R+0 C−1
Near Neighbors	R−2 C+0	R−2 C+0
	R+2 C+0	R+2 C+0
	R−1 C+2	R−1 C−2

If row and column addresses are neither both even nor both odd:

	Row Address $\leq 7F_H$	Row Address $\geq 80_H$
Nearest Neighbors	R+0 C−1	R+0 C+1
	R+2 C−1	R+2 C+1
Near Neighbors	R−2 C+0	R−2 C+0
	R+2 C+0	R+2 C+0
	R+1 C−2	R+1 C+2

Note that the algorithm changes for each half of the array due to the fact that the top half is laid out as the mirror image of the bottom half. Data in the top half of the array (as shown in Figure 1C) is stored in inverted form (absence of charge = 1), while data in the lower half is stored in true form (charge = 1). Therefore, row address bit seven is the bit which selects between true and inverted array. This may be transformed using the circuit shown in Figure 2 to compensate for this internal data inversion.

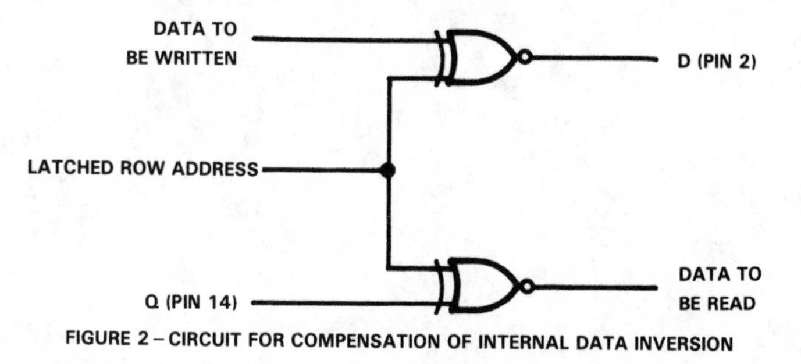

FIGURE 2 – CIRCUIT FOR COMPENSATION OF INTERNAL DATA INVERSION

MOS Memory
Applications Engineering

Applications Brief 

APRIL 1982

TMS 4416 INTERNAL TOPOLOGY

The TMS 4416 16K × 4 DRAM internal topology very closely resembles that of the TMS 4164 64K × 1 DRAM. Within the TMS 4164, the six highest order latched internal column addresses select one of 64 sense amplifier banks which activate four adjacent cells within the selected row. The data to or from these cells is carried along four I/O lines to a 4-line to 1-line multiplexer. The two lowest order latched internal column address bits select which of the four I/O lines is to be activated. The TMS 4416 differs from the TMS 4164 in that the TMS 4416 has no multiplexer circuitry on the data I/O lines. Instead all four lines are buffered and brought out to external pins. The fact that data is presented in 4-bit wide words must be taken into consideration when developing cell pattern sensitivity test algorithms. Presented here are the true binary weighting of the address lines, a bit map of the array showing cell topology, an algorithm for finding "near" and "nearest" neighbor cells, and circuit for compensating for internal data inversion.

Table 1 shows the true address bit significance for the TMS 4416. This information can be used in conjunction with Figure 1c to write various data patterns to the array.

TABLE 1 – TMS 4416 ADDRESS BIT SIGNIFICANCE

ROW		PACKAGE		COLUMN	
INTERNAL ADDRESS	BINARY WEIGHT	PIN NAME	PIN NUMBER	BINARY WEIGHT	INTERNAL ADDRESS
RA7	2^7	A7	10		
RA6	2^6	A6	6	2^5	CA5
RA5	2^5	A5	7	2^4	CA4
RA4	2^4	A4	8	2^3	CA3
RA3	2^3	A3	11	2^2	CA2
RA2	2^2	A2	12	2^1	CA1
RA1	2^1	A1	13	2^0	CA0
RA0	2^0	A0	14		

Figure 1 depicts step-by-step magnification of the TMS 4416 from a veiw of the entire package to a closeup of the array topology. The cells are arranged so as to maximize the cell size within the available area. A portion of the cell layout is shown in Figure 1d, with the address of each cell labeled as (R, CD) where R is the internal row address, and CD is the internal column/databit address.* Cells that surround any one given cell are called neighboring cells or neighbors, and are considered here for their degree of influence.

* Note that the column/databit addresses are not the same as the column addresses but rather increment four times faster. To convert from column/databit to column address simply divide the column/databit address by four and then add one to the remainder. The resulting quotient is the column address and the remainder plus one is the databit.

8

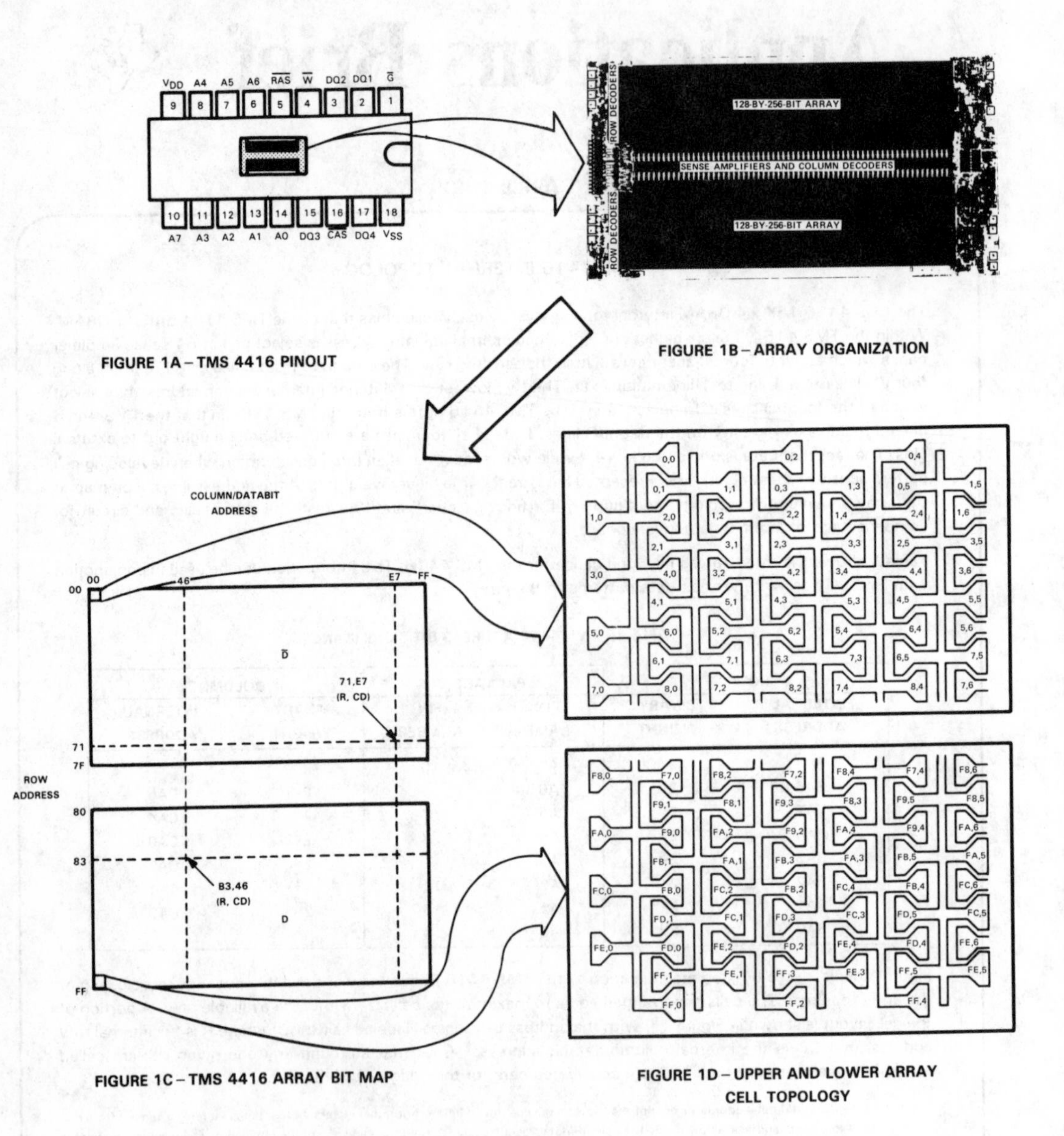

FIGURE 1A – TMS 4416 PINOUT

FIGURE 1B – ARRAY ORGANIZATION

FIGURE 1C – TMS 4416 ARRAY BIT MAP

FIGURE 1D – UPPER AND LOWER ARRAY CELL TOPOLOGY

Figure 1A shows the chip pinout, Figure 1B is a closeup of the array, Figure 1C shows the bit map for the rows and columns, and Figure 1D is a closeup of the cell topology in the array.

There are two types of neighbors – near and nearest. Nearest cells are adjacent to a given cell and are not separated by any silicon processing from that cell. For this reason, nearest neighbors have the greatest influence. Near neighbors are adjacent to but separated by the bit line diffusion from a given cell. Near neighbors have a lesser degree of influence on a given cell than do nearest neighbors.

The algorithm for finding near and nearest neighbors is given below:

Let (R, CD) represent any cell location where R = row address and CD = column/data bit address.

If row and column addresses are either both even or both odd:

	Row Address $\leq 7F_H$	Row Address $\geq 80_H$
Nearest Neighbors		
	R−2 CD+1	R−2 CD−1
	R+0 CD+1	R+0 CD−1
Near Neighbors		
	R−2 CD+0	R−2 CD+0
	R+2 CD+0	R+2 CD+0
	R−1 CD+2	R−1 CD−2

If row and column addresses are neither both even nor both odd:

	Row Address $\leq 7F_H$	Row Address $\geq 80_H$
Nearest Neighbors		
	R+0 CD−1	R+0 CD+1
	R+2 CD−1	R+2 CD+1
Near Neighbors		
	R−2 CD+0	R−2 CD+0
	R+2 CD+0	R+2 CD+0
	R+1 CD−2	R+1 CD+2

Note that the algorithm changes for each half of the array due to the fact that the top half is laid out as the mirror image of the bottom half. Data in the top half of the array (as shown in Figure 1C) is stored in inverted form (absence of charge = 1), while data in the lower half is stored in true form (charge = 1). Therefore, row address bit seven is the bit which selects between true and inverted array. This may be transformed using the circuit shown in Figure 2 to compensate for this internal data inversion.

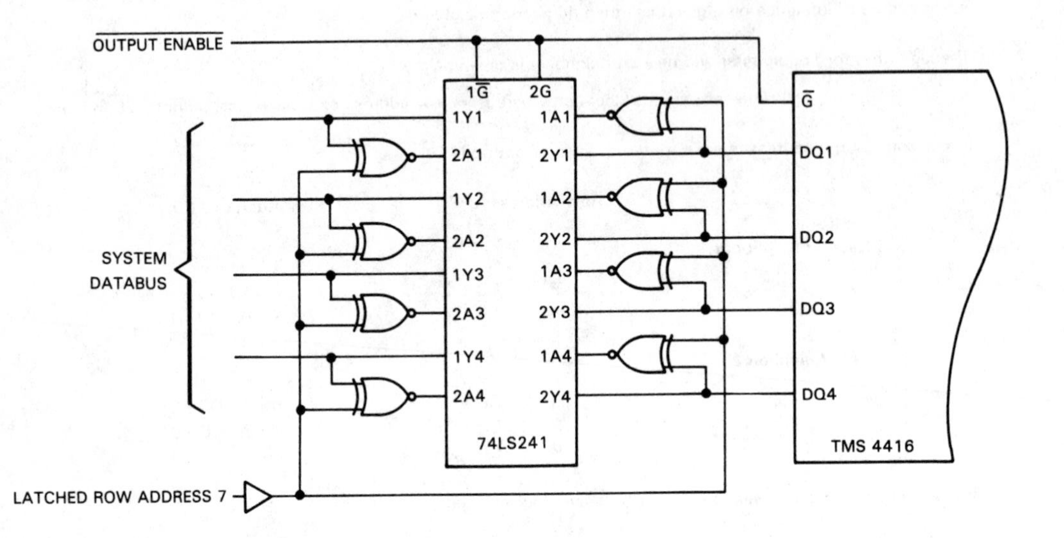

FIGURE 2 – CIRCUIT FOR COMPENSATION OF INTERNAL DATA INVERSION

When row address 7 is high, the true array is being accessed and data is passed without inversion. When row address 7 is low, the inverted array is being accessed and data is inverted as it is written to or read from the memory. In this way, true data is always presented. Also, the 74LS241 remains ready to write data to the TMS 4416 until $\overline{G}$ goes low. When this occurs, data is transferred from the TMS 4416 to the system databus for read operations.

MOS Memory
Applications Engineering

Applications Brief 

APRIL 1982

TMS 4416 16K × 4 DRAM DEVICE STRUCTURE

Upon initial inspection of the TMS 4416 pin configuration, a few departures from previous dynamic memory design formats are observed. In order to better illustrate this, the TMS 4416 pin configuration is shown below.

18-PIN PLASTIC DUAL-IN-LINE PACKAGE (TOP VIEW)

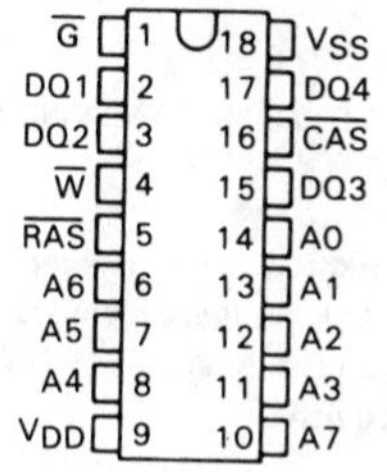

Signal	Pin	Pin	Signal
$\overline{G}$	1	18	V_{SS}
DQ1	2	17	DQ4
DQ2	3	16	$\overline{CAS}$
$\overline{W}$	4	15	DQ3
$\overline{RAS}$	5	14	A0
A6	6	13	A1
A5	7	12	A2
A4	8	11	A3
V_{DD}	9	10	A7

FIGURE 1 – TMS 4416 PIN CONFIGURATION

One feature that is noted is the inclusion of an output enable pin, $\overline{G}$. With common I/O, data must be latched in order to avoid bus conflicts unless an early write signal can be provided. Since most processors do not have this early write capability, the output enable pin has been included on the TMS 4416. This feature precludes the need for an output data latch and makes the late-write operation possible. In addition, $\overline{G}$ provides read-modify-write operation.

Another key point is the 8 × 6 (row × column)) addressing scheme. This scheme takes advantage of the reliable, time-proven architecture of TI's 64K × 1 dynamic RAM. Using half as many sense amplifiers provides lower power dissipation and reduced system cost with improved reliability. The 8 × 6 addressing scheme will also provide complete pin-for-pin upward compatibility for future intended DRAM generations. Since the pinout and refresh addressing schemes for the TMS 4416 can accommodate a 256K device, system upgrade capability is greatly simplified. The 8 × 6 format is also compatible to the 7 × 7 format as shown in the multiplex circuit below*.

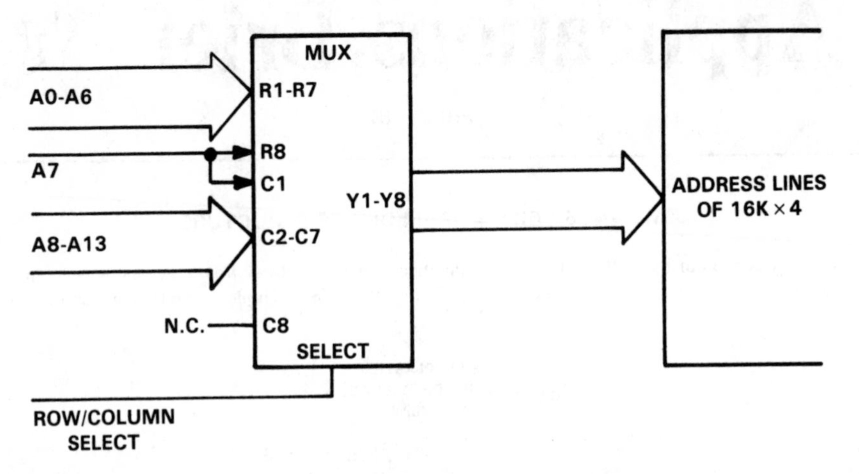

FIGURE 2 – 8 × 6 VS. 7 × 7 ADDRESSING COMPATIBILITY

The row and column address inputs are all connected to the multiplexer (MUX) in a straightforward manner except that A7 connects both to the most significant row input and the least significant column input, R8 and C1, respectively. At the multiplexer output, the most significant address line Y8 connects to pin 10 of the 16K × 4 device. This corresponds to A7 on the TMS 4416 and N.C. of the 7 × 7 addressed device.

As a final note, it should be mentioned that a 64K DRAM had previously been considered in an 8K × 8 version. This was rejected however, since this would increase the package size to 22 pins and result in an increase in board area of over 20%.

*For processors such as the Z80 which generate a 7-bit, 128 cycle refresh address see Applications Brief DR-7, "256-Cycle Refresh Conversion".

MOS Memory
Applications Engineering

Applications Brief

APRIL 1982

THE TMS 4500A IN AN ASYNCHRONOUS BUS SYSTEM

This application brief details the logic required to implement a TMS 4500A DRAM Controller within an asynchronous bus system. The particular bus to be considered has a bus protocol such that any memory must respond with an acknowledge ($\overline{ACK}$) when data is valid on the data bus. In addition to the 21 address lines ($\overline{A0}$-$\overline{A20}$) and the 16 data lines ($\overline{D0}$-$\overline{D15}$) the bus provides:

- A power-on initialization signal ($\overline{RESET}$)
- Memory read ($\overline{MEMRD}$)
- Memory write ($\overline{MEMWR}$)
- A bus clock at about 8 MHz ($\overline{BUSCLK}$).

A schematic diagram showing the TMS 4500A in an asynchronous bus system is given in Figure 1. The logic required to decode addresses in order to provide the board select and the acknowledge signals is shown in the figure. Address lines $\overline{A20}$ thru $\overline{A17}$ are buffered by a 74S240 buffer to minimize bus loading and then decoded to allow selection of any of four memory block positions by jumper J1 (each block containing up to 512K words). Using the 74S139 allows a minimum of 25 ns setup on address lines prior to the occurrence of $\overline{MEMRD}$ or $\overline{MEMWR}$ to guarantee no false triggers are placed on the $\overline{ACK}$ line. The 74LS164 8-bit shift register is held cleared until the board becomes active. The outputs are sequentially set high as $\overline{BUSCLK}$ shifts the input at A and B thru the register. Jumper J2 is placed dependent upon the frequency of $\overline{BUSCLK}$ and the access time of the memory. If this clock is not synchronous to the occurrence of $\overline{MEMRD}$ and $\overline{MEMWR}$ (as in the system under consideration), the $\overline{ACK}$ will have to be selected assuming the board becomes active just prior to shifting the 74LS164. As drawn the jumper assumes 3 to 4 clock periods are sufficient to guarantee data valid. The shift register also controls the generation of the ALE signal to be used the TMS 4500A. The clock input of the shift register is gated by the ready signals from the DRAM controllers so that if a refresh were to be in progress when the system attempted an access the $\overline{ACK}$ would be delayed until after completion of the access grant by the DRAM controller.

Two 74LS640's provide the buffering of the data bus and are controlled by the board select and read lines.

The remaining half of the 74S139 is used to select one of four TMS 4500A's each controlling 128K words of RAM. Up to four TMS 4500A's could be configured as shown in the figure. $\overline{SEL1}$ provides the chip select signal for one block of memory and RDY1 is used to gate the $\overline{ACK}$ signal onto the system bus ($\overline{SEL2}$, $\overline{SEL3}$, and $\overline{SEL4}$ provide chip select signals for memory blocks 2, 3, and 4; RDY2, RDY3, and RDY4 provide similar gating functions for blocks 2, 3, and 4). ALE, $\overline{ACR}$, $\overline{ACW}$, REN1, CLK, RA0-RA7, CA0-CA7, FS0, FS1, and TWST lines on any additional controllers would be connected in parallel with corresponding lines on the TMS 4500A. Although there are multiple connections, the load on those lines is minimal since the input current required by the TMS 4500A is less than 10 μA in either the low or high state.

Although a single board containing one megabyte of memory may be ambitious in most systems the same design can be adapted to a variety of asynchronous bus structures and memory requirements.

Mos Memory
Applications Engineering

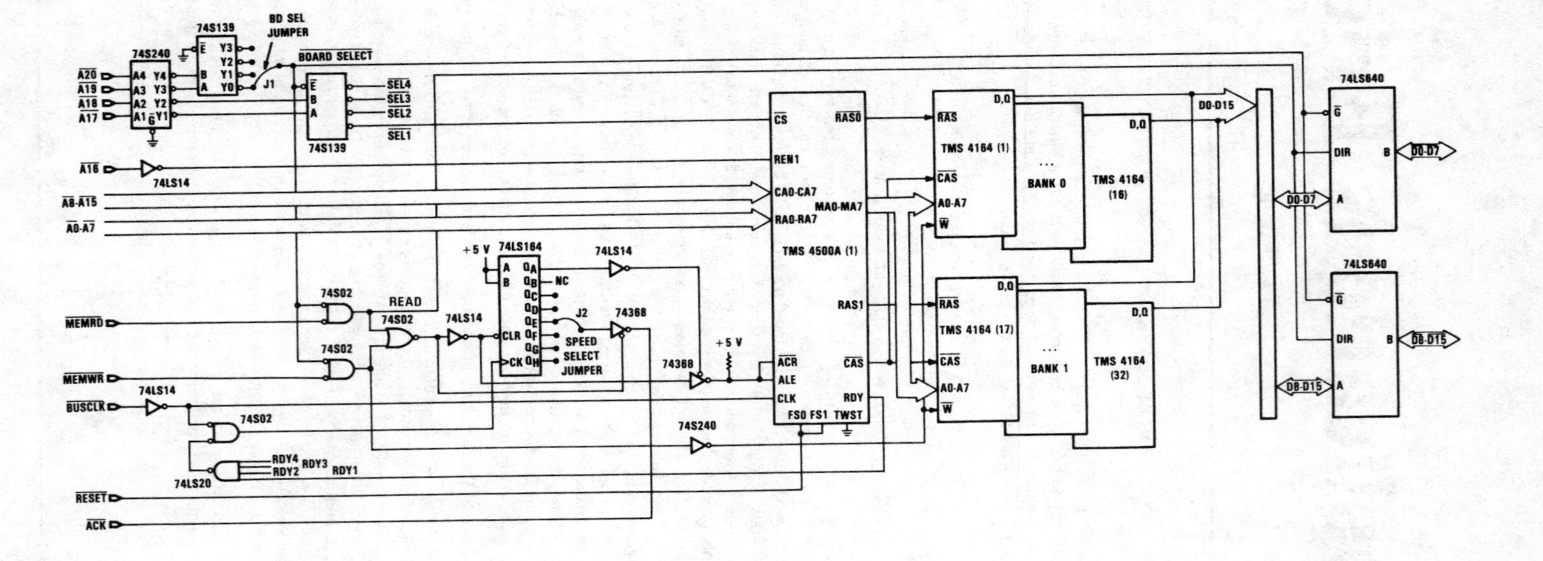

FIGURE 1 – TMS 4500A IN AN ASYNCHRONOUS BUS SYSTEM

Applications Brief

APRIL 1982

TMS 4500A CLOCK SYNCHRONIZATION

A simple clock synchronization circuit for use in asynchronous systems utilizing the TMS 4500A DRAM controller is described in this brief. Synchronization is accomplished by gating whatever signal is used to drive ALE with the clock signal. The first example assumes that an asynchronous clock with a frequency of less than 10 MHz is available (see Figure 1).

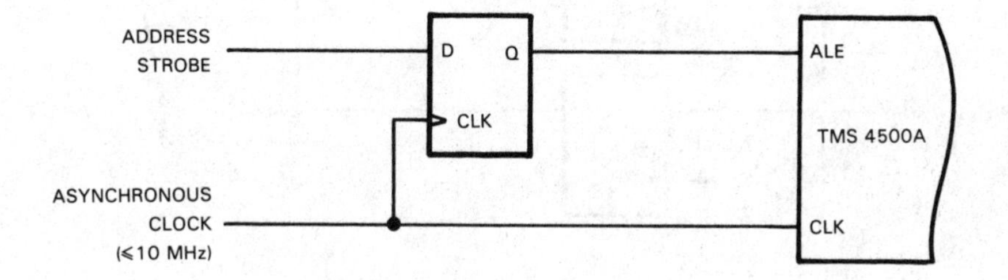

FIGURE 1 – TMS 4500A CLOCK SYNCHRONIZATION INTERFACE

The critical synchronization point is where the trailing edges of ALE and CLK occur. The falling edges of these two signals must not coincide within 10 nanoseconds of each other. The above circuit gates ALE in on a rising clock edge, thus eliminating any conflicts.

For systems that supply an asynchronous clock of frequencies greater than 10 MHz, the same principle is applied. This time however, (in addition to synchronizing ALE and CLK), the clock must be divided down to derive a signal less than 10 MHz to drive the TMS 4500A. Two possible configurations are illustrated in Figure 2.

8

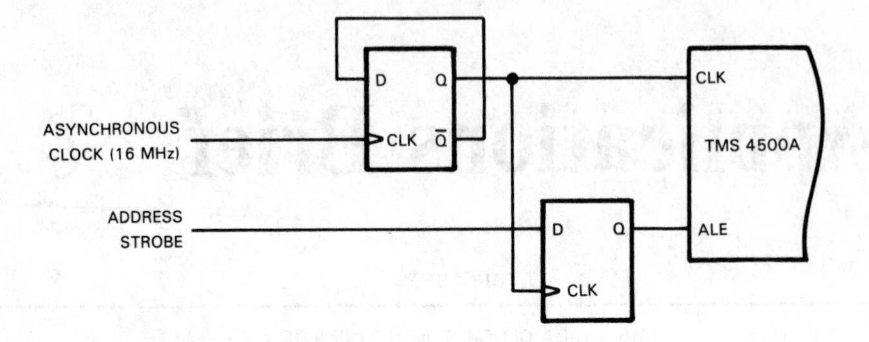

FIGURE 2a – TMS 4500A FAST CLOCK (>10 MHz) SYNCHRONIZATION INTERFACE

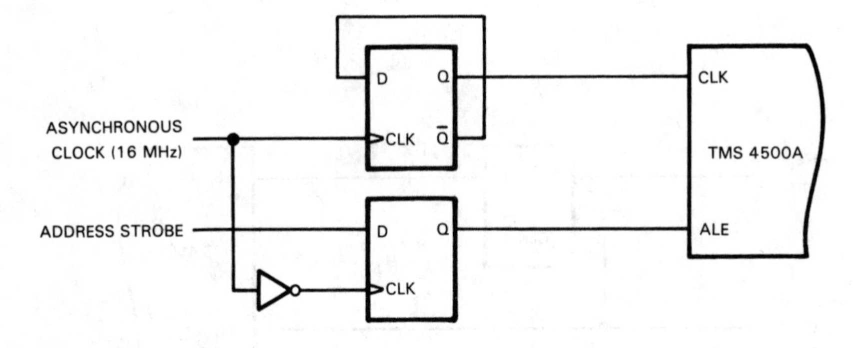

FIGURE 2b – TMS 4500A FAST CLOCK FAST SYNCHRONIZATION INTERFACE

In Figure 2a, a 16 MHz clock signal is divided down to 8 MHz by a D-type flip-flop. This signal is used both as the CLK input to the TMS 4500A and as a gating signal for the ALE input. This circuit has the advantage of lower complexity but can delay ALE by more than 125 nanoseconds. With the circuit of Figure 2b, this delay is reduced by nearly half with the addition of an inverter. Either schottky or advanced schottky devices should be used to minimize propagation delays and skewing.

Finally, it should be noted that the circuits in Figure 2 can be used in cases where a clock signal must be generated exclusively for the TMS 4500A. In this case, higher clock frequencies can be generated and appropriately divided down in order to minimize the synchronization delay.

MOS Memory
Applications Engineering

Applications Brief

MEMORY SYSTEM UPGRADE: 16K EPROM TO 32K EPROM AND 32K EPROM TO 32K ROM

This application brief describes, in detail, how to upgrade from the TMS 2516 (as well as other, compatible 16K 5 V EPROMs such as the Intel 2716) to the TMS 2532 32K 5 V EPROM. Also described is the upgrade from the TMS 2532 to the TMS 4732 32K ROM.

By designing or modifying a system to enable the upgrade from the 16K to the 32K EPROM, time and money can be saved when increasing EPROM memory size. How? Costly redesign is eliminated since devices can be plugged in using one or more jumpers. Also extra board space is not needed in order to double board density. Additionally, TMS 2532's can be replaced with cost-effective TMS 4732s (or any other compatible 32K 5 V ROMs).

This last point especially deserves elaboration. By being able to replace the TMS 2532 with the TMS 4732 the user has the option of prototyping using 32K EPROMs and then plugging in cost-effective, production proven 32K ROMs. The EPROMs, once replaced by ROMs, can be erased and programmed again, and then used to prototype other systems thus providing additional benefits. And, to make matters easy, no modifications are needed to allow this upgrade. The TMS 4732 is directly plug-in compatible.

It can be seen that there are significant savings to be made by planning ahead for upgrades, both from 16K to 32K EPROMs and from 32K EPROMs to 32K ROMs. In this brief you will find a number of schemes which facilitate these upgrades. They are presented as follows:

I. System Upgrades Involving Read Mode Only
   - A. Upgrade: TMS 2516 to TMS 2532
   - B. Upgrade: TMS 2532 to TMS 4732

II. System Upgrades Involving Read and Program Modes, TMS 2516 to TMS 2532
   - A. When Both Modes are Available on the TMS 2516 Circuit But No In-System Programming is Intended on the TMS 2532 Upgrade
   - B. When No In-System Programming is Intended on the TMS 2516 Circuit But Both Modes are to be Available on the TMS 2532 Upgrade
   - C. When Both Modes are to be Available on the TMS 2516 Circuit and the TMS 2532 Upgrade
     1. Scheme 1
     2. Scheme 2

I. SYSTEM UPGRADES INVOLVING READ MODE ONLY

A. Upgrade: TMS 2516 To TMS 2532

Most users will probably program both TMS 2516s and TMS 2532s on 5 V EPROM programmers. A list of suggested programmers which can be used for these devices is given at the end of the brief. If, however, you do presently program or plan to program your 16K 5 V EPROMs in system and/or would like the TMS 2532 upgrade circuit to be conducive to in-system programming, please read the upgrade schemes discussed later.

As long as no in-system programming is intended, though, the TMS 2516 circuit can be arranged as follows:

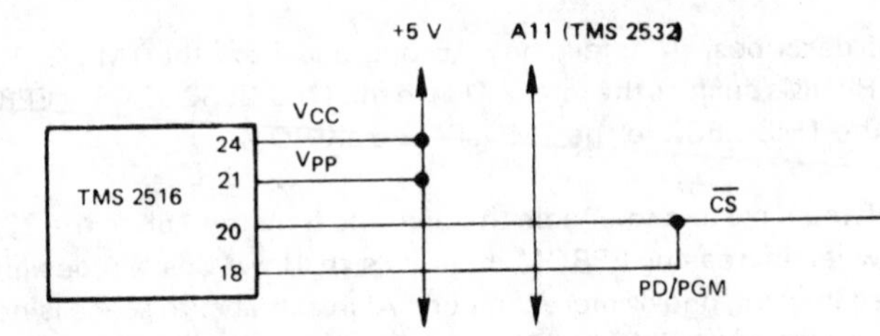

Note that Pin 21, V_{PP} can be commoned to the V_{CC} +5 V supply since no programming (which requires V_{PP} to be +25 V) is planned. Also Pin 18, PD/PGM, can be commoned to the $\overline{CS}$ PC run. This latter connection causes the device to be powered down whenever it is deselected. (More specifically, for any device not being read (i.e., CS high), PD/PGM will also be high.) The result is a significant savings in power.

In order to replace some or all of the TMS 2516s with TMS 2532s only one small modification needs to be made, that is to have an A11 bus and jumpers near pin 18. To upgrade simply disconnect pin 18 from the $\overline{CS}$ PC run and reconnect to the A11 bus. That's it! The PD/PGM signal applied to the TMS 2532 during the non programming modes is functionally identical to the $\overline{CS}$ signal applied to the TMS 2516 thus maintaining timing compatibility. All other pins on the TMS 2516 and TMS 2532 are identical. The resultant TMS 2532 upgrade circuit is as follows:

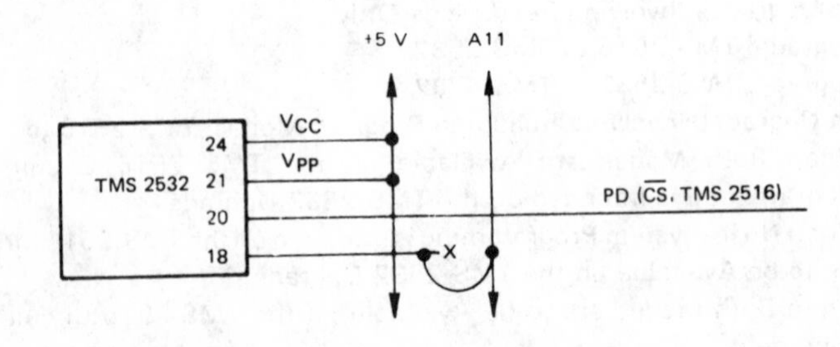

B. Upgrade: TMS 2532 To TMS 4732

When upgrading to the TMS 4732 it would be most economical to start from the TMS 2532 circuit just discussed where the VPP and VCC share a common supply of +5 V. However, regardless of the external circuitry, the TMS 4732 can be directly plugged into the TMS 2532 socket with no modifications. Both devices have 600 mil packages and both have the same pinout (pins 20 and 21 are capable of compatible signal levels though appearing different between devices). For direct plug-in compatibility all that is required is to select the ROM chip selects so that pin 20 chip select 1 is active low ($\overline{CS1}$) and pin 21 chip select 2 is active high (CS2).

	TMS 2532		TMS 4732	
Pin 20	FCN:	PD/$\overline{PGM}$ (Performs $\overline{CS}$ Function) $t_{a(PD)} = 450$ ns Max. $t_{a(A)} = 450$ ns Max. V_{IL} to Read $V_{IL} = 0.65$ V Max.	FCN:	$\overline{CS1}$ $t_{a(CS)} = 200$ ns Max. $t_{a(A)} = 450$ ns Max. V_{IL} to Read $V_{IL} = 0.65$ V Max.
Pin 21	FCN:	V_{PP} V_{CC} to Read $V_{CC} = 5$ V ±5% (4.75 V to 5.25 V)	FCN:	CS2 V_{IH} to Read $V_{IH} = 2.0$ V to 5.25 V

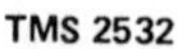

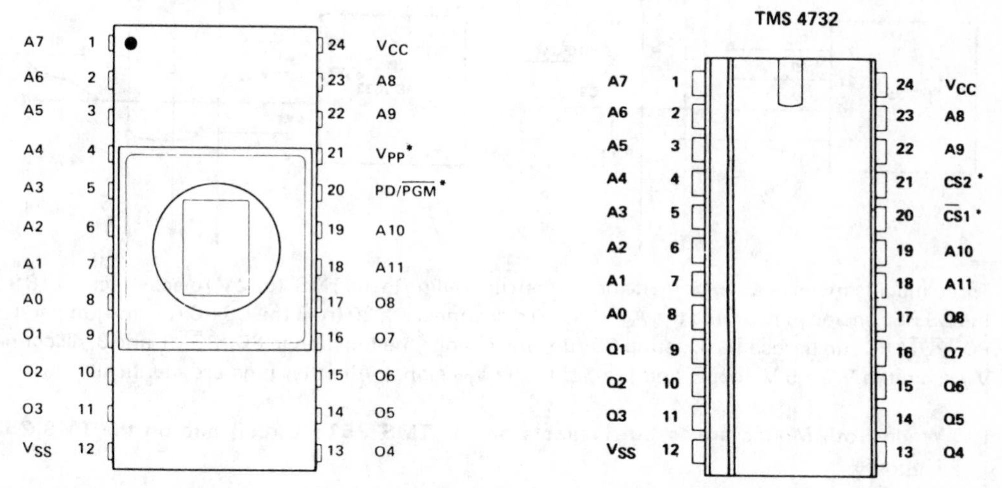

*These pins though functionally different are compatible when upgrading to the TMS 4732. ($\overline{CS1}$ must be active low and CS2 active high).

II. SYSTEM UPGRADES INVOLVING READ AND PROGRAM MODES, TMS 2516 TO TMS 2532

A. When Both Modes are Available on the TMS 2516 Circuit But No In-System Programming is Intended on the TMS 2532 Upgrade

If only reading is to be done from the upgraded TMS 2532 circuit, but the present or planned TMS 2516 (or other compatible 16K 5 V EPROM) circuit also offers in-system programming only one modification is needed to upgrade and it is the same as in the scheme just discussed, i.e., disconnect pin 18 from the PD/PGM run and jumper it to the A11 bus. The only difference between the upgrade here and the upgrade already discussed is that V_{PP} and V_{CC} are now not commoned to the same +5 V bus. Instead, V_{PP} will now be commoned to its own bus to allow it to vary from +5 V to +25 V needed for in system programming.

B. When No In-System Programming is Intended on the TMS 2516 Circuit But Both Modes are to be Available on the TMS 2532 Upgrade

If your TMS 2516 (or other compatible 16K 5 V EPROM) circuit is or is planned to be set up only for reading (i.e., programming is external) but you would like the upgraded TMS 2532 circuit to allow in-system programming as well as reading, then the following upgrade applies:

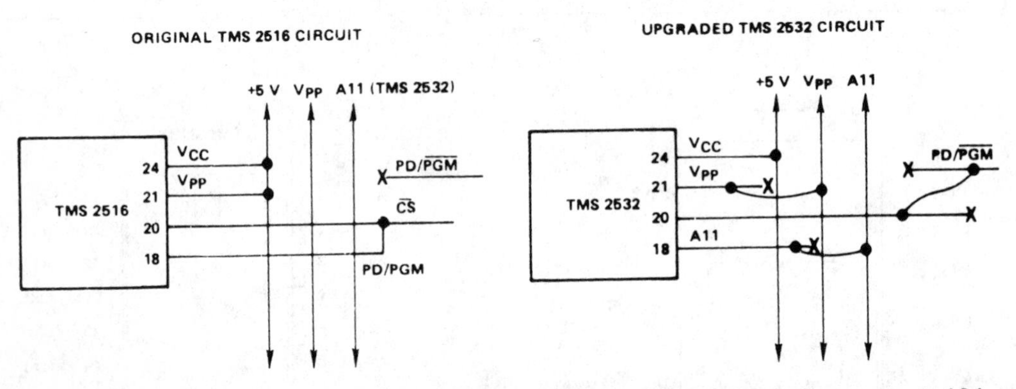

Three modifications need to be made here when upgrading to the TMS 2532: 1) disconnect pin 18 from the $\overline{CS}$ PC run and jumper it to the A11 bus; 2) disconnect pin 20 from the $\overline{CS}$ PC run and jumper it to a PD/$\overline{PGM}$ PC run (or use the existing $\overline{CS}$ run, and change the function to PD/$\overline{PGM}$); and 3) disconnect V_{PP} from the V_{CC} +5 V supply and jumper to the V_{PP} signal. All other pins are identical.

C. When Both Modes are to be Available on the TMS 2516 Circuit and on the TMS 2532 Upgrade

If you would like to be able to upgrade to a TMS 2532 circuit which allows both reading and programming from a TMS 2516 (or compatible 5 V EPROM) circuit also set up for both reading and programming, there are two upgrade schemes which will be of interest.

The first upgrade scheme looks as follows:

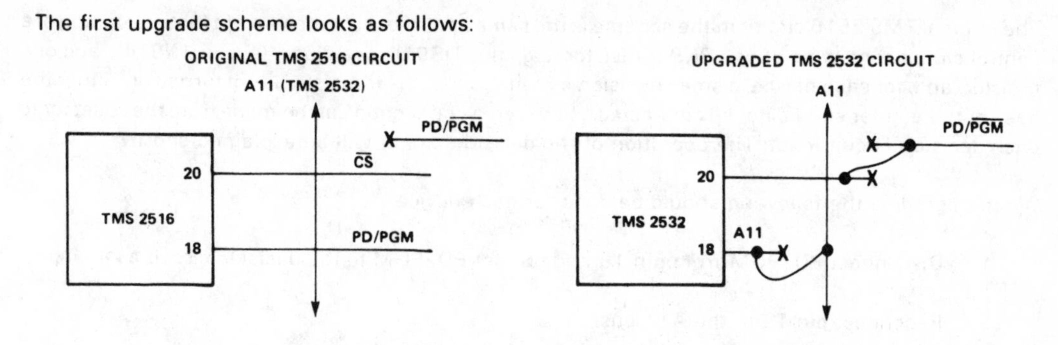

NOTE: For both the original and upgrade circuit V_{PP} (pin 20) and V_{CC} (pin 21) are not commoned.

This scheme requires only two modifications in order to upgrade: 1) disconnect pin 18 from the PD/PGM run and jumper to the A11 bus and 2) disconnect pin 20 from the $\overline{CS}$ run and jumper it to a PD/$\overline{PGM}$ run (or use the existing $\overline{CS}$ run, and change the function to PD/$\overline{PGM}$).

The second possible upgrade (below) is more complicated but generates the full PD/$\overline{PGM}$ function automatically from the PD/PGM signal.

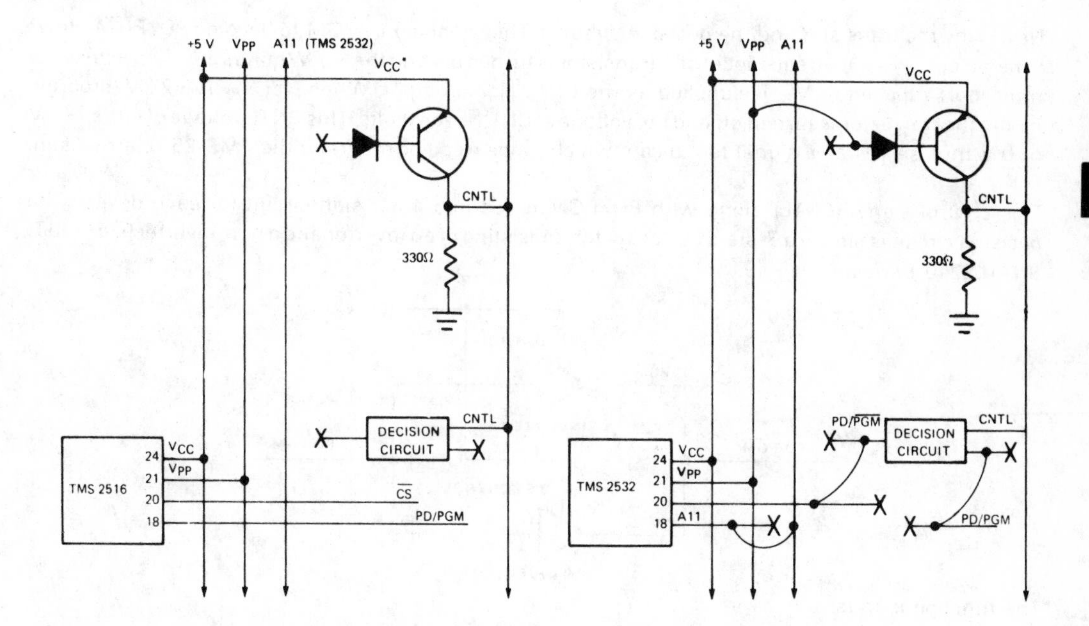

*In the original circuitry before upgrade one of the power supplies, either V_{CC} or V_{PP}, should not be connected to the transistor. (In above example we show V_{CC} connected.) In the upgraded TMS 2532 circuit, both supplies are connected to the transistor.

8

The original TMS 2516 circuit in the scheme is the same as earlier schemes except for the addition of a control circuit consisting of one PNP transistor (e.g., the TIS91), one diode (e.g., a 1N914), and one resistor; and, for each device, a small decision circuit. So that the transistor is not turned on, either the base or the emitter should not be connected. However, a connection can be made from the collector to each decision circuit input. The operation of the decision circuit will be explained shortly.

When upgrading the following should be done for each device:

1. Disconnect PD/PGM from pin 18 and jumper PD/PGM to the decision circuit as an input.

2. Reconnect pin 18 to the A11 bus.

3. Disconnect pin 20 from $\overline{CS}$ and jumper pin 20 to the output of the decision circuit (PD/$\overline{PGM}$).

4. Connect the V_{PP} supply (or the V_{CC} supply if it was originally disconnected) to the transistor.

Note that only one transistor/diode/resistor control circuit is needed for each array of TMS 2532s being used together. A decision circuit, though, is needed for each TMS 2532 unless a 16-bit word configuration is being used. In this case just one decision circuit is required per each pair of devices whose pin 20s are commoned.

Now, why the transistor and the decision circuits? They enable PD/PGM to become PD/$\overline{PGM}$. How? Whenever V_{PP} is +5 V (read mode), the transistor is turned on, and the +5 V (minus any drop across the transistor) supplied by V_{CC} is supplied as the CNTL signal output. Whenever V_{PP} is +25 V (program mode) the transistor is turned off and the voltage at CNTL is zero volts. This CNTL voltage of either +5 V or 0 V thus serves as a signal to indicate which mode (read or program) the TMS 2532 array is in.

This control signal (CNTL) along with PD/PGM is fed into a decision circuit for each device. The decision circuit is simply a 3-state buffer switch consisting of an inverter and a non-inverter (e.g., the TI SN74LS240 Series).

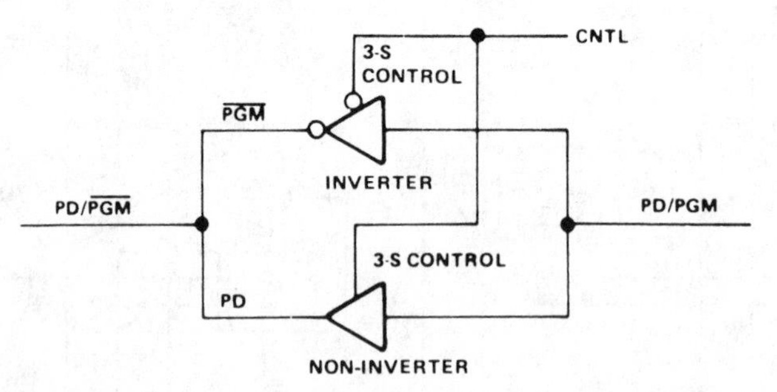

The function is thus:

If CNTL is +5 V (read mode), the inverter buffer is disabled and the non-inverter buffer is enabled. This provides the PD function at the circuits output. If CNTL is 0 V (program mode), the inverter buffer is enabled and the non-inverter buffer is disabled, thus providing the inverted program function, $\overline{PGM}$ at the circuit's output. The net result is the PD/$\overline{PGM}$ signal required for the TMS 2532.

Timing:

For programming upgrade compatibility (regarding this last upgrade) V_{PP} for the TMS 2516 must adhere to the same timing requirements as it does for the TMS 2532. Typically V_{PP} remains at +25 V for the TMS 2516 during all programming modes. However it also can be at +5 V during the read/verify mode. This option must be taken to satisfy the TMS 2532 $\overline{V_{PP}}$ timing requirements (V_{PP} has to be +5 V during the read mode). Note that as a result the desired PD/$\overline{\text{PGM}}$ signal is generated through the buffer switch – PD/PGM is inverted with V_{PP} at +25 V (programming) and left as is with V_{PP} at +5 V (reading). (There will be about a 20 ns delay for PD/PGM to return to V_{IL} once V_{PP} is at +5 V).

Program Cycle Timing

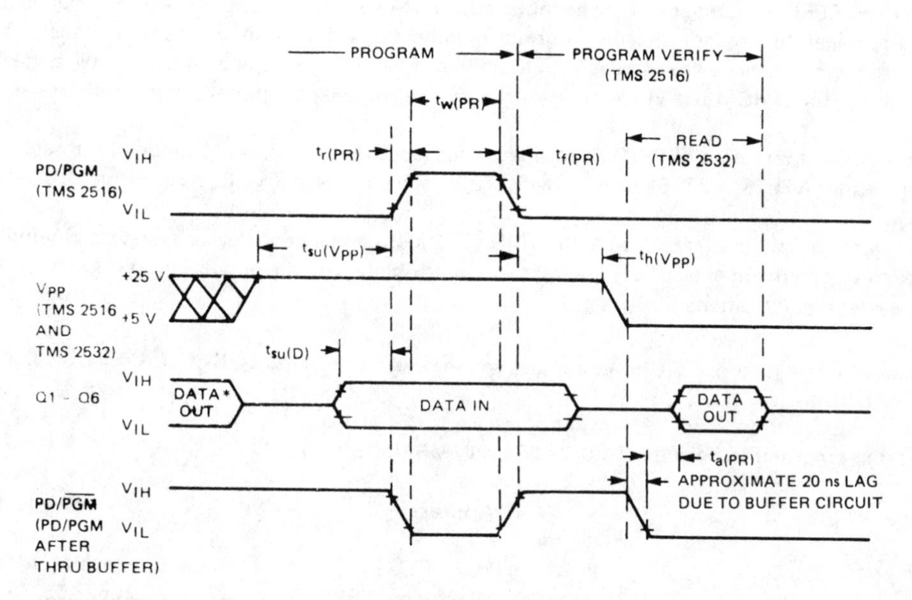

*HI-Z for TMS 2532

NOTE: For complete program cycle timing diagram and timing specifications refer to the TMS 2516/TMS 2532 Data Sheet.

SUMMARY

This application brief has described how to design or modify your system so that TMS 2516s (or other industry compatible 16K 5 V EPROMs) can be directly replaced by TMS 2532 32K 5 V EPROMs. It has also discussed how the TMS 4732 32K EPROM can be directly plugged into the TMS 2532 socket with no modifications whatsoever.

For upgrading from the TMS 2516 to the TMS 2532 five methods were presented. The method requiring least system modification was described first as it is applicable to most systems where memories are programmed externally. The other upgrades described later allowed in-system programming in either the original 16K EPROM or the upgraded 32K EPROM circuit or both.

It should be mentioned that for all upgrades from the TMS 2516 to the TMS 2532, each device in the original 16K EPROM system need not be replaced by a TMS 2532 32K EPROM. Any number of devices can be replaced (upgraded) and the upgrade modifications need only be made for those devices. Therefore memory size can be increased in multiples of 2K or 4K bytes at a time up to twice the size of the original 16K EPROM based memory. And it can be increased whenever the user desires.

Also of note is that, in any EPROM upgrade discussed, pin 20s can be commoned for each pair of devices (either two TMS 2716s or two TMS 2532s) which together yield a 16-bit word.

The conversion from the TMS 2532 to the TMS 4732 also can be done device by device whenever the user desires. The result of all this is greater system flexibility with no increase in board space and little if any system modifications.

For data sheets on these devices, please contact your nearest TI Field Sales Office or Authorized Distributor.

Suggested programmers for the TMS 2516 and TMS 2532*

Company	Address	Contact
DATA I/O	P. O. Box 308 Issaquah, Washington 98027	Steve Montgomery 206/455-3990
PRO-LOG	2411 Garden Road Monterrey, California 93940	Stan Noble 408/372-4593
SHEPARDSON MICRO SYSTEMS	Bldg. C-4 20823 Stevens Creek Blvd. Cupertino, California 95014	Bob Shepardson 408/257-9900

Company	Address	Contact
OLIVER ADVANCED ENGINEERING	676 West Wilson Avenue Glendale, California 91203	Doug Oliver 213/468-8080
TEXAS MICRO SYSTEMS	3320 Bering Drive Houston, Texas 77057	Michael Loeb 713-789-9820
MICRO PRO	424 Oak Mead Parkway Sunnyvale, California 94086	Jim Moon 408/737-0500

*Information on programmers is provided only for user convenience and does not indicate any preference by TI.

MOS Memory
Applications Engineering

8

Applications Brief

MEMORY SYSTEM UPGRADE
16K AND 32K EPROM TO 64K EPROM
64K EPROM TO 64K ROM

The TMS 2564 64K EPROM offers the user the opportunity to increase the density of a system that uses 16K and 32K EPROMS by a factor of 2 or 4. In addition, the TMS 2564 is pin compatible with 64K ROMs from at least eight competitive sources, allowing eventual replacement by low-cost fixed storage.

16K EPROM/32K EPROM TO 64K EPROM UPGRADE

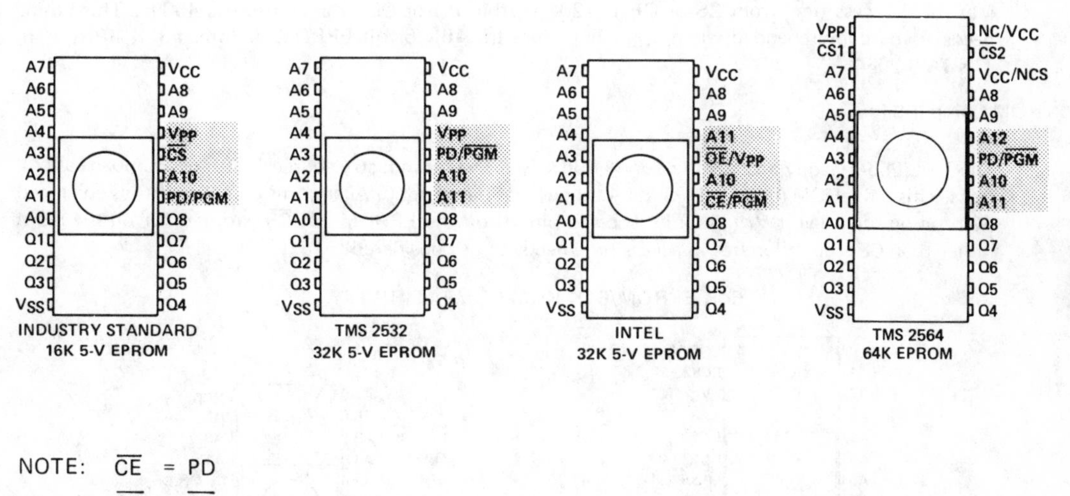

NOTE: $\overline{\text{CE}}$ = PD
$\overline{\text{OE}}$ = $\overline{\text{CS}}$

- Assuming 28-pin socket not reserved in advance:

  Four 16K 5-volt EPROMs can be replaced by a single TMS 2564 with the same operation plus a second chip select by providing additional address decoding for A11 and A12.

  The method for switching from TMS 2532 or Intel 2732 is the same. The decoding logic to two of the 32K EPROMs is replaced by A12, the next order address. The TMS 2564 offers the user two additional chip selects over the TMS 2532, and one over the Intel 2732.

• Assuming 28-pin socket is reserved in advance:

Upgrade is easy once the 28-pin socket is reserved. When reserving this socket only three pins need to be considered.

1) A jumper is reserved on pin 21 for A12
2) V_{CC} (+5 V) must be applied to pin 26 (pins 1 and 28 can be wired together as pin 26 will be the only pin requiring a supply)
3) The PD trace to the TMS 2516 is properly included in the address decoder as it becomes the next order address.

With this socket designed in, upgrade from 16K to 32K to 64K is straightforward.

This same 28-pin socket can also be prepared to accept an Intel 32K EPROM. $\overline{CE}$ (equivalent to PD) on the I2732 will be included in the address decoder to be A11, $\overline{OE}$ is replaced by PD, and A11 by A12. Since addresses are arbitrarily labeled, changing address numbers should not pose problems if this upgrade is planned in advance.

Timing Parameters

Maximum access time from $\overline{CS}$ or $\overline{OE}$ is 120 ns; from PD or $\overline{CE}$ or an address is 450 ns. This timing needs to be comprehended when upgrading from the 16K 5-volt EPROM or Intel's 32K EPROM to TI's TMS 2564.

ROM Compatibility

The TMS 2564 readily accepts the popular 64K ROM (eight industry sources) with no PC board alterations. This is the ROM that the TMS 2564 pinout was based upon. Again, this compatibility is enhanced by running the +5-volt V_{CC} supply trace to pin 26 on the EPROM. (If $\overline{CS}$ is opted for on the ROM rather than CE, the difference in access time needs to be considered).

64K EPROM/64K ROM COMPATIBILITY

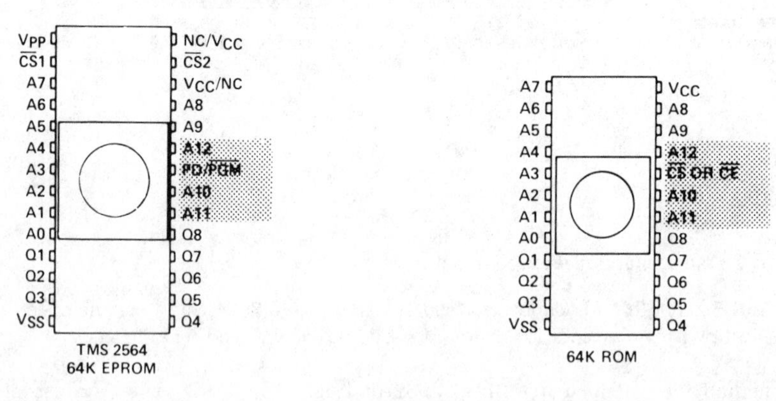

For further information on TI's flexible EPROM family, please contact your nearest sales office or authorized distributor or write Texas Instruments Incorporated, P.O. Box 1443, M/S 6946, Houston, Texas 77001.

MOS Memory
Applications Engineering

Applications Brief

RAM – EPROM COMPATIBILITY
TMS 4016 – TMS 2516

TMS 4016
2K X 8 STATIC RAM

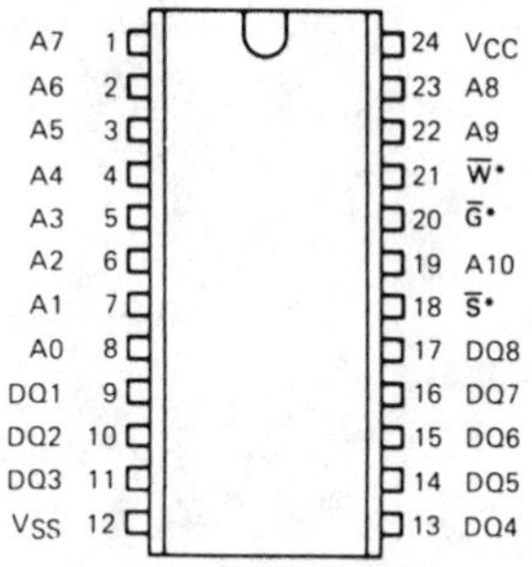

TMS 2516
2K X 8 5V EPROM

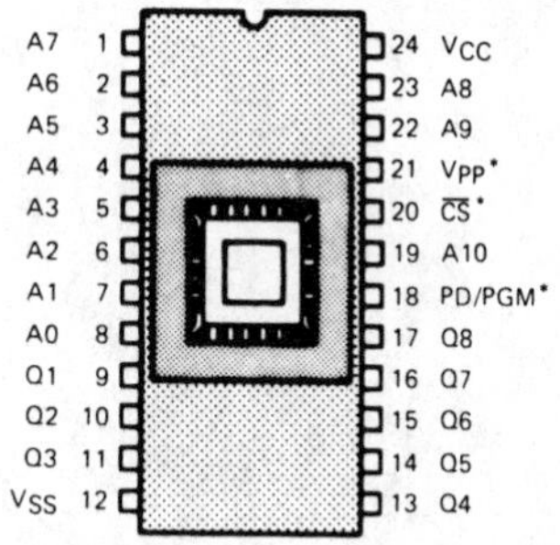

*These pins though apparently different are compatible when switching to the TMS 2516.

Memory boards may now be designed with essentially one pinout type, leaving read/write versus read-only partitioning decisions until later. TI's TMS 4016 static RAM is plug-compatible with 16K 5 V EPROMs. Extensive compatibility exists between the TMS 4016 and TMS 2516. Both the TMS 4016 and TMS 2516 have a 2K X 8 organization. Both come in 600 mil, 24-pin DIP packages. As can be seen above all addresses, data-in/data-out, and V_{SS} and V_{CC} are on the same pins on both devices. The select functions on pins 18 and 20 are also compatible. Provisions must be made for pin 21 since $\overline{W}$ on the TMS 4016 is a MOS input and V_{PP} on the TMS 2516 draws several milliamperes. To find out more please write: Texas Instruments, P. O. Box 1443, M/S 6946, Houston, Texas 77001, for the RAM-EPROM Compatibility Application Brief. Or call or write your nearest TI sales office or authorized distributor.

Mechanical Data

general

Electrical characteristics presented in this catalog, unless otherwise noted, apply to device type(s) listed in the page heading, regardless of package. Factory orders for devices described should include the complete part-type numbers listed on each page.

MOS NUMBERING SYSTEM

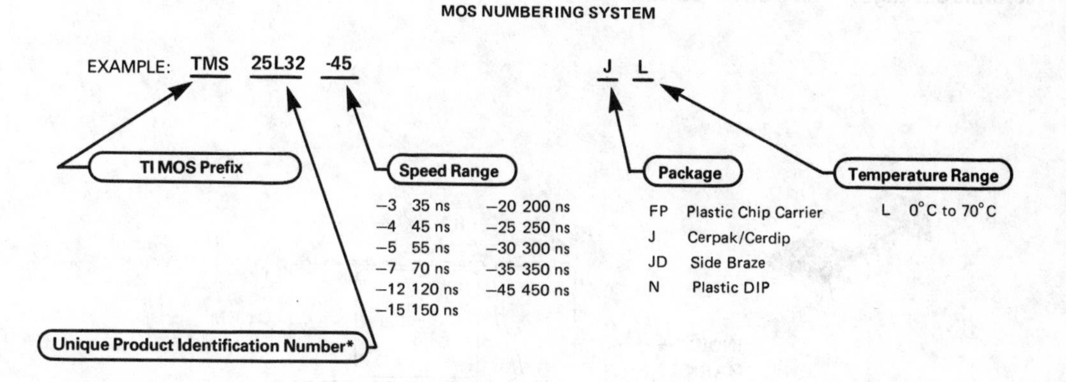

manufacturing information

Die-attach is by standard gold silicon eutectic or by conductive polymer.

Thermal compression gold wire bonding is used on plastic packaged circuits. Typical bond strength is 5 grams. Bond strength is monitored on a lot-to-lot basis. Any preseal bond strength of less than 2 grams causes rejection of the entire lot of devices. On hermetic devices either thermal compression or ultrasonic wire bonding is used. All hermetic MOS LSI devices produced by TI are capable of withstanding 5×10^{-7} atm cc/sec inspection and may be screened to 5×10^{-8} atm cc/sec fine leak, if desired by the customer, for special applications.

All packages are capable of withstanding a shock of 3000 g. All packages are capable of passing a 20,000 g acceleration (centrifuge) test in the Y-axis. Pin strength is measured by a pin-shearing test. All pins are able to withstand the application of a force of 6 pounds at 45° in the peel-off direction.

dual-in-line packages

A pin-to-pin spacing of 2.54 mm (100 mils) has been selected for standard dual-in-line packages (both plastic and ceramic).

TI uses three types of hermetically sealed ceramic dual-in-line packages: cerdip, cerpak, and sidebrazed. The cerdip and cerpak packages have tin-plated leads. The sidebraze package has gold-plated leads.

* Inclusion of an "L" in the product identification indicates the device operates at low power.

MECHANICAL DATA

All measurements are given using both metric and English systems. Under the metric system, the measurements are given in millimeters; under the English system, the measurements are given in inches. The English system measurements are indicated in parentheses next to the metric.

ceramic packages — side braze (JD suffix)

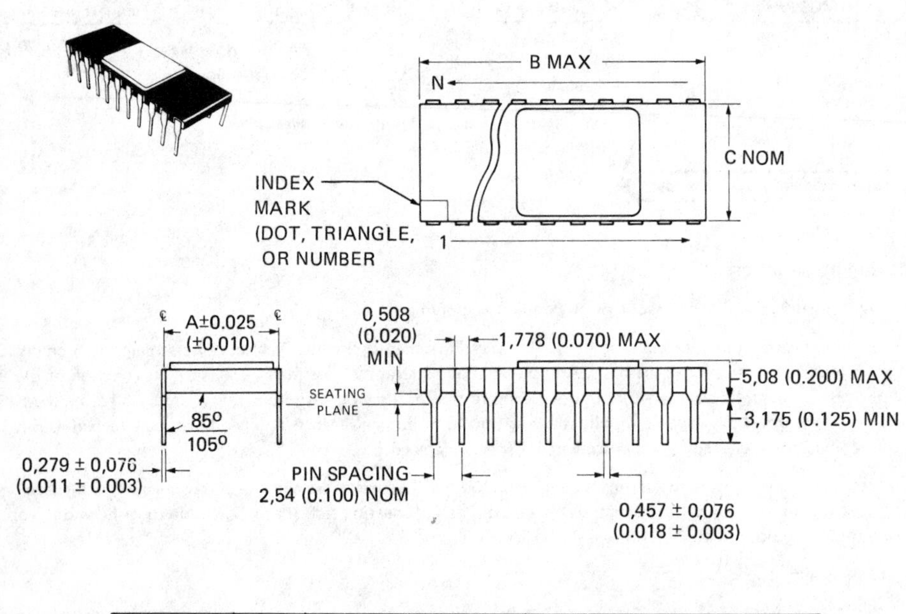

DIM. \ PINS	16	18	20	22	24	24	28	40
A ± 0,025 (±0.010)	7,62 (0.300)	7,62 (0.300)	7,62 (0.300)	10,16 (0.400)	7,62 (0.300)	15,24 (0.600)	15,24 (0.600)	15,24 (0.600)
B(MAX)	20,57 (0.810)	23,11 (0.910)	25,65 (1.010)	27,94 (1.100)	30,86 (1.215)	32,77 (1.290)	35,94 (1.415)	51,31 (2.020)
C(NOM)	7,493 (0.295)	7,493 (0.295)	7,493 (0.295)	10,03 (0.395)	7,493 (0.295)	15,11 (0.595)	15,11 (0.595)	15,11 (0.595)

TEXAS INSTRUMENTS
INCORPORATED
POST OFFICE BOX 225012 • DALLAS, TEXAS 75265

ceramic packages — cerdip/300 mil cerpak (J suffix)

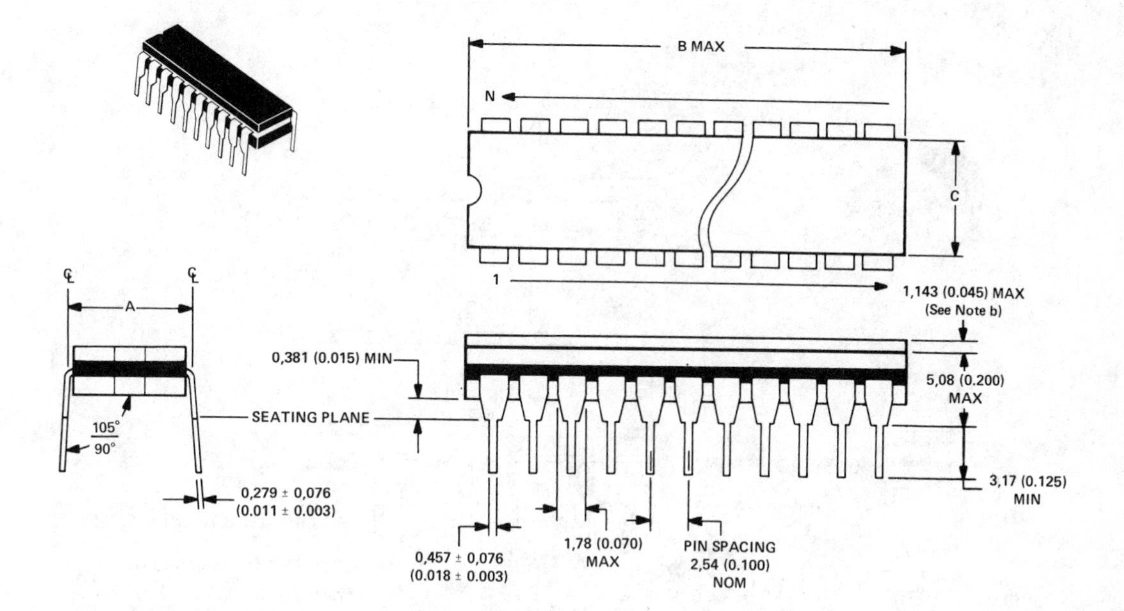

DIM. \ PINS	16*	18	20	24
A(MAX)	8,255 (0.325)	8,255 (0.325)	8,255 (0.325)	8,255 (0.325)
B(MAX)	19,56 (0.770)	22,86 (0.900)	24,38 (0.960)	32,00 (1.260)
C(MAX)	7,645 (0.301)	7,645 (0.301)	7,645 (0.301)	7,645 (0.301)

* Dimensions A, B, and C are applicable for both 16-pin cerdip and cerpak.

NOTES: a. All dimensions are shown in millimeters and parenthetically in inches. Millimeter dimensions govern.

b. Cerpak only.

9

MECHANICAL DATA

ceramic packages — 600 mil cerpak (J suffix)

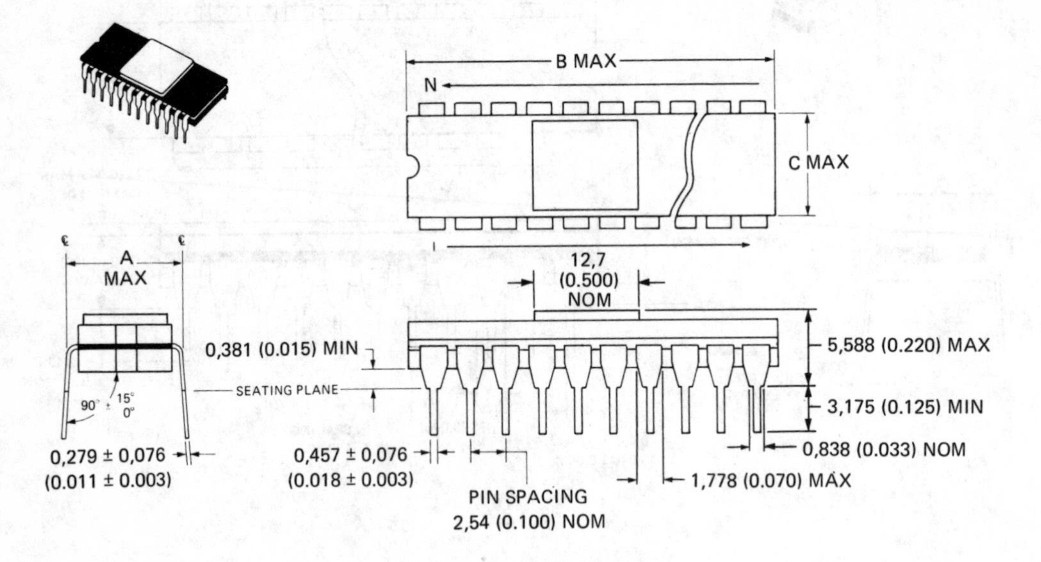

DIM. \ PINS	24	28
A (MAX)	15,88 (0.625)	15,88 (0.625)
B (MAX)	32,77 (1.290)	37,85 (1.490)
C (MAX)	15,24 (0.600)	15,24 (0.600)

TEXAS INSTRUMENTS
INCORPORATED
POST OFFICE BOX 225012 • DALLAS, TEXAS 75265

plastic packages (N suffix)

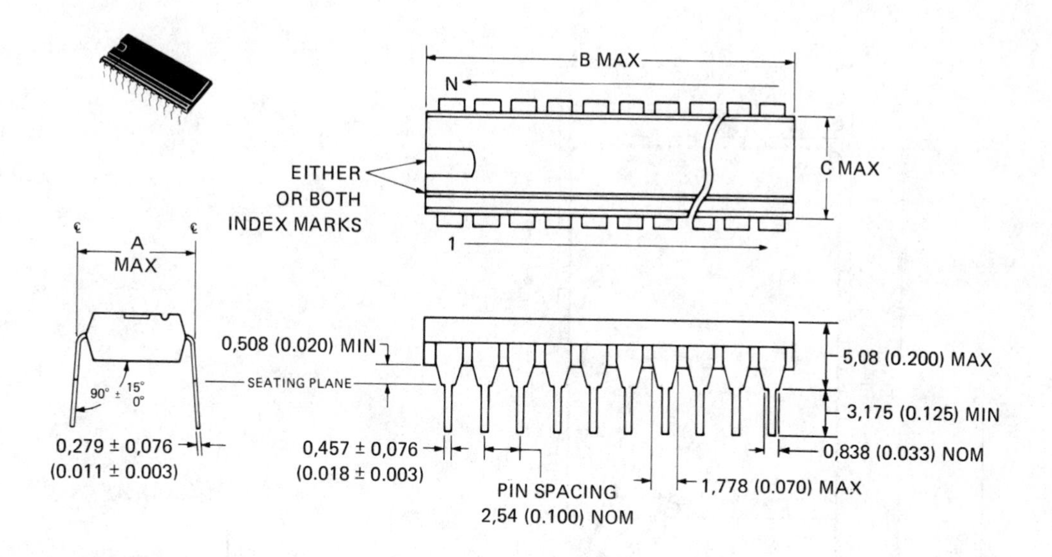

DIM. \ PINS	16	18	20	22	24	28
A (MAX)	8,255 (0.325)	8,255 (0.325)	8,255 (0.325)	10,80 (0.425)	15,88 (0.625)	15,88 (0.625)
B (MAX)	22,1 (0.870)	23,37 (0.920)	27,18 (1.070)	28,45 (1.120)	32,26 (1.270)	36,58 (1.440)
C (MAX)	6,858 (0.270)	6,858 (0.270)	6,858 (0.270)	9,017 (0.355)	13,97 (0.550)	13,97 (0.550)

MECHANICAL DATA

plastic chip carrier package (FP suffix)

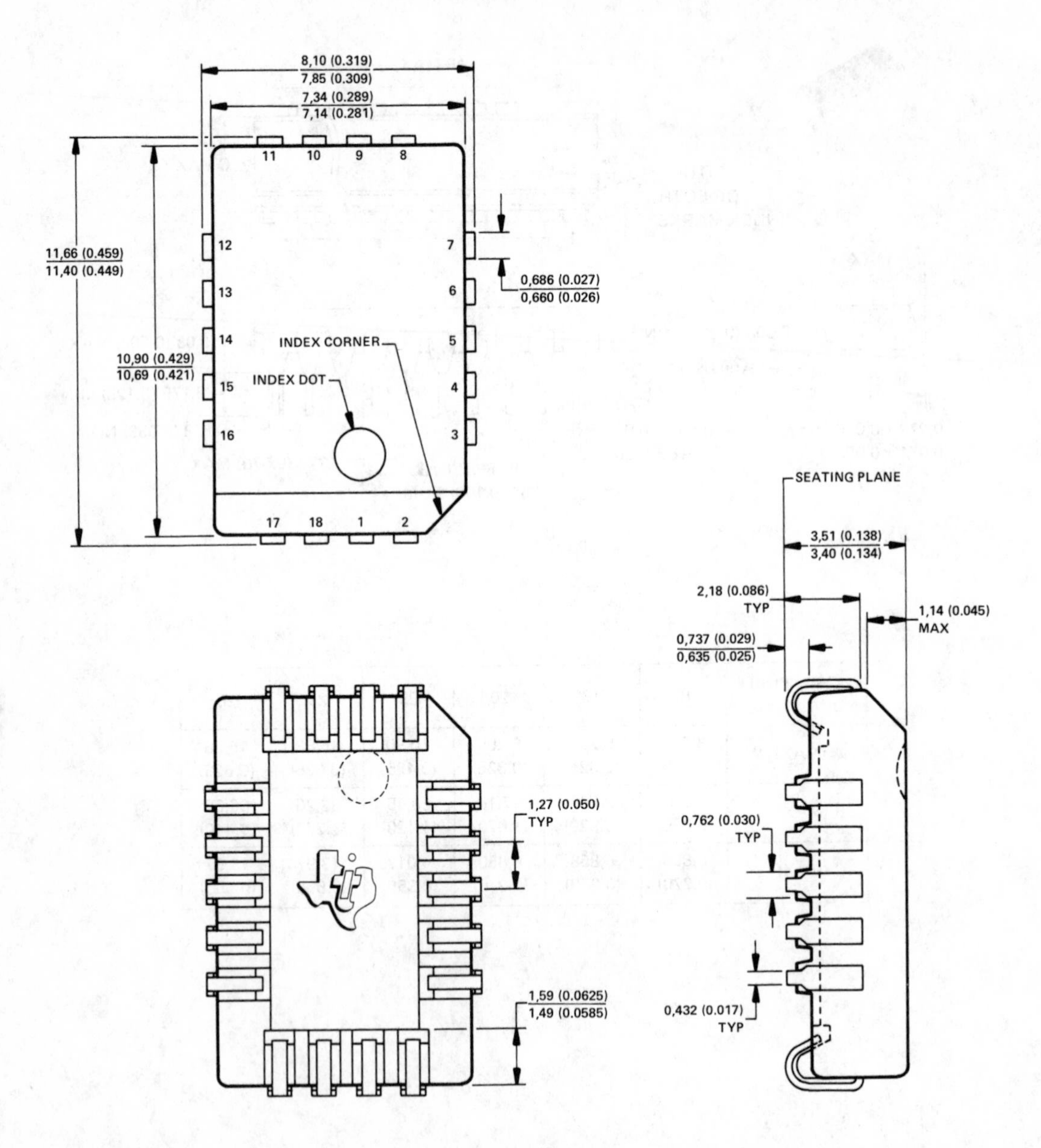

TEXAS INSTRUMENTS
INCORPORATED
POST OFFICE BOX 225012 • DALLAS, TEXAS 75265

Manufacturing Flow

Standard Hermetic Processing Flow*

Slice Sawing
↓
Chip Inspection
↓
Chip Alloyed into Header
↓
Ultrasonic, Thermocompression, or Thermosonic Bond
↓
Preseal Inspection
↓
Seal
↓
Temperature Cycle (10 cycles at –65°C to 150°C)
↓
Tin Plate
↓
Fine Leak (5 X 10^{-8} atm cc/sec)
↓
Gross Leak (C_2 modified)
↓
Lead Tie-bar Sheared
↓
Testing
↓
Q/A Inspection
↓
Shipment

* For cerdip, cerpak, and sidebraze ceramic packages.

TEXAS INSTRUMENTS
INCORPORATED
POST OFFICE BOX 225012 • DALLAS, TEXAS 75265

Standard Plastic Processing Flow

Slice Sawing
↓
Chip Inspection
↓
Chip Epoxied onto Leadframe
↓
Thermocompression or Thermosonic Bond
↓
Premold Inspection
↓
Mold
↓
Tie-bar Sheared
↓
Testing
↓
Q/A Inspection
↓
Shipment

TEXAS INSTRUMENTS
INCORPORATED
POST OFFICE BOX 225012 • DALLAS, TEXAS 75265

Testing/Reliability

In order to ensure the highest in quality and performance, each and every MOS memory device manufactured by the TI MOS Memory Division is thoroughly tested before being shipped. Testing is done during assembly by process engineering (indicated on the manufacturing flow in the previous section); after assembly by product engineering (final test); and after final test by quality and reliability assurance engineering. Every device is tested during the first two stages after which they are received by QRA for random screening for reliability. Outlines of the final test procedure and QRA screening process by family type are included in this section.

TMS 4164 DYNAMIC RAM – FINAL TEST

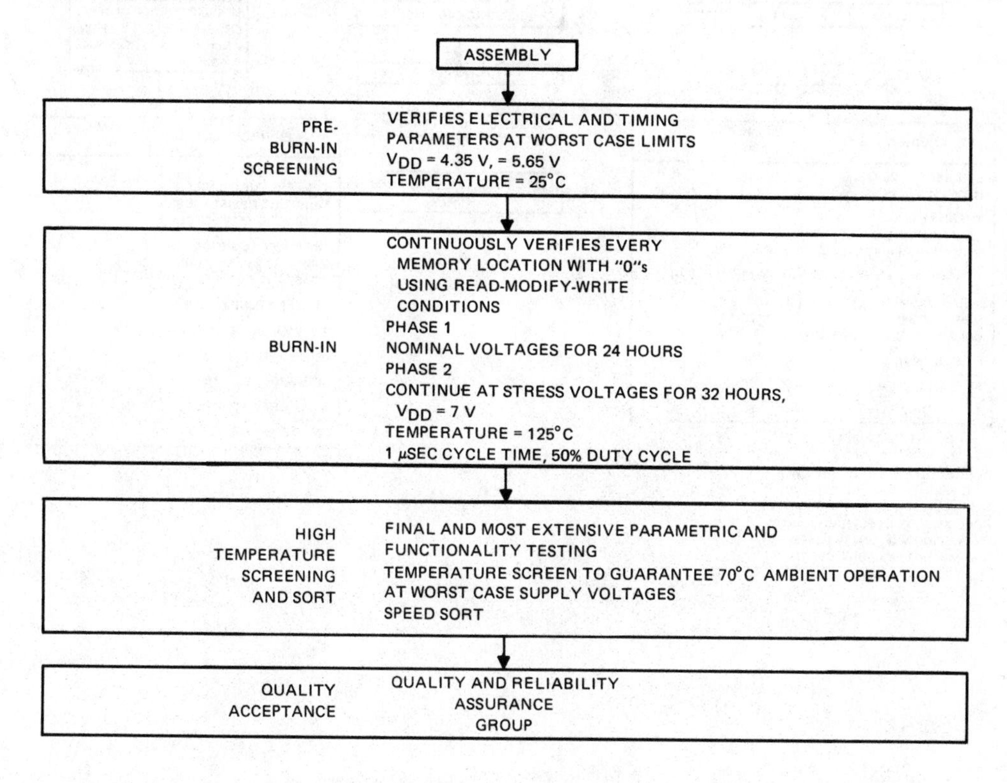

TMS 4164 DYNAMIC RAM — QRA FLOW

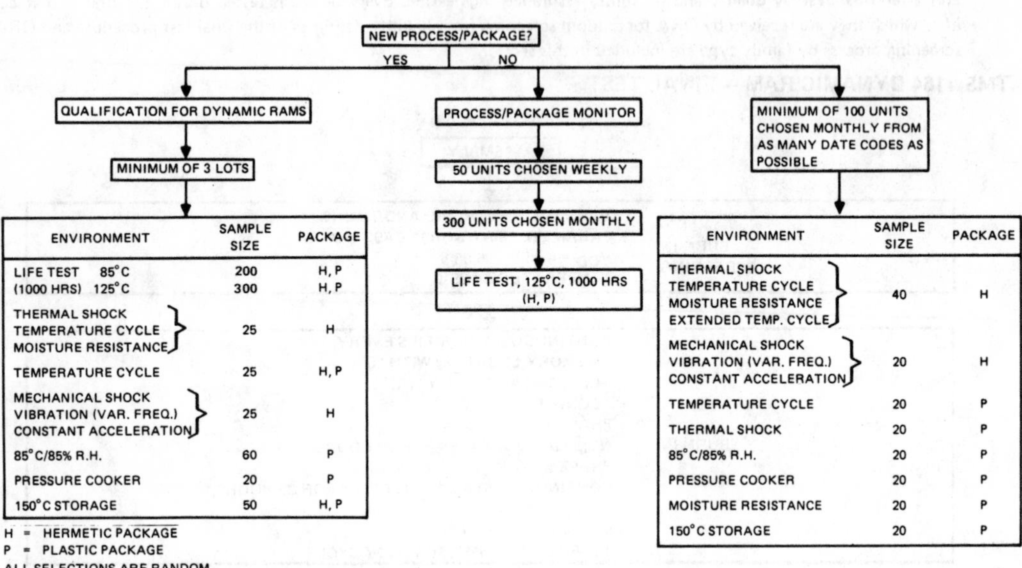

ENVIRONMENT	SAMPLE SIZE	PACKAGE
LIFE TEST 85°C	200	H, P
(1000 HRS) 125°C	300	H, P
THERMAL SHOCK TEMPERATURE CYCLE MOISTURE RESISTANCE	25	H
TEMPERATURE CYCLE	25	H, P
MECHANICAL SHOCK VIBRATION (VAR. FREQ.) CONSTANT ACCELERATION	25	H
85°C/85% R.H.	60	P
PRESSURE COOKER	20	P
150°C STORAGE	50	H, P

H = HERMETIC PACKAGE
P = PLASTIC PACKAGE
ALL SELECTIONS ARE RANDOM
LIFE TEST UNITS ELECTRICALLY TESTED; PACKAGE ENVIRONMENTS FOLLOWED BY ELECTRICAL AND HERMETICITY (H ONLY) TESTS (SEE MECHANICAL DATA SECTION)

ENVIRONMENT	SAMPLE SIZE	PACKAGE
THERMAL SHOCK TEMPERATURE CYCLE MOISTURE RESISTANCE EXTENDED TEMP. CYCLE	40	H
MECHANICAL SHOCK VIBRATION (VAR. FREQ.) CONSTANT ACCELERATION	20	H
TEMPERATURE CYCLE	20	P
THERMAL SHOCK	20	P
85°C/85% R.H.	20	P
PRESSURE COOKER	20	P
MOISTURE RESISTANCE	20	P
150°C STORAGE	20	P

TEXAS INSTRUMENTS
INCORPORATED
POST OFFICE BOX 225012 • DALLAS, TEXAS 75265

TMS 4116 DYNAMIC RAM – FINAL TEST

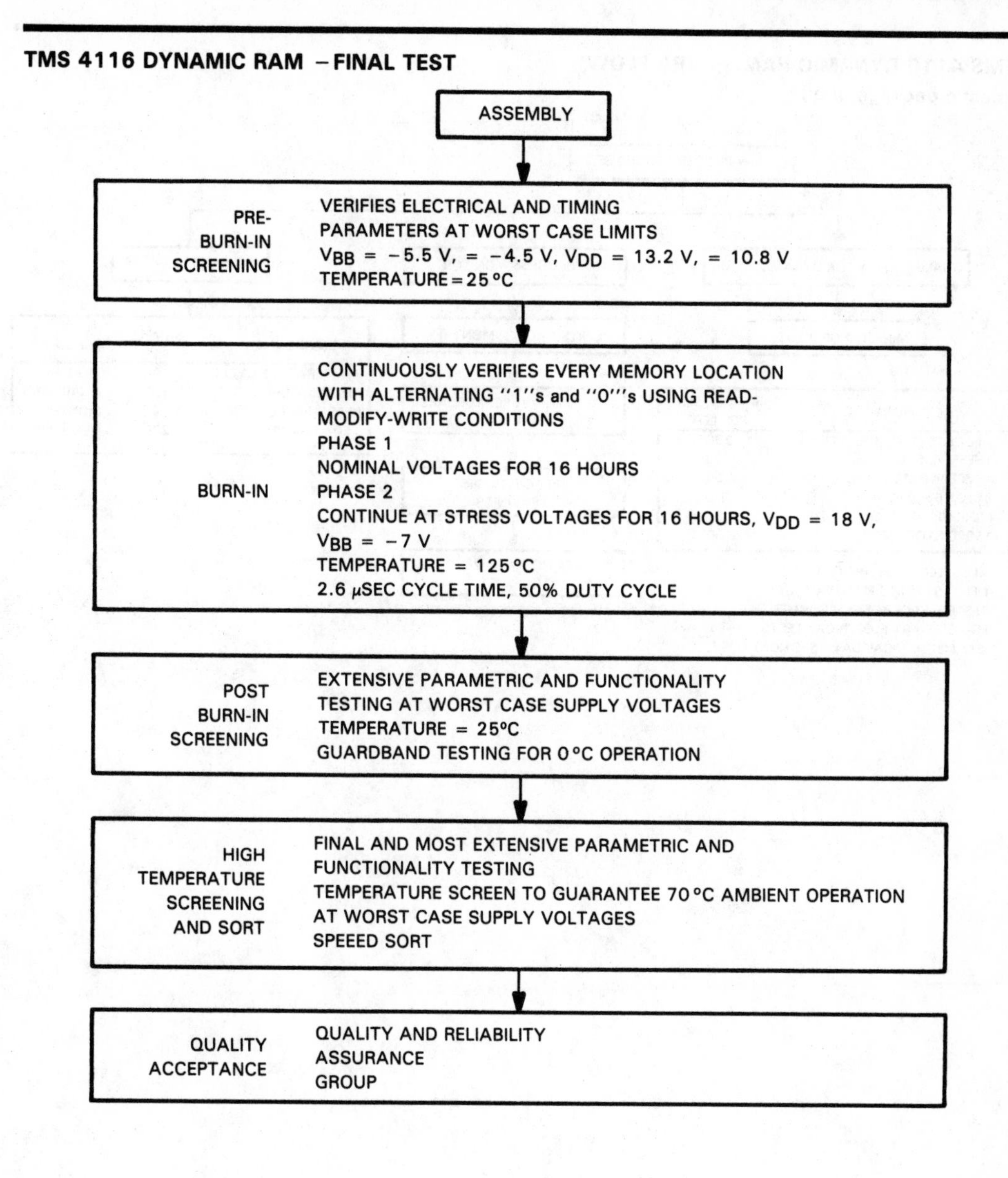

ASSEMBLY
PRE-BURN-IN SCREENING
VERIFIES ELECTRICAL AND TIMING PARAMETERS AT WORST CASE LIMITS
V_{BB} = −5.5 V, = −4.5 V, V_{DD} = 13.2 V, = 10.8 V
TEMPERATURE = 25 °C
BURN-IN
CONTINUOUSLY VERIFIES EVERY MEMORY LOCATION WITH ALTERNATING "1"s and "0"s USING READ-MODIFY-WRITE CONDITIONS
PHASE 1
NOMINAL VOLTAGES FOR 16 HOURS
PHASE 2
CONTINUE AT STRESS VOLTAGES FOR 16 HOURS, V_{DD} = 18 V, V_{BB} = −7 V
TEMPERATURE = 125 °C
2.6 μSEC CYCLE TIME, 50% DUTY CYCLE
POST BURN-IN SCREENING
EXTENSIVE PARAMETRIC AND FUNCTIONALITY TESTING AT WORST CASE SUPPLY VOLTAGES
TEMPERATURE = 25°C
GUARDBAND TESTING FOR 0 °C OPERATION
HIGH TEMPERATURE SCREENING AND SORT
FINAL AND MOST EXTENSIVE PARAMETRIC AND FUNCTIONALITY TESTING
TEMPERATURE SCREEN TO GUARANTEE 70 °C AMBIENT OPERATION AT WORST CASE SUPPLY VOLTAGES
SPEEED SORT
QUALITY ACCEPTANCE
QUALITY AND RELIABILITY ASSURANCE GROUP

TMS 4116 DYNAMIC RAM – QRA FLOW
(plastic package only)

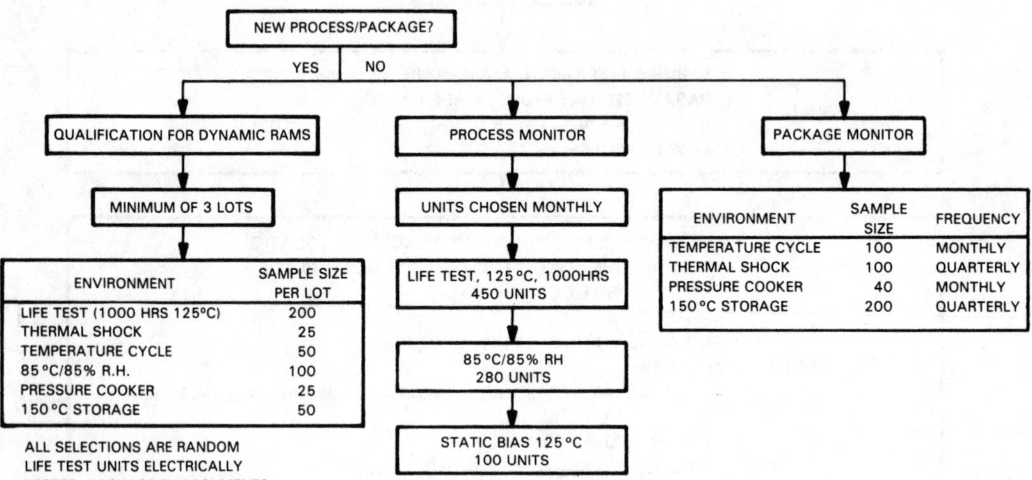

ALL SELECTIONS ARE RANDOM
LIFE TEST UNITS ELECTRICALLY TESTED; PACKAGE ENVIRONMENTS FOLLOWED BY ELECTRICAL TESTS
(SEE MECHANICAL DATA SECTION)

STATIC RAMs – FINAL TEST

STATIC RAMs – QRA FLOW

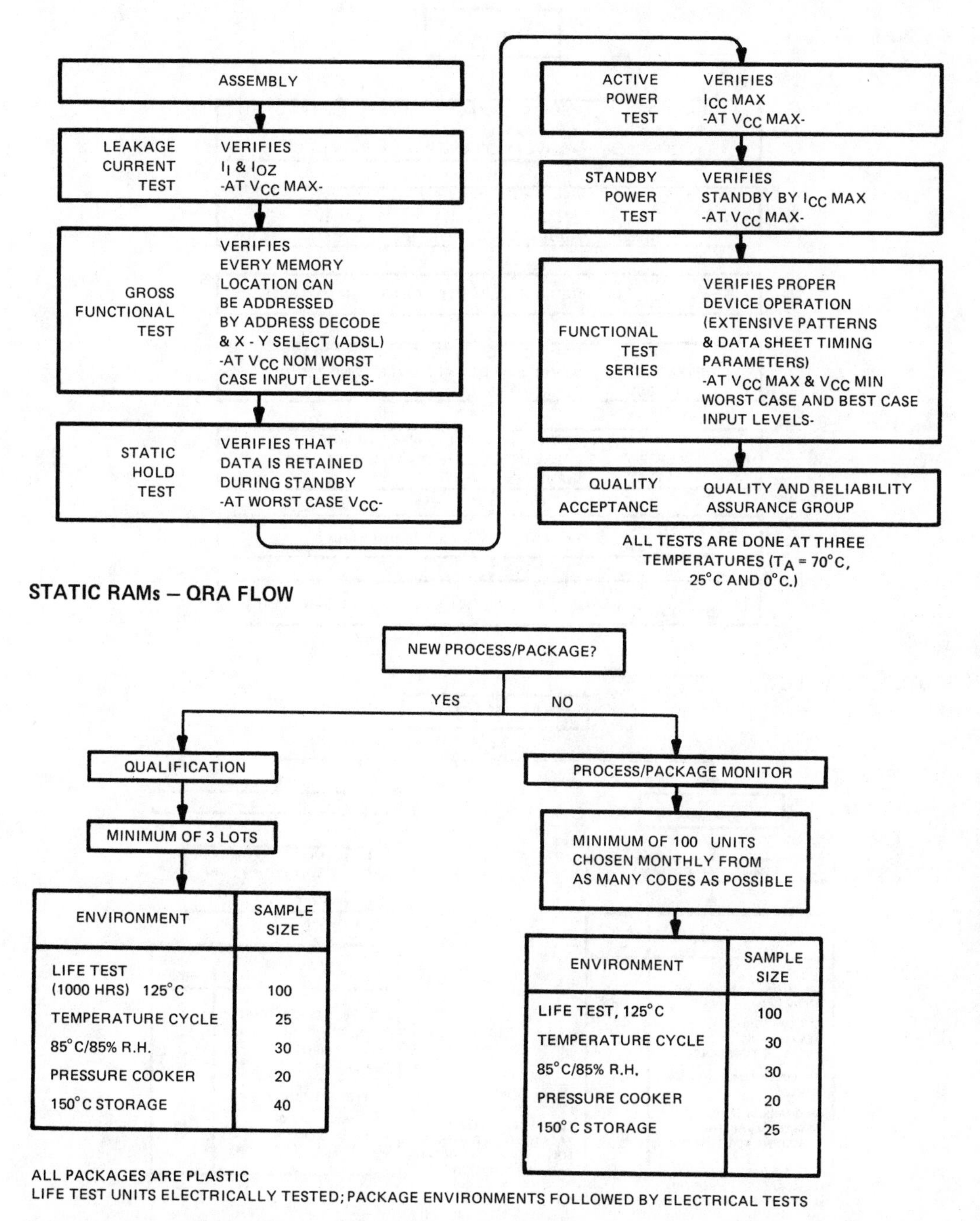
ASSEMBLY
LEAKAGE CURRENT TEST
VERIFIES I_I & I_{OZ} -AT V_{CC} MAX-
GROSS FUNCTIONAL TEST
VERIFIES EVERY MEMORY LOCATION CAN BE ADDRESSED BY ADDRESS DECODE & X - Y SELECT (ADSL) -AT V_{CC} NOM WORST CASE INPUT LEVELS-
STATIC HOLD TEST
VERIFIES THAT DATA IS RETAINED DURING STANDBY -AT WORST CASE V_{CC}-
ACTIVE POWER TEST
VERIFIES I_{CC} MAX -AT V_{CC} MAX-
STANDBY POWER TEST
VERIFIES STANDBY BY I_{CC} MAX -AT V_{CC} MAX-
FUNCTIONAL TEST SERIES
VERIFIES PROPER DEVICE OPERATION (EXTENSIVE PATTERNS & DATA SHEET TIMING PARAMETERS) -AT V_{CC} MAX & V_{CC} MIN WORST CASE AND BEST CASE INPUT LEVELS-
QUALITY ACCEPTANCE
QUALITY AND RELIABILITY ASSURANCE GROUP
ALL TESTS ARE DONE AT THREE TEMPERATURES (T_A = 70°C, 25°C AND 0°C.)
NEW PROCESS/PACKAGE?
YES
NO
QUALIFICATION
MINIMUM OF 3 LOTS
PROCESS/PACKAGE MONITOR
MINIMUM OF 100 UNITS CHOSEN MONTHLY FROM AS MANY CODES AS POSSIBLE

ENVIRONMENT	SAMPLE SIZE
LIFE TEST (1000 HRS) 125°C	100
TEMPERATURE CYCLE	25
85°C/85% R.H.	30
PRESSURE COOKER	20
150°C STORAGE	40

ENVIRONMENT	SAMPLE SIZE
LIFE TEST, 125°C	100
TEMPERATURE CYCLE	30
85°C/85% R.H.	30
PRESSURE COOKER	20
150°C STORAGE	25

ALL PACKAGES ARE PLASTIC
LIFE TEST UNITS ELECTRICALLY TESTED; PACKAGE ENVIRONMENTS FOLLOWED BY ELECTRICAL TESTS

EPROMs – FINAL TEST

FINAL TEST

ASSEMBLY

PARAMETRIC TEST SERIES	VERIFY LEAKAGE CURRENT – INPUT/OUTPUT, HIGH/LOW; POWER SUPPLY CURRENT – ACTIVE MODE AND POWER DOWN MODE
FUNCTIONAL TEST SERIES	VERIFY ERASEABILITY (ALL ONES); PROGRAMMABILITY (WORST CASE); ADDRESS DECODERS
PROGRAMMING TEST	VERIFIES ALL BITS ARE PROGRAMMABLE
DATA RETENTION TEST	VERIFIES LONG LIFE CHARGE STORAGE – DEVICES ARE BAKED OVER LONG PERIOD TO INSURE SUCH
FUNCTIONAL SPEED TEST	VERIFIES MAXIMUM SPEED AT WORST CASE SUPPLY LEVELS AND WORST CASE INPUT LEVELS
ERASEABILITY TEST	VERIFIES ALL BITS ARE ERASEABLE
QUALITY ACCEPTANCE	QAULITY AND RELIABILITY ASSURANCE GROUP

EPROMs – QRA FLOW

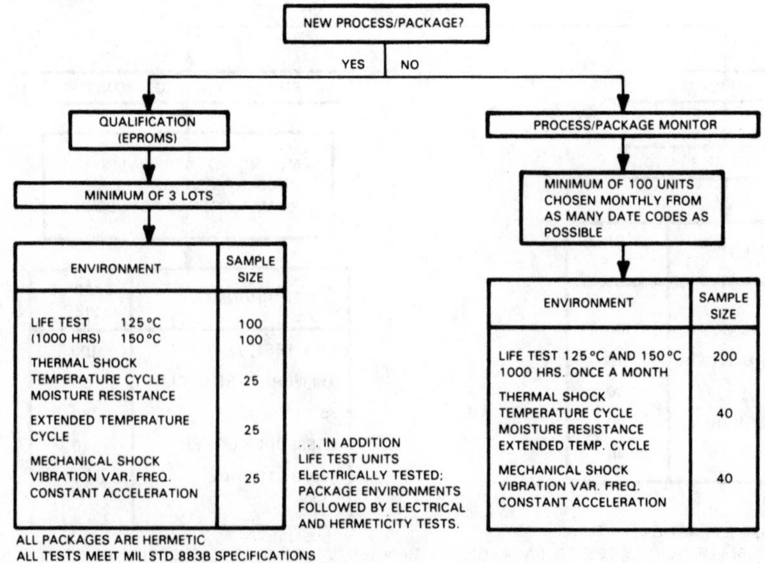

TEXAS INSTRUMENTS
INCORPORATED
POST OFFICE BOX 225012 • DALLAS, TEXAS 75265

Glossary/Timing Conventions/ Data Sheet Structure

PART I – GENERAL CONCEPTS AND TYPES OF MEMORIES

Address – Any given memory location in which data can be stored or from which it can be retrieved.

Automatic Chip-Select/Power Down – (see Chip Enable Input)

Bit – Contraction of Binary digIT, i.e., a 1 or a 0; in electrical terms the value of a bit may be represented by the presence or absence of charge, voltage, or current.

Byte – A word of 8 bits (see word)

Chip Enable Input – A control input to an integrated circuit that when active permits operation of the integrated circuit for input, internal transfer, manipulation, refreshing, and/or output of data and when inactive causes the integrated circuit to be in a reduced power standby mode.

Chip Select Input – Chip select inputs are gating inputs that control the input to and output from the memory. They may be of two kinds:

1. Synchronous – Clocked/latched with the memory clock. Affects the inputs and outputs for the duration of that memory cycle.
2. Asynchronous – Has direct asynchronous control of inputs and outputs. In the read mode, an asynchronous chip select functions like an output enable.

Column Address Strobe (CAS) – A clock used in dynamic RAMs to control the input of column addresses. It can be active high (CAS) or active low ($\overline{\text{CAS}}$).

Data – Any information stored or retrieved from a memory device.

Dynamic (Read/Write) Memory (DRAM) – A read/write memory in which the cells require the repetitive application of control signals in order to retain the stored data.

NOTES:
1. The words "read/write" may be omitted from the term when no misunderstanding will result.
2. Such repetitive application of the control signals is normally called a refresh operation.
3. A dynamic memory may use static addressing or sensing circuits.
4. This definition applies whether the control signals are generated inside or outside the integrated circuit.

Electrically Alterable Read-Only Memory (EAROM) – A nonvolatile memory that can be field-programmed like a PROM or EPROM, but that can be electrically erased by a combination of electrical signals at its inputs.

Erasable and Programmable Read-Only Memory (EPROM)/Reprogrammable Read-Only Memory – A field-programmable read-only memory that can have the data content of each memory cell altered more than once.

Erase – Typically associated with EPROMs and EAROMs. The procedure whereby programmed data is removed and the device returns to its unprogrammed state.

Field-Programmable Read-Only Memory – A read-only memory that after being manufactured, can have the data content of each memory cell altered.

Fixed Memory – A common term for ROMs, EPROMs, EAROMs, etc., containing data that is not normally changed. A more precise term for EPROMs and EAROMs is nonvolatile since their data may be easily changed.

Fully Static RAM – In a fully static RAM, the periphery as well as the memory array is fully static. The periphery is thus always active and ready to respond to input changes without the need for clocks. There is no precharge required for static periphery.

K – When used in the context of specifying a given number of bits of information, 1K = 2^{10} = 1024 bits. Thus, 64K = 64 X 1024 = 65,536 bits.

Large-Scale Integration (LSI) – The description of any IC technology that enables condensing more than 100 gates onto a single chip.

Mask-Programmed Read-Only Memory — A read-only memory in which the data content of each cell is determined during manufacture by the use of a mask, the data content thereafter being unalterable.

Memory — A medium capable of storage of information from which the information can be retrieved.

Memory Cell — The smallest subdivision of a memory into which a unit of data has been or can be entered, in which it is or can be stored, and from which it can be retrieved.

Metal-Oxide Semiconductor (MOS) — The technology involving photolithographic layering of metal and oxide to produce a semiconductor device.

NMOS — A type of MOS technology in which the basic conduction mechanism is governed by electrons. (Short for N-channel MOS)

Nonvolatile Memory — A memory in which the data content is maintained whether the power supply is connected or not.

Output Enable — A control input that, when true, permits data to appear at the memory output, and when false, causes the output to assume a high-impedance state. (See also chip select)

PMOS — A type of MOS technology in which the basic conduction mechanism is governed by holes. (Short for P-channel MOS)

Parallel Access — A feature of a memory by which all the bits of a byte or word are entered simultaneously at several inputs or retrieved simultaneously from several outputs.

Power Down — A mode of a memory device during which the device is operating in a low-power or standby mode. Normally read or write operations of the memory are not possible under this condition.

Program — Typically associated with EPROM memories, the procedure whereby logical 0's (or 1's) are stored into various desired locations in a previously erased device.

Program Enable — An input signal that when true, puts a programmable memory device into the program mode.

Programmable Read-Only Memory (PROM) — A memory that permits access to any of its address locations in any desired sequence with similar access time to each location.
NOTE: The term as commonly used denotes a read/write memory.

Read — A memory operation whereby data is output from a desired address location.

Read-Only Memory (ROM) — A memory in which the contents are not intended to be altered during normal operation.
NOTE: Unless otherwise qualified, the term "read-only memory" implies that the content is determined by its structure and is unalterable.

Read/Write Memory — A memory in which each cell may be selected by applying appropriate electrical input signals and the stored data may be either (a) sensed at appropriate output terminals, or (b) changed in response to other similar electrical input signals.

Row Address Strobe (RAS) — A clock used in dynamic RAMs to control the input of the row addressed. It can be active high (RAS) or active low ($\overline{RAS}$).

Scaled-MOS (SMOS) — MOS technology under which the device is scaled down in size in three dimensions and in operating voltages allowing improved performance.

Semi-Static (Quasi-Static, Pseudo-Static) RAM — In a semi-static RAM, the periphery is clock-activated (i.e., dynamic). Thus the periphery is inactive until clocked, and only one memory cycle is permitted per clock. The peripheral circuitry must be allowed to reset after each active memory cycle for a minimum precharge time. No refresh is required.

Serial Access — A feature of a memory by which all the bits are entered sequentially at a single input or retrieved sequentially form a single output.

Static RAM (SRAM) — A read/write random-access device within which information is stored as latched voltage levels. The memory cell is a static latch that retains data as long as power is applied to the memory array. No refresh is required. The type of periphery circuitry sub-categorizes static RAMs.

TEXAS INSTRUMENTS
INCORPORATED
POST OFFICE BOX 225012 • DALLAS, TEXAS 75265

Very-Large-Scale Integration (VLSI) — The description of any IC technology that is much more complex than large-scale integration (LSI), and involves a much higher equivalent gate count. At this time an exact definition including a minimum gate count has not been standardized by JEDEC or the IEEE.

Volatile Memory — A memory in which the data content is lost when power supplied is disconnected.

Word — A series of one or more bits that occupy a given address location and that can be stored and retrieved in parallel.

Write — A memory operation whereby data is written into a desired address location.

Write Enable — A control signal that when true causes the memory to assume the write mode, and when false causes it to assume the read mode.

PART II – OPERATING CONDITIONS AND CHARACTERISTICS (INCLUDING LETTER SYMBOLS)

Capacitance

The inherent capacitance on every pin, which can vary with various inputs and outputs.

Example symbology:

C_i Input capacitance

C_o Output capacitance

$C_{i(D)}$ Input capacitance, data input

Current

High-level input current, I_{IH}
The current into an input when a high-level voltage is applied to that input.

High-level output current, I_{OH}
The current into* an output with input conditions applied that according to the product specification will establish a high level at the output.

Low-level input current, I_{IL}
The current into an input when a low-level voltage is applied to that input.

Low-level output current, I_{OL}
The current into* an output with input conditions applied that according to the product specification will establish a low level at the output.

Off-state (high-impedance-state) output current (of a three-state output), I_{OZ}
The current into* an output having three-state capability with input conditions applied that according to the product specification will establish the high-impedance state at the output.

Short-circuit output current, I_{OS}
The current into* an output when the output is short-circuited to ground (or other specified potential) with input conditions applied to establish the output logic level farthest from ground potential (or other specified potential).

Supply current I_{BB}, I_{CC}, I_{DD}, I_{PP}
The current into, respectively, the V_{BB}, V_{CC}, V_{DD}, V_{PP} supply terminals.

Operating Free-Air Temperature

The temperature (T_A) range over which the device will operate and meet the specified electrical characteristics.

* Current out of a terminal is given as a negative value.

Voltage

High-level input voltage, V_{IH}

An input voltage within the more positive (less negative) of the two ranges of values used to represent the binary variables.

NOTE: A minimum is specified that is the least positive value of high-level input voltage for which operation of the logic element within specification limits is guaranteed.

High-level output voltage, V_{OH}

The voltage at an output terminal with input conditions applied that according to the product specification will establish a high level at the output.

Low-level input voltage, V_{IL}

An input voltage level within the less positive (more negative) of the two ranges of values used to represent the binary variables.

NOTE: A maximum is specified that is the most positive value of low-level input voltage for which operation of the logic element within specification limits is guaranteed.

Low-level output voltage, V_{OL}

The voltage at an output terminal with input conditions applied that according to the product specification will establish a low level at the output.

Supply Voltages, V_{BB}, V_{CC}, V_{DD}, V_{PP}

The voltages supplied to the corresponding voltage pins that are required for the device to function. From one to four of these supplies may be necessary, along with ground, V_{SS}.

Time Intervals

New or revised data sheets in this book use letter symbols in accordance with standards recently adopted by JEDEC, the IEEE, and the IEC. Two basic forms are used. The first form is usually used in this book when intervals can easily be classified as access, cycle, disable, enable, hold, refresh, setup, transition, or valid times and for pulse durations. The second form can be used generally but in this book is used primarily for time intervals not easily classifiable. The second (unclassified) form will be described first. Since some manufacturers use this form for all time intervals, symbols in the unclassified form are given with the examples for most of the classified time intervals.

Unclassified time intervals

Generalized letter symbols can be used to identify almost any time interval without classifying it using traditional or contrived definitions. Symbols for unclassified time intervals identify two signal events listed in from-to sequence using the format:

$$t_{AB\text{-}CD}$$

Subscripts A and C indicate the names of the signals for which changes of state or level or establishment of state or level constitute signal events assumed to occur first and last, respectively, that is, at the beginning and end of the time interval. Every effort is made to keep the A and C subscript length down to one letter, if possible (e.g., R for $\overline{RAS}$ and C for $\overline{CAS}$ of TMS 4116).

Subscripts B and D indicate the direction of the transitions and/or the final states or levels of the signals represented by A and C, respectively. One or two of the following is used:

H = high or transition to high
L = low or transition to low
V = a valid steady-state level
X = unknown, changing, or "don't care" level
Z = high-impedance (off) state

The hyphen between the B and C subscripts is omitted when no confusion is likely to occur.

For examples of symbols of this type, see TMS 4116 (e.g., t_{PLCL}).

Classified time intervals (general comments, specific times follow)

Because of the information contained in the definitions, frequently the identification of one or both of the two signal events that begin and end the intervals can be significantly shortened compared to the unclassified forms. For example, it is not necessary to indicate in the symbol that an access time ends with valid data at the output. However, if both signals are named (e.g., in a hold time), the from-to sequence is maintained.

Access time

The time interval between the application of a specific input pulse and the availability of valid signals at an output.

Example symbology:

Classified	Unclassified	Description
$t_{a(A)}$	t_{AVQV}	Access time from address
$t_{a(S)}$, $t_{a(CS)}$	t_{SLQV}	Access time from chip select (low)

Cycle time

The time interval between the start and end of a cycle.

NOTE: The cycle time is the actual time interval between two signal events and is determined by the system in which the digital circuit operates. A minimum value is specified that is the shortest interval that must be allowed for the digital circuit to perform a specified function (e.g., read, write, etc.) correctly.

Example symbology:

Classified	Unclassified	Description
$t_{c(R)}$, $t_{c(rd)}$	$t_{AVAV(R)}$	Read cycle time
$t_{c(W)}$	$t_{AVAV(W)}$	Write cycle time

NOTE: R is usually used as the abbreviation for "read"; however, in the case of dynamic memories, "rd" is used to permit R to stand for RAS.

Disable time (of a three-state output)

The time interval between the specified reference points on the input and output voltage waveforms, with the three-state output changing from either of the defined active levels (high or low) to a high-impedance (off) state.

Example symbology:

Classified	Unclassified	Description
$t_{dis(S)}$	t_{SHQZ}	Output disable time after chip select (high)
$t_{dis(W)}$	t_{WLQZ}	Output disable time after write enable (low)

These symbols supersede the older forms t_{PVZ} or t_{PXZ}.

Enable time (of a three-state output)

The time interval between the specified reference points on the input and output voltage waveforms, with the three-state output changing from a high-impedance (off) state to either of the defined active levels (high or low).

NOTE: For memories these intervals are often classified as access times.

Example symbology:

Classified	Unclassified	Description
$t_{en(SL)}$	t_{SLQV}	Output enable time after chip select low

These symbols supercede the older form t_{PZV}.

Hold time

The time interval during which a signal is retained at a specified input terminal after an active transition occurs at another specified input terminal.

NOTES: 1. The hold time is the actual time interval between two signal events and is determined by the system in which the digital circuit operates. A minimum value is specified that is the shortest interval for which correct operation of the digital circuit is guaranteed.
2. The hold time may have a negative value in which case the minimum limit defines the longest interval (between the release of the signal and the active transition) for which correct operation of the digital circuit is guaranteed.

Example symbology:

Classified	Unclassified	Description
$t_{h(D)}$	t_{WHDX}	Data hold time (after write high)
$t_{h(RHrd)}$	t_{RHWH}	Read (write enable high) hold time after $\overline{RAS}$ high)
$t_{h(CHrd)}$	t_{CHWH}	Read (write enable high) hold time after $\overline{CAS}$ high)
$t_{h(CLCA)}$	$t_{CL\text{-}CAX}$	Column address hold time after $\overline{CAS}$ low
$t_{h(RLCA)}$	$t_{RL\text{-}CAX}$	Column address hold time after $\overline{RAS}$ low
$t_{h(RA)}$	$t_{RL\text{-}RAX}$	Row address hold time (after $\overline{RAS}$ low)

These last three symbols supersede the older forms:

NEW FORM	OLD FORM
$t_{h(CLCA)}$	$t_{h(ACL)}$
$t_{h(RLCA)}$	$t_{h(ARL)}$
$t_{h(RA)}$	$t_{h(AR)}$

NOTE: The from-to sequence in the order of subscripts in the unclassified form is maintained in the classified form. In the case of hold times, this causes the order to seem reversed from what would be suggested by the terms.

Pulse duration (width)

The time interval between specified reference points on the leading and trailing edges of the pulse waveform.

Example symnbology:

Classified	Unclassified	Description
$t_{w(W)}$	t_{WLWH}	Write pulse duration
$t_{w(RL)}$	t_{RLRH}	Pulse duration, $\overline{RAS}$ low

Refresh time interval

The time interval between the beginnings of successive signals that are intended to restore the level in a dynamic memory cell to its original level.

NOTE: The refresh time interval is the actual time interval between two refresh operations and is determined by the system in which the digital circuit operates. A maximum value is specified that is the longest interval for which correct operation of the digital circuit is guaranteed.

Example symbology:

Classified	Unclassified	Description
t_{rf}		Refresh time interval

Setup time

The time interval between the application of a signal at a specified input terminal and a subsequent active transition at another specified input terminal.

NOTES:
1. The setup time is the actual time interval between two signal events and is determined by the system in which the digital circuit operates. A minimum value is specified that is the shortest interval for which correct operation of the digital circuit is guaranteed.
2. The setup time may have a negative value in which case the minimum limit defines the longest interval (between the active transition and the application of the other signal) for which correct operation of the digital circuit is guaranteed.

Example symbology:

Classified	Unclassified	Description
$t_{su(D)}$	t_{DVWH}	Data setup time (before write high)
$t_{su(CA)}$	t_{CAV-CL}	Column address setup time (before $\overline{CAS}$ low)
$t_{su(RA)}$	t_{RAV-RL}	Row address setup time (before $\overline{RAS}$ low)

Transition times (also called rise and fall times)

The time interval between two reference points (10% and 90% unless otherwise specified) on the same waveform that is changing from the defined low level to the defined high level (rise time) or from the defined high level to the defined low level (fall time).

Example symbology:

Classified	Unclassified	Description
t_t		Transition time (general)
$t_{t(CH)}$	t_{CHCH}	Low-to-high transition time of $\overline{CAS}$
$t_{r(C)}$	t_{CHCH}	$\overline{CAS}$ rise time
$t_{f(C)}$	t_{CLCL}	$\overline{CAS}$ fall time

Valid time

(a) General

The time interval during which a signal is (or should be) valid.

(b) Output data-valid time

The time interval in which output data contines to be valid following a change of input conditions that could cause the output data to change at the end of the interval.

Example symbology:

Classified	Unclassified	Description
$t_{v(A)}$	t_{AXQX}	Output data valid time after change of address.

This supersedes the older form t_{PVX}.

12

PART III – TIMING DIAGRAMS CONVENTIONS

TIMING DIAGRAM SYMBOL	MEANING: INPUT FORCING FUNCTIONS	MEANING: OUTPUT RESPONSE FUNCTIONS
	Must be steady high or low	Will be steady high or low
	High-to-low changes permitted	Will be changing from high to low some time during designated interval
	Low-to-high changes permitted	Will be changing from low to high sometime during designated interval
	Don't Care	State unknown or changing
	(Does not apply)	Centerline represents high-impedance (off) state.

PART IV – BASIC DATA SHEET STRUCTURE

The front page of the data sheet begins with a list of key *features* such as organization, interface, compatibility, operation (static or dynamic), access and cycle times, technology (N or P channel, silicon or metal oxide gate), and power. In addition, the top view of the device is shown with the *pinout* provided. Next a general *description* of the device, system interface considerations, and elaboration on other device chracteristics are presented. The next section is an explanation of the device's *operation* which includes the function of each pin (i.e., the relationship between each input (output) and a given type of memory). The functions basically involve starting, achieving, and ending a given type of memory cycle (e.g., programming or erasing EPROMs, or reading a memory location).

Augmenting the descriptive text there appears a *logic symbol* prepared in accordance with forthcoming IEEE and IEC standards and explained in the section of this book following this one. Following the symbol is usually a *functional block diagram,* a flow chart of the basic internal structure of the device showing the signal paths for data, addresses, and control signals, as well as the internal architecture. Usually the next few pages contain the absolute maximum ratings (e.g., voltage supplies, input voltage, and temperature) applicable over the *operating free-air temperature range.* If the device is used outside of these values, it may be permanently destroyed or at least it would not function as intended. Next, typically, are the *recommended operating conditions* (e.g., supply voltages, input voltages, and operating temperature). The memory device is guaranteed to work reliably and to meet all data sheet parameters when operated in accord with the recommended operating conditions and within the specified timing. If the device is operated outside of these limits (minimum/maximum), the device's operation is no longer guaranteed to meet the data sheet parameters. Operation beyond the absolute maximum ratings as just described can result in catastrophic failures.

The next section provides a table of *electrical characteristics over full ranges of recommended operating conditions* (e.g., input and output currents, output voltages, etc.). These are presented as minimum, typical, and maximum values. Typical values are representative of operation at an ambient temperature of $T_A = 25°C$ with all power supply voltages at nominal value. Next, input and output capacitances are presented. Each pin has a capacitance (whether an input, an output, or control pin). Minimum capacitances are not given, as the typical and maximum values are the most crucial.

The next few tables involve the device timing characteristics. The parameters are presented as minimum, typical (or nominal), and maximum. The *timing requirements over recommended supply voltage range and operating free-air temperature* indicate the device control requirements such as hold times, setup times, and transition times. These values are referenced to the relative positioning of signals on the timing diagrams, which follow. The *switching characteristics over recommended supply voltage range* are device performance characteristics inherent to device

TEXAS INSTRUMENTS
INCORPORATED
POST OFFICE BOX 225012 • DALLAS, TEXAS 75265

operation once the inputs are applied. These parameters are guaranteed for the test conditions given. The interrelationship of the timing requirements to the switching characteristics is illustrated in *timing diagrams* for each type of memory cycle (e.g., read, write, program).

At the end of a data sheet additional *applications information* may be provided such as how to use the device, graphs of electrical characteristics, or other data on electrical characteristics.

Logic Symbols

EXPLANATION OF NEW LOGIC SYMBOLS FOR MEMORIES

1. INTRODUCTION

The International Electrotechnical Commission (IEC) has been developing a very powerful symbolic language that can show the relationship of each input of a digital logic circuit to each output without showing explicitly the internal logic. At the heart of the system is dependency notation, which will be partially explained below.

The system was introduced in the USA in a rudimentary form in IEEE/ANSI Standard Y32.14-1973. Lacking at that time a complete development of dependency notation, it offered little more than a substitution of rectangular shapes for the familiar distinctive shapes for representing the basic functions of AND, OR, negation, etc. This is no longer the case.

Internationally, IEC Technical Committee TC-3 has prepared a new document (Publication 617-12) that will consolidate the original work started in the mid 1960's and published in 1972 (Publication 117-15) and the amendments and supplements that have followed. Similarly for the USA, IEEE Committee SCC 11.9 has revised the publication IEEE Std 91/ANSI Y32.14. Texas Instruments participated in the work of both organizations and this 1982 Edition of the MOS Memory Data Book introduces new logic symbols in anticipation of the new standards. When changes are made as the standards develop, future editions of this book will take those changes into account.

The following explanation of the new symbolic language is necessarily brief and greatly condensed from what the standards publications will finally contain. This is not intended to be sufficient for those people who will be developing symbols for new devices. It is primarily intended to make possible the understanding of the symbols used in this book.

2. EXPLANATION OF A TYPICAL SYMBOL FOR A STATIC MEMORY

The TMS 2114 symbol will be explained in detail. This symbol includes almost all the features found in the others. Section 4, Diagramatic Summary, should be referred to while reading this explanation.

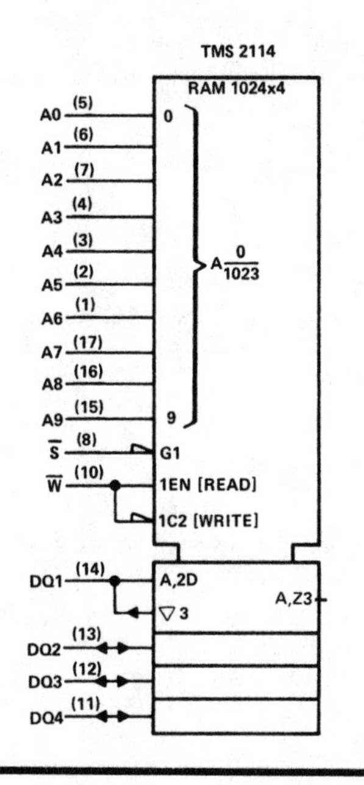

By convention all input lines are located on the left and output lines are located on the right. When an exception is made, an arrowhead shows reverse signal flow. The input/output lines (DQ1 through DQ4) illustrate this.

The polarity indicator ⊳ indicates that the external low level causes the internal 1 state (the active state) at an input or that the internal 1 state causes the external low level at an output. The effect is similar to specifying positive logic and using the negation symbol o .

The rest of this discussion concerns features inside the symbol outline. The address inputs are arranged in the order of their assigned binary weights and the range of the addresses are shown as $A\frac{m}{n}$ where m is the decimal equivalent of the lowest address and n is the highest. The inputs and outputs affected by these addresses are designated by the letter A.

The letter Z followed by a number is used to transfer a signal from one point in a symbol to another. Here the signal at output A,Z3 transfers to the 3 at the left side of the symbol in order to form an input/output port. The A means the output comes from the storage location selected by the address inputs.

The ▽ symbol designates a three-state output. Three-state outputs will always be controlled by an EN function. When EN stands at its internal 1 state, the outputs are enabled. When EN stands at its internal 0 state, the three-state outputs stand at their high-impedance states.

Since the boxes associated with DQ2, DQ3, and DQ4 have no internal qualifying symbols, it is to be understood that these boxes are identical to the box associated with DQ1.

Any D input is associated with storage. Whatever internal state is taken on by the D input is stored. The letter A (in A,Z3) indicates that the state of the D input will be stored in a cell selected by the A inputs. If the D input is disabled, the storage element retains its content.

Various types of relationships between ports can be indicated by what is called dependency notation. A letter indicating the type of dependency (e.g., C, G, Z) is placed at the affecting input (or output) and this is followed by a number. Each affected input (or output) is labeled with that same number. The Z symbol explained above is one form of dependency notation. Several other types of dependency have been defined but their use has not been anticipated in this book.

The numeral 2 at the D input indicates that the D input is affected by another input, in this case a C input (i.e., 1C2). When a C input stands at its internal 1 state, it enables the affected D input(s). When the C input stands at its internal 0 state, it disables the D input(s) so that it (they) can no longer alter the contents of the storage element(s).

The C input is itself affected by another input. The numeral 1 in front of the C shows that a dependency relationship exists with a G input. The letter G indicates an AND relationship. When a G input stands at its internal 1 state (low in this case), the affected inputs (EN and C2 here) are enabled. When the G input stands at its internal 0 state, it imposes the 0 state on the affected inputs.

Pin 10 has two functions. Its function as a C input has just been explained. Note that for the C input function to stand at its 1 state, pin 10 must be low and pin 8 must also be low. The other function of pin 10 is as an EN input. This controls the 3-state outputs. This EN input is also affected by the AND relationship with pin 8 so for the EN function to stand at its internal 1 state (enabling the outputs), pin 10 must be high and pin 8 must be low.

Labels within square brackets are merely supplementary and should be self-explanatory.

3. CACHE ADDRESS COMPARATOR

The block diagram for the TMS 2150 uses the RAM symbol (explained in Section 2) and also the following:

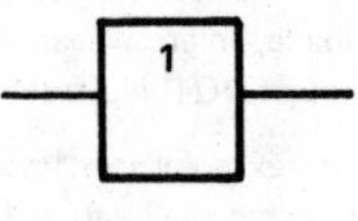

Buffer without special amplification. If special amplification is included, the numeral 1 is replaced by ▷.

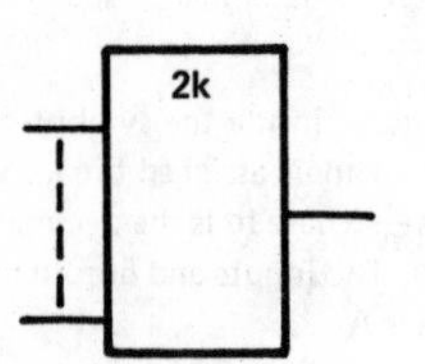

Even-parity element. The output stands at its 1-state if an even number of inputs stand at their 1-states.

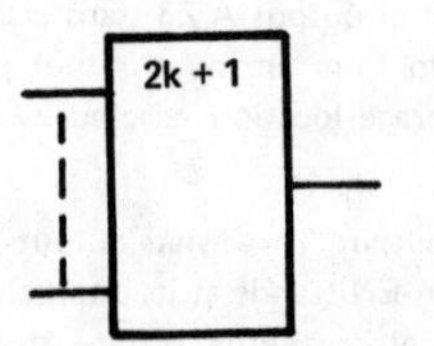

Odd-parity element. The output stands at its 1-state if an odd number of inputs stand at their 1-states.

NOTE: TMS 2150 uses one of these to generate even parity by adding the output as a ninth bit.

TEXAS INSTRUMENTS
INCORPORATED
POST OFFICE BOX 225012 • DALLAS, TEXAS 75265

4. DIAGRAMATIC SUMMARY

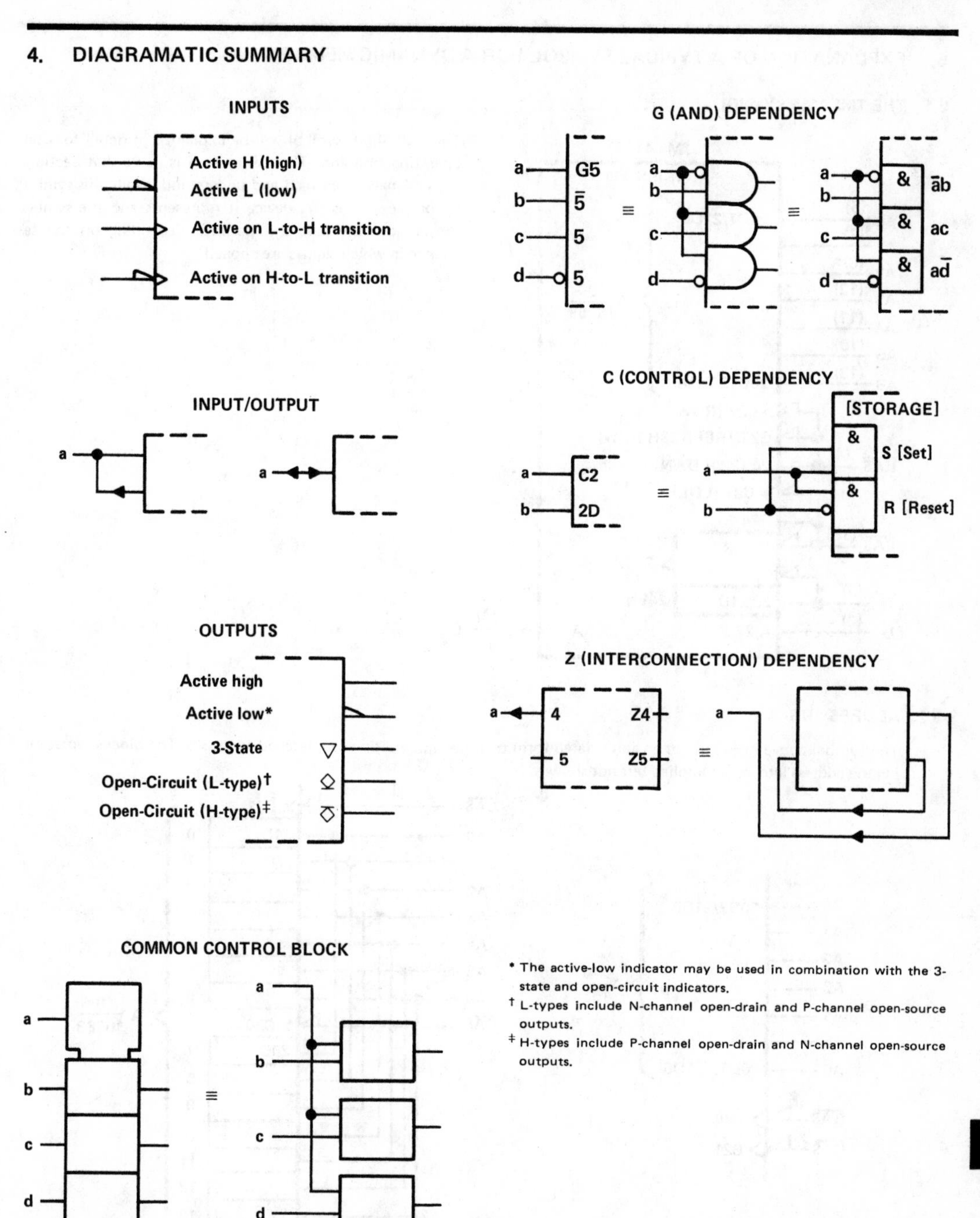

* The active-low indicator may be used in combination with the 3-state and open-circuit indicators.

† L-types include N-channel open-drain and P-channel open-source outputs.

‡ H-types include P-channel open-drain and N-channel open-source outputs.

5. EXPLANATION OF A TYPICAL SYMBOL FOR A DYNAMIC MEMORY

5.1 THE TMS 4116 SYMBOL

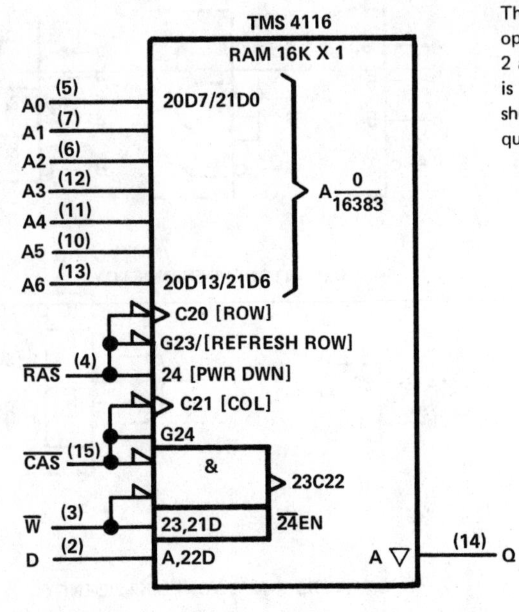

The TMS 4116 symbol will be explained in detail for each operating function. The assumption is made that Sections 2 and 4 have been read and understood. While this symbol is complex, so is the device it represents and the symbol shows how the part will perform depending on the sequence in which signals are applied.

5.2 ADDRESSING

The symbol above makes use of an abbreviated form to show the multiplexed, latched addresses. The blocks representing the address latches are implied but not shown.

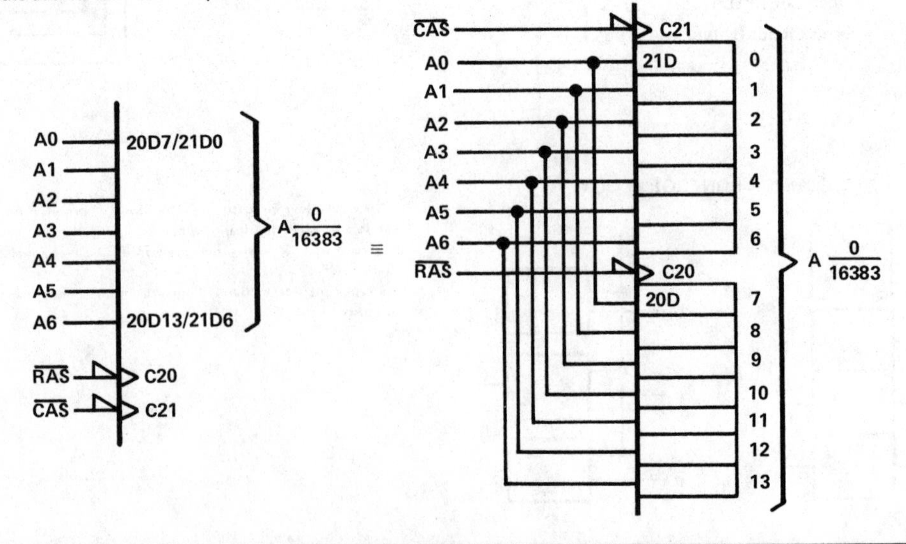

TEXAS INSTRUMENTS
INCORPORATED
POST OFFICE BOX 225012 • DALLAS, TEXAS 75265

When $\overline{RAS}$ goes low, it momentarily enables (through C20, ▷ indicates a dynamic input) the D inputs of the seven address registers 7 through 13. When $\overline{CAS}$ goes low, it momentarily enables (through C21) the D inputs of the seven address registers 0 through 6. The outputs of the address registers are the 14 internal address lines that select 1 of 16,384 cells.

5.3 REFRESH

When $\overline{RAS}$ goes low, row refresh starts. It ends when $\overline{RAS}$ goes high. The other input signals required to carry out refreshing are not indicated by the symbol.

5.4 POWER DOWN

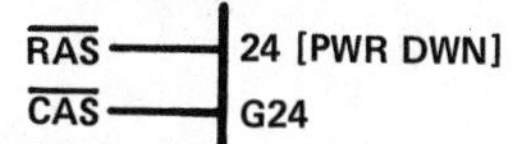

$\overline{CAS}$ is AND'ed with $\overline{RAS}$ (through G24) so when $\overline{RAS}$ and $\overline{CAS}$ are both high, the device is powered down.

5.5 WRITE

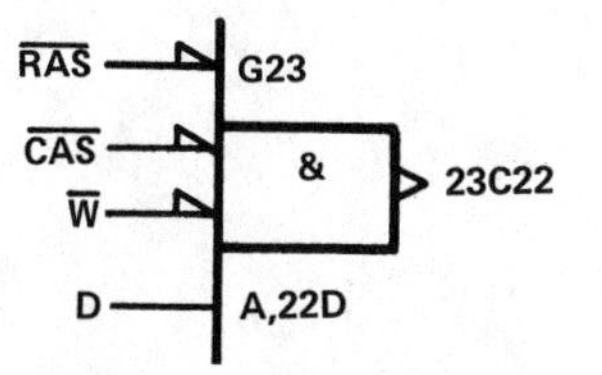

By virtue of the AND relationship between $\overline{CAS}$ and $\overline{W}$ (explicitly shown), when either one of these inputs goes low with the other one and $\overline{RAS}$ already low ($\overline{RAS}$ is AND'ed by G23), the D input is momentarily enabled (through C22). In an "early-write" cycle it is $\overline{W}$ that goes low first; this causes the output to remain off as explained below.

5.6 READ

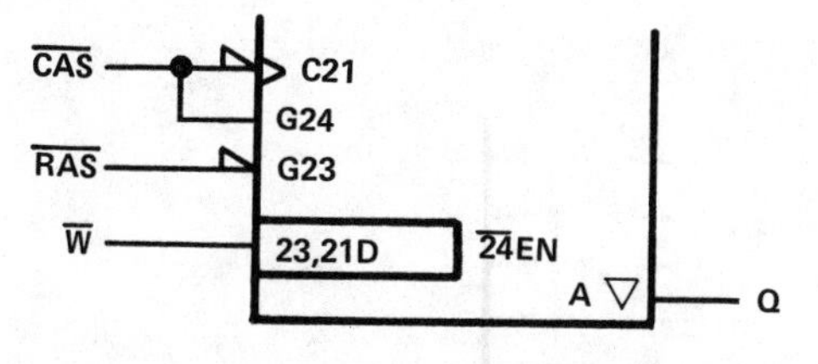

The AND'ed result of $\overline{RAS}$ and $\overline{W}$ (produced by G23) is clocked into a latch (through C21) at the instant $\overline{CAS}$ goes low. This result will be a "1" if $\overline{RAS}$ is low and $\overline{W}$ is high. The complement of $\overline{CAS}$ is shown to be AND'ed with the output of the latch (by G24 and $\overline{24}$). Therefore, as long as $\overline{CAS}$ stays low, the output is enabled. In the "early-write" cycle referred to above, a "0" was stored in the latch by $\overline{W}$ being low when $\overline{CAS}$ went low, so the output remained disabled.

If you have questions on this Explanation of New Logic Symbols, please contact:

F.A. Mann MS 84
Texas Instruments Incorporated
P.O. Box 225012
Dallas, Texas 75265
Telephone (214) 995-3746

IEEE Standards may be purchased from:

Institute of Electrical and Electronics Engineers, Inc.
345 East 47th Street
New York, N.Y. 10017

International Electrotechnical Commission (IEC) publications may be purchased from:

American National Standards Institute, Inc.
1430 Broadway
New York, N.Y. 10018

If you're not already on it and want to keep up to date, get on our mailing list. Complete and return the card below.

NAME ______________________ TITLE ______________________
DESIGN RESPONSIBILITY ______________________
COMPANY ______________________
COMPANY STREET ADDRESS ______________________
CITY __________ STATE __________ ZIP __________ PHONE (___) __________

1. For which of the following applications do you have an influence on the design? (Please circle only one letter but as many numbers as applicable)

A. MAINFRAME COMPUTERS
1. Main memory
2. Control store
3. Cache memory
4. I/O buffers
5. Other __________

B. MINICOMPUTERS
1. Main memory
2. Control store
3. Cache memory
4. I/O buffers
5. Other __________

C. MICROCOMPUTERS
1. Systems
2. Boards

D. COMPUTER PERIPHERAL EQUIPMENT
1. Dumb terminals
2. Intelligent terminals
3. Graphics terminals
4. Printers
5. Disc storage
6. High-speed peripheral controllers
7. Other __________

E. TELECOMMUNICATIONS SYSTEMS

F. TEST, MEASUREMENT AND INSTRUMENTATION

G. MEDICAL ELECTRONICS

H. MILITARY/GOVERNMENT ELECTRONICS

I. INDUSTRIAL CONTROLS

J. CONSUMER ELECTRONICS

K. WORD PROCESSING

L. OTHER __________

2. What MOS Memory architecture is best suited for your needs? (circle one)

A. ×1
B. ×4
C. ×8
D. Other __________

3. Which MOS Memory products do you utilize in your designs? What is your estimated annual usage of each for 1982? (Circle appropriate letters)

	in 1,000 units		
	<10	10-100	100-500
16K DRAM	A	B	C
64K × 1 DRAM	A	B	C
16K × 4 DRAM	A	B	C
4K × 1 High-Speed Static RAM	A	B	C
1K × 4 High-Speed Static RAM	A	B	C
2K × 8 Medium Performance Static RAM	A	B	C
16K EPROM	A	B	C
32K EPROM	A	B	C
64K EPROM	A	B	C
32K ROM	A	B	C
64K ROM	A	B	C
Cache Address Comparator (TMS 2150)	A	B	C
DRAM Controller (TMS 4500A)	A	B	C
Other __________	A	B	C
Other __________			

4. Who are your preferred vendors for MOS Memory Products?

DRAMs __________ REASON: __________
STATIC RAMs __________ __________
NONVOLATILE __________ __________

TEXAS INSTRUMENTS
INCORPORATED

NO POSTAGE
NECESSARY
IF MAILED
IN THE
UNITED STATES

BUSINESS REPLY MAIL

FIRST CLASS PERMIT NO. 6189 HOUSTON, TX

POSTAGE WILL BE PAID BY ADDRESSEE

Texas Instruments Incorporated
M/S 6946
P.O. Box 1443
Houston, Texas 77001